LIFE WORLDS OF MIDDLE EASTERN OIL

LIFE WORLDS OF MIDDLE EASTERN OIL

Histories and Ethnographies of Black Gold

Edited by Nelida Fuccaro and Mandana E. Limbert

EDINBURGH
University Press

Edinburgh University Press is one of the leading university presses in the UK. We publish academic books and journals in our selected subject areas across the humanities and social sciences, combining cutting-edge scholarship with high editorial and production values to produce academic works of lasting importance. For more information visit our website: edinburghuniversitypress.com

Edinburgh University Press Ltd
The Tun – Holyrood Road
12 (2f) Jackson's Entry
Edinburgh EH8 8PJ

Typeset in 11/15pt Adobe Garamond Pro by
Cheshire Typesetting Ltd, Cuddington, Cheshire, and
printed and bound in Great Britain

A CIP record for this book is available from the British Library

ISBN 978 1 3995 0614 4 (hardback)
ISBN 978 1 3995 0616 8 (webready PDF)
ISBN 978 1 3995 0617 5 (epub)

CONTENTS

FIGURES

NOTES ON CONTRIBUTORS

Mattin Biglari is a Postdoctoral Fellow at SOAS, University of London. His research focuses on the intersection of labour, infrastructure and environment, especially in Iran and the Middle East. He has wider interests in connecting science and technology studies, energy/environmental humanities, and postcolonial studies. His monograph, titled *Refining Knowledge: Labour, Politics and Oil Nationalisation in Iran, 1933–51,* will be published in 2023. This focuses on labour and expertise in the history of Iran's oil nationalisation in 1951, especially everyday experiences in the refinery town of Abadan. It illuminates how anti-colonialism, urban politics and labour activism coalesced in opposition to the Anglo-Iranian Oil Company, and how this process culminated in the reproduction of colonial epistemologies even as the Iranian government expelled the foreign oil company.

Dawn Chatty is Emeritus Professor in Anthropology and Forced Migration and former Director of the Refugee Studies Centre, University of Oxford, United Kingdom. She was elected Fellow of the British Academy in 2015. Her research interests include refugee youth in protracted refugee crises, conservation and development, pastoral society and forced settlement She is the author of *Displacement and Dispossession in the Modern Middle East* (2010), *From Camel to Truck* (2013) and *Syria: The Making and Unmaking of a Refuge State* (2018).

Nathan J. Citino is the Barbara Kirkland Chiles Professor of History at Rice University. He is the author of *From Arab Nationalism to OPEC: Eisenhower, King Saʿud, and the Making of US–Saudi Relations* (2002, 2d ed. 2010). His second book, *Envisioning the Arab Future: Modernization in US–Arab Relations, 1945–1967* (2017), was awarded the Robert H. Ferrell Book Prize by the Society for Historians of American Foreign Relations. In 2020, he received the Scholar's Award from the Truman Library Institute to support a new project about US empire in the Middle East.

Nelida Fuccaro is Professor of Middle Eastern History at NYU Abu Dhabi. Formerly based at SOAS University of London, she has written on cities, public violence, and on the interplay between ethnicity and nationalism in frontier societies. In the last few years her research has focused on the social, material and visual cultures of the oil in the Persian Gulf, Iraq and the Arabian Peninsula. Among her publications, she is the author of *Histories of City and State in the Persian Gulf: Manama since 1800* (2009, paperback 2011), the guest editor of the thematic contribution 'Histories of Oil and Urban Modernity in the Middle East' in *Comparative Studies in South Asia, Africa and the Middle East* (2013) and the editor of *Violence and the City in the Modern Middle East* (2016).

Rania Ghosn is Associate Professor of Architecture and Urbanism at the Massachusetts Institute of Technology and founding partner of Design Earth. Her practice engages design as a speculative medium for making visible and public the geographies of the climate crisis. She is editor of *New Geographies 2: Landscapes of Energy* (2009), and co-author of *Geographies of Trash* (2015), *Geostories: Another Architecture for the Environment* (2nd ed. 2020) and *The Planet After Geoengineering* (2021).

Laura Hindelang is a post-doc researcher at the University of Bern, Institute of Art History. Her book *Iridescent Kuwait: Petro-Modernity and Urban Visual Culture since the Mid-Twentieth Century* (2022) is a transdisciplinary study of Kuwait's urban visual culture and the (in)visibilities of petroleum. She co-edited the bilingual anthology *Into the Wild. Art and Architecture in a Global Context* (2018) and has published articles on the magazine culture, architecture, and contemporary art of the Gulf States. Her current research interests are nostalgia, glass in art and architecture and gender in pre-modern architectural history. She is a committee member of Manazir – Swiss Platform for the Study of Visual Arts, Architecture and Heritage in the MENA Region and an editorial member of *Manazir Journal.*

Mandana E. Limbert is Associate Professor of Anthropology at Queens College and the Graduate Center, City University of New York. She is a historical anthropologists interested in state-building in the Sultanate of Oman and has written about resources, everyday experiences of infrastructural transformation, and the politics of history. Most recently, she has focused on processes of territorialisation and racialisation in the Western Indian Ocean. She is a co-editor of *Timely Assets: The Politics of Resources and their Temporalities* (2005) and the author of *In the Time of Oil: Piety, Memory, and Social Life in an Omani Town* (2010). Her work has also appeared in journals such as *Social Text, Comparative Studies in South Asia, Africa and the Middle East*, and *The Journal of the Royal Anthropological Institute.*

Matthew MacLean is an independent scholar based in New York City. He completed his Ph.D. in History and Middle Eastern Studies at NYU in 2017. In the course of his research he was a Humanities Research Fellow and Associate Lecturer at NYU Abu Dhabi, a Visiting Doctoral Scholar at the Al-Qasimi Foundation in Ras Al Khaimah, and a Fulbright Scholar at Zayed University in Dubai.

Arthur Mason is Associate Professor in Social Anthropology at the Norwegian University of Science and Technology. He holds a Ph.D. in cultural anthropology from the University of California at Berkeley. His previous edited volumes include *Arctic Abstractive Industry: Assembling the Valuable and Vulnerable North* (2022) and *Subterranean Estates: Life Worlds of Oil and Gas*, with coeditors Hanna Appel and Michael Watts (2015). Mason studies energy consultants involved in oil and gas development.

Farah Al-Nakib is Associate Professor of History at California Polytechnic State University in San Luis Obispo. She received her Ph.D. (2011) and MA

(2006) in History from SOAS University of London. Al-Nakib's research focuses on the urban social history of Kuwait before and after oil, on which she has published numerous articles as well as her award-winning book *Kuwait Transformed: A History of Oil and Urban Life* (2016). She also writes on collective memory, forgetting, and nostalgia in relation to urban modernity in the Arab Gulf. Her current research focuses on untold histories of the 1990–1 Iraqi invasion and occupation of Kuwait. As a founding member of Cal Poly's Public Humanities Collaborative, Al-Nakib is also currently involved in developing oral history and storytelling projects among underrepresented communities living on California's central coast.

Zeynep Oguz is a political anthropologist of resources, energy, and the environment. She is currently a Senior Postdoctoral Researcher at the Laboratory of Social and Cultural Anthropology at the University of Lausanne. Between 2019 and 2021, Zeynep was a Postdoctoral Fellow in Environmental Humanities at Northwestern University with a joint appointment at the Department of Anthropology. She received her Ph.D. in Anthropology in September 2019 at the Graduate Center, the City University of New York. Zeynep's essays and articles on the politics of oil, geology, and coloniality in Turkey have appeared or are forthcoming in *Political Geography, Cultural Anthropology, Journal of Cultural Economy.* She is the co-editor (with Jerome Whitington) of an *Environmental Humanities* special journal issue, titled 'Earth as Praxis: Geology, Power, and the Planetary' (2023).

Douglas Rogers is Professor of Anthropology at Yale University and author of *The Depths of Russia: Oil, Power, and Culture After Socialism* (2015). His current research projects concern the history of hydrocarbon biotechnology and geology, both in and beyond the Soviet Union.

Andrea Wright is an Assistant Professor of Anthropology and Asian & Middle Eastern Studies at William & Mary. Her work examines the histories of capitalism and its contemporary expressions in South Asia and the Arabian Peninsula. In her first book, *Between Dreams and Ghosts: Indian Migration and Middle Eastern Oil* (2021), Wright uses ethnographic and archival materials to explore labour migration as a social process that shapes global capitalism. Currently, Wright is finishing her second book, *Producing Labor Hierarchies: A History of Oil in the Arabian Sea*. Focusing on the relationships among governments, oil companies, and mobile workforces, *Producing Labor Hierarchies* uncovers the process by which the lines between citizens and noncitizens were drawn and enforced in South Asia and the Middle East over the course of the twentieth century.

Ala Younis is an artist and curator. She seeks instances where historical and political events collapse into personal ones. Her work also looks into how the archive plays on predilections and how its lacunas and mishaps manipulate the imagination. She curated Kuwait's first pavilion at the Venice Biennale (2013). She is co-founder of the publishing initiative Kayfa ta, co-head of Berlinale's Forum Expanded, member of the Academy of the Arts of the World (Cologne), and co-artistic director of Singapore Biennale 2022. She is also a research scholar at al Mawrid Arab Center for the Study of Art at New York University Abu Dhabi.

ACKNOWLEDGEMENTS

This volume arises from a workshop, 'The Life Worlds of Middle Eastern Oil', held at NYU Abu Dhabi in April 2019. We were fortunate to have been able to capitalise on the endowment of place, that is to organise this oil event in a region that – thanks to its vast petroleum resources – has played such an important part in shaping oil lives and cultures, both locally and globally. We would like to thank the NYU Abu Dhabi Institute who has generously sponsored the workshop, particularly Reinerdt Falkenburg and Martin Klimke. We are also much indebted to the Institute Staff who has helped us to organise the event, particularly Gila Bessarat Waels, Nahed Ahmad, Nora Hanaa Yousif Toma, Manal Demaghlatrous, Omar Hindawi, Antoine Jean El Khayat. Dean Rober Young, Jesusita Santillan, Tina Galanopoulos-Papavasileiou and Caitlin Newsom from NYU Abu Dhabi Arts and Humanities Division have also offered their unfailing support, alongside Aleksandra Markov who was the most efficient and delightful student assistant we could possibly have.

This volume has benefited from the discussions and insightful ideas of the workshop participants. We wish to thank Frauke Heard-Bey, David Heard, and Muhamad al-Mubarak for their vivid and personal accounts of past oil lives in the United Arab Emirates, and Finn E. Krogh and Murtaza Vali for having shared with us their knowledge of oil heritage and art. A heartfelt thanks also to Douglas Rogers who has provided most generous and useful comments to the opening chapter of the volume; to the two anonymous readers who have enthusiastically endorsed this project; to Julio Moreno Cirujano who worked tirelessly as our editorial assistant; and to Ala Younis for help with the image for the book cover. Finally, we are indebted to the wonderful editorial team at Edinburgh University Press whose unfailing support has made the publication of this book possible. Our sincerest thanks to Nicola Ramsey, Emma Rees and Louise Hutton, and to Michael Ayton, our freelance copy-editor – it was a pleasure working with you!

PART I

EXPOSING OIL

1

SCOUTING FOR OIL IN THE MIDDLE EAST

Nelida Fuccaro and Mandana E. Limbert

Rivers of plastic waste float down streams and alleys in Beirut; an art gallery in Dubai showcases a pile of plastic sandals; geologists in Arabia and Turkey take photographs and rock samples; asphalt roads in the United Arab Emirates cross and frame regional and national borders; family-oriented compounds and labour camps house engineers and rig operators; and refinery fires in Kuwait light the sky with toxins. These are just a few of the ways in which petroleum and petroleum by-products appear in and shape everyday life in the Middle East. Some of these manifestations are recognised as involving petroleum, others less so; some involve specialised knowledge, others are more public. There is so much more; petroleum and petroleum by-products permeate life.

This volume explores how oil has shaped and mediated everyday life in the modern and contemporary Middle East. We aim to highlight the diversity of this experience as a corrective to what we see as the 'downstream effect' that we believe has dominated readings and public perceptions of the Middle East's petroleum resources. This 'downstream effect', coloured by the interruptions of Middle East oil supplies to European and American consumers, the pernicious effects of petro-dollars flooding financial markets and the global reverberation of the regional geopolitics of oil conflict, has shaped attention on Middle East oil almost exclusively as a site of geopolitics and production. Indeed, we aim to highlight how a range of experiences of oil in the Middle East are shaped by the simultaneity of upstream and downstream processes and dynamics and are embedded in multiple political-economic scales, from the local to the global.

In the chapters that follow, oil is approached as a natural resource explored by geologists, engineers and biologists; as an extractive industry that has triggered contestation over labour and borders as well as environmental degradation; as a commodity forging new national subjectivities and spatial imaginaries, including those shaped by enhanced mobilities and automobility; as potential wealth that structures financial projections, anticipation, anxieties and disappointments; and as a substance and source of energy framed by images in advertising, corporate literature, graphs and aerial photographs, as well as film, art, architecture and design.

This approach challenges ideas of oil as an industry and commodity with singular effects, or shaping modular political communities, social hierarchies, nations, citizens and consumers. We recognise instead an array of oil's material forms, geographies and temporalities, and the variety of ways oil has channelled and produced people's life experiences. To be clear, our attention to the simultaneity of upstream and downstream markets and the multi-layered scales of oil's political-economic circulations in shaping oil's multifocality – and differing visibilities – recognises the search for profit and power that motivates and structures the processes we examine and the economic and political inequalities that have ensued.

As this volume highlights, if oil feels 'hidden in plain sight' in much of the world, it is particularly so in the Middle East where production, distribution and consumption are integral to local labour markets and evident in direct employment with oil companies, national bureaucracies and transnational migrations. While not quite isolated to a 'zone of confinement', concealed from – though sometimes integral to – the operations of everyday life, oil is nevertheless somewhat invisible, despite its ubiquitousness. It is precisely the shifting visibility of oil that has drawn our attention and that has encouraged us to consider when, how and for whom oil gains visibility as it shapes people's lives. Indeed, several of the contributors emphasise the significance of other senses – smell and sound in particular – that signal oil. It is in this sense that this volume scouts for Middle Eastern petroleum.

Throughout the Middle East, oil's numerous circles of reach have formed porous, extendable and often overlapping spheres of influence, experience and visibility not necessarily coterminous with political and geographical boundaries, or with the constraints posed and possibilities created by the international oil industry. Attuned to the varying scales of oil's presence and effect, this volume therefore recognises tensions between the scales of enormity and detail, of global markets and technological exactitude, that structure many of the everyday experiences of oil explored here. We suggest that an emphasis on global oil, as has been the primary approach of the scholarship on the Middle East, obfuscates the multifaceted aspects of 'oil's life worlds', an approach explored by Hannah Appel, Arthur Mason and Michael Watts in their influential volume *Subterranean Estates*.[1] For us, highlighting the local geographies and temporalities of oil production in particular brings into focus the specificities of oil's workings, performances and incarnations as it manifests as a geological resource and an object of local expertise and labour and as it shapes and is framed by social life, images and politics.

The Middle Eastern Oil Experience

Despite petroleum's importance in shaping the twentieth and twenty-first centuries, its effects on the texture and struggles of human experience have been largely unexplored, especially in the Middle East. It is as if academic literature and public perceptions of Middle East oil have cast a spell on the granular experiences of petroleum existence. When we shift attention to the everyday – including the experiences of engineers and geologists or artists and labourers – the limits and over-simplification of the 'resource curse' paradigm with its phantasmagoric wealth narratives and apocalyptic scenarios become more evident.

Indeed, while US oil has long been recognised as the substance organising modern spaces and American life – from cities, highways and car mobility to plastics and material cultures of consumption – its Middle Eastern counterpart has been the subject of much 'resource curse' speculation as the main architect of authoritarian regimes and failed states, and as an appendix to the global oil industry. Although, in the last decade or so, development economics has started to paint the 'curse' of oil outside the USA and Europe as elusive – easing what Terry Lynn Karl has called 'the paradox of plenty' – the axis oil resources–poor governance still perpetuates the Western dominance of global capitalism. Moreover, the geopolitical importance of the Persian Gulf and Saudi Arabia has also sealed the fate of Middle Eastern oil, attracting the attention of policy-oriented disciplines such as economics, political science and strategic/international studies, averse to digging deep into the micro level of oil experiences.[2]

As oil scholars with humanist inclinations, we appreciate that it is often difficult to translate oil experience into an academic text at all. As Stephanie LeMenager puts it, 'Living in oil, through injury and pleasure, is personal and not easily transmissible as a story.'[3] While sociologists, economists and political scientists have often approached the experience of fossil fuels as the 'semantics' of modern capitalism or as generalised projections about political regimes, cultural theorists have wondered how to write about oil encounters in a way that is not so reductive, or reproductive of our world's ultimate obsession with oil capital.[4] Middle Eastern petro-writing – both academic and fictional – has similarly had a long history of identification with Big Oil – the foreign corporations and governments that have controlled the industry – and with the 'resource curse' and 'paradox of plenty' which transfix oil's power as corporate outreach, surplus wealth and dystopian governance. Despite a recent focus on less abstract and more grounded manifestations of oil – particularly urbanism, labour and infrastructure – the truth of the matter is that the effects and affect generated by many encounters with petroleum throughout the region remain rather modular.[5]

In this volume, we go back to basics – in LeMenager's words, to 'evoke the scale of the human'[6] – by reaching beyond the level of institutions, technological zones, and the technopolitics that have made up the oil assemblage. We recognise, in other words, petroleum's inducing entanglements at different scales, from the local and the individual to the global and the collective. We share with Timothy Mitchell and others a concern with oil's material politics and technopolitics, and recognise the co-constitutive nature of petroleum infrastructure, capital and societies. We also acknowledge the importance of the 'oil complex' paradigm whose focus on the intersecting global and national architectures of oil's political economy has helped to locate important modes and nodes of ethical and environmental responsibility that are shaping Big Oil and petro-states throughout the Middle East and globally.[7] Yet these ways of examining the worlds of oil tend to abstract petroleum encounters by concentrating on the reverberations of the macro-level interconnections between the guts of the industry, global finance and politics. Treating the industry as a modular ensemble tends to conceal the specificity of oil's action, and the interaction and power struggles between humans and technology, not only the industry's infrastructure but also machines that are moved by oil or are made with materials that are oil derivatives such as cars, cameras, computers and many others.

For instance, oil is often said to have been the fossil fuel industry able to seal the triumph of technology over labour. Insightfully, Kaveh Ehsani has argued that oil's staggering power of 'occlusion' has also concealed the agency of workers, partly because of a disproportionate focus within the literature on less labour-intensive extraction sites such as fields, offshore platforms and pipelines. Some of the contributions to this volume speak to the centrality of the human factor – the industry's workforce – in shaping oil's technological zones, also signposting the influence of machines, technology and expertise in framing experience and forming new subjectivities.[8] They document how the industry shaped professional trajectories that reverberated far beyond oil fields, refineries and workshops by servicing new fields of company outreach: the new world of media, as in the case of the photographers and film makers in Fuccaro's chapter; and that of sustainable global development, as exemplified by the microbiologists on the payroll of BP discussed by Rogers who strived to feed the world with newly-discovered petroprotein.

Examining the life worlds of Middle Eastern oil prompts us, therefore, not only to grapple with the more ephemeral aspects of oil experiences, but also to reflect on how to read petroleum encounters from the perspective of their past and present 'protagonists' within the oil complex: the labourers, scientists, company executives and consumers. This reflection on the 'protagonists', however, also brings into sharp relief the ethics of problematisation that we share as scholars, arising from the tensions and ambiguities of petroleum experiences. It is what Lakoff and Collier have called 'regimes of living' as ethical configurations formed in relation to technology and biopolitics that 'are brought into alignment in problematic and uncertain situations'.[9] The chapter on Kuwait's oil fires by Al-Nakib tests the complexity of oil's ethical and cognitive reach both by attending to the view of oil engineers and by emphasising the incomprehensibility of the environmental and human disaster that unfolded after the withdrawal of the Iraqi army. Indeed, in this chapter, the unintelligibility of the disaster is mitigated by the actions and stories of Kuwaitis who have claimed the oil fires back from outside intervention and global media attention: the petroleum engineer Sara Ahmad with her intimate knowledge of the oil wells, Al-Nakib's own experience of the fires and as a researcher in the archives of the Iraq Memory Foundation, and the artist/filmmaker Monira al-Qadiri, whose work brings back context to the more opaque and abstract cinematic representation of Kuwait's oil disaster in Werner Herzog's acclaimed documentary *Lesson of Darkness*. Al-Nakib's chapter brings into focus the ambiguities of oil's experience and visibility as against the grip of personal lives exercised by the geopolitics of oil and the many supposed certainties popularised by the industry before the Iraqi invasion.

The history of Middle Eastern oil offers an in-depth diagnostic as well of the making and un-making of what Douglas Rogers has called the 'spectacles and dramas' of oil modernity.[10] In recognition of the hyper-technological futures of a world *after* oil simulated and projected by some of the large producers in the Gulf States, we might also add oil's post-modernity.[11] The virtual reality tour of Dubai in the Future Dubai Gallery is a case in point. It envisions the post-oil city as a hub for innovation and a testing ground for future technologies, a robotic futuristic society with no individual and collective agency.[12] While the dazzling quality and glitziness of Dubai's (and Gulf) petro-futures might monopolise global attention and serve to elide labour contestation, uneven wealth distribution and environmental damage, the glitz is also not necessarily representative of past and present petro-'spectacles and dramas', which remain haunted by the history of oil exploration.

At the level of individual Middle Eastern nations, it should be remembered, oil development has had mixed results and has unfolded unevenly; it has by no means been modular, unilinear or synchronic. Moreover, not all countries in the region have experienced oil booms. Some places live in constant expectation and speculation about their underground petroleum resources as in the case of Turkey, as Oguz analyses in this volume. And, in others, as Mandana Limbert has examined, contemporary expectations of depletion puncture projections of teleological progress and developmentalist trajectories from the start.[13] Nevertheless, with few exceptions, all Middle Eastern nations have extracted (or attempted to extract) oil from their national territories and have accommodated oil infrastructure: pipelines dotted by pumping stations that ensured the smooth flow of crude from oil fields to the Mediterranean and Persian Gulf, refineries located in coastal areas, and fuel distribution centres which since the 1950s epitomised by iconic petrol stations. Many individuals and communities in the Middle East have literally 'sniffed' oil without necessarily having been the direct recipients of oil bonanzas.

In addition, and at the regional scale, we might ask whether the Middle Eastern love affair with petroleum can be compared to that which characterised the twentieth century as oil's American century. The short answer perhaps is no. Despite the love affair, there is, as the chapters in this volume also testify, a deep ambivalence about oil's development. Geological abundance, geographical disconnection, national boundaries and foreign exploitation come to mind when thinking about the history of Middle Eastern oil. As early as 1958, soon after the creation of the first Oil Committee by the Arab League, an American scholar-cum-oil executive noted that oil prosperity in the Middle East was already developing as 'parochial rather than regional'.[14] This sentiment has persisted despite (or perhaps because of) the copiousness with which oil has continued to gush out of the ground in its crude form. Yet the largest reserves were confined to sparsely populated frontier regions bordering the Persian Gulf, Iran, Iraq and Turkey, and the vast deserts of North Africa. Moreover, post-World War I Arab regional powers such as Egypt and Syria had negligible oil reserves that did not match their political influence. After extraction, oil entered the life cycles of the Middle East through networks of global hydrocarbon capitalism and statism, controlled by foreign oil companies and, after nationalisation, by the governments of major oil producers. Before becoming a state enterprise, the highly sophisticated industry that built the infrastructure to extract and transport crude, and the chemical plants that transformed it into refined products, had dangerous attachments to European neo-colonial mercantilism and after World War II to American capitalism. A British oil company with large government participation such as British Petroleum controlled oil concessions in Iran and Iraq and in many Persian Gulf States, including Oman. Aramco in Saudi Arabia and Bapco in Bahrain were established as subsidiaries of the powerful Standard Oil of California.

While corporate, the history of oil in the Middle East has not, as in the USA, been driven by nationally-based capitalist expansion and nurtured by the myth of individual success against the odds, free entrepreneurship and the expansion of hydrocarbon frontiers. Oil in the Middle East has a varied colonial and postcolonial history, tainted by economic dependency, political subjugation and fragmentation and post-nationalisation corruption, but also defined more recently by the glitzy economic 'success' and projections of stewardship promoted by the various rulers of smaller Gulf States. Indeed, at present there is still some discomfort with equating the origin story of large oil-producing countries solely with oil development, and little sign of narratives (including from environmentalists) such as those that declared the twentieth century petroleum's American century. And perhaps it is for all these reasons that oil's circles of reach in the Middle East are often still opaque and difficult to identify despite (or because of) petroleum's status as the most important natural resource and industry in the region.

Acting 'with a force suggestive of a form of life',[15] petroleum has thus penetrated the existence of Middle Eastern individuals, collectivities and nations in unprecedented, albeit ephemeral and multiple, ways, and in forms that elide modular analysis. The outcomes of oil's actions as natural substance, infrastructure and commodity in the one hundred years or so of its regional history are by no means fixed, as much of the 'resource curse' literature depicts.[16] While oil appears to shape history uniformly by establishing seemingly consistent trends of social and political development, it is, as Antti Salminen and Tere Vadén have noted, multifocal, producing a 'forests of foci'. And, by detecting the often slippery and bituminous traces of oil we recognise its instability, ambiguity, and what Salminen and Vadén also call its 'con-distancing' effects: its ability to produce both intimacy and alienation, appearing as a unifier of experience and social action while creating a sense of removal and alienation.[17]

Uncovering and Disentangling Oil

This volume, thus, 'scouts' for oil, noting the dissonance between oil's simultaneous ubiquitousness and its taken-for-granted near-invisibility. In so doing, and in recognising the more ephemeral forms of its presence, the chapters presented here note the ways in which oil both mediates modern life and serves as a metonym for wealth or potential wealth, not to mention inequalities and dangers.[18]

While the concept of 'mediation' often refers to the processes by which technologies of media, from newspapers or photography to electronic audiovisual technologies and practices (with particular communicational structures), shape 'social envisioning', as William Mazzarella describes, we recognise oil as mediating – even while its visibility is uneven – as well.[19] It is a natural substance that has entered the life cycles of human and natural environments as energy and through geological, technological and techno-scientific expertise and infrastructure, as well as through the discourses and images of the industry, art, and its material 'by-products', such as plastic. The possibilities and limits of oil – and oil institutions and infrastructures – have become central to people's understandings of national history and their political futures.

As Oguz describes in this volume, for example, Turkey's oil scarcity has shaped the knowledge world of Faruk, the geologist of the state-owned Turkish Petroleum, and has structured memories and discourses of Turkey's geopolitical formations: from the early Republican era and the violent state of the 1990s to the current populist AKP-led government. In this case, for example, the fluidity and elusiveness of oil as an underground/underwater liquid defies nation-state border politics and practice, but also becomes a critical node for speculation and conspiracy. Training to become oil 'experts' in the Abadan Technical Institute functioned as a catalyst for the formation of anti-colonial nationalism among its young Iranian students, as Biglari points out in his chapter. Finally, as Rogers highlights, the oil microbes that in the 1980s triggered a rush to produce petroprotein in Kuwait and Iran shaped the failed global food revolution championed by national and international organisations in the previous decade.

This volume also exhibits art as a field of oil's workmanship, both mediating further interpretations and drawing on oil's varying metonymic qualities. Younis's photo essay, for example, is structured around a series of nested histories of Iraqi oil that are extracted from her multi-media installation *Al-Bahithun* (The Re-Searchers). The installation assembles original documentation and historical cinematic footage with sculpted plastic and ceramic pieces. Besides contemporary works such as *Al-Bahithun*, we can locate material and metaphysical traces left by oil in past artwork by insiders to the oil industry. Without dealing directly with petroleum subjects (but rekindling memories and milieux of earlier oil eras) the sophisticated black and white photography by Latif Al-ʿAni – an employee of the Iraq Petroleum Company – offers a striking portrayal of working lives and urban environments in 1950s and 1960s Iraq. The modernist brushstrokes of Anglo-Iranian Oil Company illustrator Houshang Pezeshknia are more explicit in their depiction of Iran's oil fields and its workers. His canvasses present the viewer with a corrosive political critique of AIOC's exploitation tinged with a stint of early environmental activism.[20]

As representative of petroleum craftmanship, art is also increasingly mediating knowledge of oil's histories, locations and strategies, whether it is in the executive boardrooms in the form of graphs, or in art works in the form of abstraction, as Mason examines in this volume. Contemporary oil art is contributing, as well, to enriching the visual and material oil archive, adding collages, sculptures and multi-media installations to the bureaucratic, technical and material evidence left by the industry and corporations in physical, electronic and paper form.[21] As its curator Murtaza Vali has noted, the *Crude* exhibition held at the Jameel Art Centre in Dubai in 2018–19, showing *inter alia* Al-ʿAni, Pezeshknia and Younis's work, was this kind of oil-driven creative enterprise. It looked back at the petroleum pasts of the Middle East as a conscious critical intervention into and retrieval of material, visual and documentary evidence to build a future oil archive.[22]

The combination of oil as a metonym and a medium – often a catalyst, causal agent and interposer of sorts – has greatly contributed to its con-distancing powers. The life cycles of oil shape its role in mediating society, culture and environment as oil's transformed state triggers dialectics that oscillate between engagement and alienation: intimacy but also detachment characterises relations with oil objects and substances that shape everyday life; oil's industrial infrastructure inspires opportunity but disengagement; the spectacle of oil's many environmental disasters instils awe and fear. And yet, its presence in everyday life, as many scholars have noted, is not only often taken for granted, but oil also often seems invisible to contemporary consumers – except perhaps when people pump gasoline into their cars.[23] Furthermore, plastic, the most widespread of petroleum's by-products, seems rarely recognised as linked to oil. Oil's relative 'absence' outside a relatively small number of labourers and an elite circle of technocrats, economists and investment managers feels particularly stark here as oil production and prices have determined so much of everyday life, from jobs and lifestyles to national wealth.

Of course, for those working within the oil industry – whether in the upstream or downstream sector, in exploration and extraction or in refining and distribution – oil has not been quite as invisible as it might be for consumers outside the industry. The oil industry, from executives and traders to engineers, explorers, rig labourers and gas station employees, certainly *see* oil, as do those who live near the industry's pumping stations or refineries. In this volume the 'guts' of the industry are exposed in different modalities: Al-Nakib's analysis of the spectacles and horrors of burning oil wells are spectacles of toxicity and environmental disaster; Ghosn's portraits of oil infrastructures speculate on oil's future lifestyles; Biglari's Abadan refinery comes into plain sight through the disciplining of Iranian trainees in the Abadan Technical Institute. However, as Timothy Mitchell has noted, oil presents itself as a standardised form, rather than recognised as 'engineered', as it flows into and through contemporary life as a necessary source of energy. Indeed, and as Fernando Coronil once pointed out, wealth itself has often appeared in petro-states even more miraculously, as though neither oil nor the labour that is required to transform it into energy and wealth is recognised as transfiguring nature for profit.[24]

Thus, as oil serves as metonym and mediates human lives, whether as a source of energy enabling speed, as contentious fantasies of wealth unequally distributed, as a catalyst for border formation and confrontation, or in the form of plastic as it has permeated human life, it is also itself framed by image worlds that render it unevenly visible. In addressing oil's 'forest of foci', this volume thus emphasises two specific aspects of its presence that we recognise as particularly salient: the forms and unevenness of its visibility in Part II, and its profound influence on shaping subjecthood in Part III.

Oil Images

The thematic overview of this introduction and Al-Nakib's chapter on Kuwait's oil fires (Part I, 'Exposing Oil') set out the two main themes running through Parts II and III of this volume: images and subjecthood. Al-Nakib's chapter on the fires, which dissects their spectacle through the triad of vision, mediation and representation, focuses both on the image worlds produced by the fires and on the questions of oil subjecthood they raise. Part II ('Image Worlds') highlights tension between oil's ubiquitousness and its seeming-invisible, or unevenly visible, nature by attending directly to oil images as they have been produced and circulated. In our focus on these images, we wish to draw attention to the visual forms through which oil has circulated and to ask how, when and for whom oil becomes visible. We pose these questions as they arise both within the oil industry and other corporate sources and on the part of artists and designers who have engaged with it. Thus, while we extend the work of mediation beyond technologies of *media*, we recognise too the particular power in images.

Public attention to fossil fuel and plastic contamination of ecosystems, including within the Middle East, has certainly been compounded by broadly circulating images. Images of plastic bottle seas are hard to ignore or forget, as are the flames and toxins from refineries in the background of Iraqi villages, recently (16 July 2020) reported by the *New York Times*. And yet, while these moments of public recognition have sometimes helped to draw attention to the pervasive presence of oil and its by-products or to alter consumption, they also often seem fleeting as they enter a complex ecology, similar to the processes examined by Karen Strassler in Indonesia, of reverberating social media with shifting relationships to authenticity and formed within particular histories and practices of image production and circulation.[25] In Part II of this volume, we wish to emphasise these histories and practices of image production by also situating oil's visual forms in the specific temporal horizons that they have shaped and that have shaped them: those of historical experience (Section 1: Visualising Petroleum Pasts) and of expectation (Section 2: Projecting Futures).

In Part II of this volume, we highlight a historical shift towards an increased use of petroleum images, mostly photographs but also films, in the 1950s and 1960s, particularly in areas with proven large oil reserves.[26] The first professional photographers and film makers entered the region on the payroll of business ventures, and eventually of oil companies, that prospected for mineral resources. Initially, the work of explorers like Arnold

Heim projected new frontiers, less through sight and more through smell, of 'crude in the wild', as Hindelang examines in this volume. By the 1950s, as Fuccaro explores, the manicured corporate still and moving images of workers were not only deployed to illustrate social happiness and civilisational achievement, but were also used as metonymic evidence that objectified the companies' benevolent values and world-views. Fuccaro's chapter traces, in particular, the emergence of the oil propaganda machine in Bahrain, Iraq, Kuwait and Saudi Arabia targeting increasingly dissatisfied local labour. Not surprisingly, a highly politicised workforce and the middle classes were among the first consumers of company magazines, films and newspapers in the Middle East after World War II. While deployed to pacify subversive instincts, they often rekindled antagonism, as much of the oil-related industrial unrest unfolding throughout the region in the 1950s and 1960s attests. Mason's chapter in this volume locates the emergence too of a global culture of graphs in a drive by policymakers and the corporate world to produce reassuring yet mystifying energy forecasts in the aftermath of the 1973 oil crisis. Indeed, as he argues, the graph itself – through its simultaneous transparency and opaqueness, its apparent accuracy and abstractness – has become a vehicle of specialised knowledge distinction and appreciation among oil executives while appealing to demands for public accountability and accounting.

Part II of this volume also draws attention to the work of image production both from *within* the oil industry and corporate world and *beyond* it, with a particular focus on the creative scene in the visual arts and design. Although, certainly, a sharp distinction between inside and outside the industry is untenable, especially in the circulation of images in the electronically mediated contemporary world, this distinction is helpful for outlining the vast diversity of oil images and their publics. Image production from within the oil industry, so this volume illustrates, has been oriented at different audiences: geologists exploring new fields (Hindelang), executives and shareholders analysing and projecting profit (Mason), publics whose opinions of the industry could be swayed to be sympathetic, or employees who – similarly – may be attracted by the various fantasies and potentials of working for an oil company (Fuccaro). By examining the corporate sources of their production, we attend, nevertheless, to the processes by which oil's transformations of daily life – as well as the tensions within it – have been naturalised and made commonsensical, mediating, in other words, our relations with oil as it is extracted and enters the global market and consumed.

As these authors make evident too, and as we wish to highlight in this volume, there are also significant differences in the forms of images as well, not to mention tensions within them: from the still and moving propaganda images from the 1950s and 1960s in Fuccaro's chapter, with their own dynamics of labour politics, nationalisation potentials or fears, and the production of a distorted suburban domesticity, to the geologist's photographic eye, which came to reaffirm Orientalist frontier fantasies of wild nature from which oil appears, as described by Hindelang. And finally, the image world of the industry itself, with a kaleidoscope of different image forms (from the graphs to the salon, and the artistic), provides yet another set of tensions. As Mason describes it, 'the energy image mediates between ignorance and knowledge in complex ways, sometimes propagandising while other times capturing, assembling and performing the complexity of the system's various forms. The energy image works recursively to diagnose the system, but also to depict the interactions that continually dissolve industry while demonstrating those coming into being.'[27] We get the sense, in all these cases, of the constant instability of the images, among forces that aim to stabilise them and their referents.

Image worlds produced outside the industry are also varied, both by form and by audience: film, design, sculpture, mixed media – and all with a keen appreciation of the politics and economics of the history (and prospects) of oil extraction and wealth in the region. This volume, therefore, recognises both the tension between oil's ubiquity and invisibility and the power of oil imagery in shaping the moments and modalities of its visibility. As Al-Nakib examines here, Werner Herzog's cinematic appropriation of the Kuwait disaster in his internationally acclaimed *Lessons of Darkness* presents oil as a sort of Orientalist currency that erases the local context and people. In its cinematic form, oil has a history of both obfuscating and sensationalising image-making. Bernardo Bertolucci's 1965–6 documentary *La Via del Petrolio* is an early example of petroleum's cinematic wizardry. Commissioned by the multinational Ente Nazionale Idrocarburi (ENI), it projects the image of

Zoroastrian fire onto Iranian oil in order to instil a sense of marvel and miracle-working in audiences.[28]

Indeed, for some artists and designers, oil motifs and themes have moved them and viewers to go beyond the aesthetic qualities of artistic production and to get 'to the crunch of things', as expressed by an oil art curator. Installations, sculptures and collages have prompted political, social and environmental reflections about the past and future, as Younis and Ghosn's works in this volume demonstrate. Particularly in the Gulf States, such artistic production has made visible a critique of the present condition of neoliberal consumerism defined by commodity landscapes and an aesthetics of luxury that is often not explicitly articulated.[29]

While there is no single common quality to oil images examined in this volume, especially as different images have different intended audiences and as practices of image production and circulation have shifted over the years, the tensions and oscillations, in particular, between closeness and distance that these images produce help to reinforce a sense of 'vertigo' with regard to the industry and commodity – a 'vertigo' that is not confined to intellectual engagement with oil.[30] Mason's chapter in this volume, for example, illustrates the allure of the oil/energy graph, its powers to foster abstraction and self-interpretation, creating an 'empathetic engagement' in viewers but at the same time detaching them from the factual energy messages they seek to convey. Fuccaro highlights the oil companies' self-projection of modernist fantasies through their public relations efforts, along with the power of images to alienate publics, while Hindelang emphasises the simultaneous ability of oil to focus and expose, all the while obscuring existing landscapes. Hindelang emphasises too the limits of visibility and the significance of the other sense, *smell*, in early geological exploration. The same goes for the work of artists and designers, who employ that vertigo in their work. In her collages, Younis draws together the enormity of historical events with details such as intimate gestures captured on postage stamps, or of machinery such as tractors. Also, Younis emphasises *sound* as a corollary of sight in oil exploration. Ghosn presents an overall structural and infrastructural architecture of oil's futures that emphasises an enormity of scale, while simultaneously emphasising the painstaking detail manifested in engineering.

Oil images, the contributions to this volume therefore highlight (whether as graphs, art or design; whether of gushers, homes, machinery, cars or geological terrain), seem to oscillate, navigating the magnitude of economies and subterranean flows and the finitude of detail. The artistic practice of *pointillism* becomes a useful metaphor for thinking of the ways images in the oil industry work, reflecting the tensions embodied in oil's con-distancing powers discussed above. Nevertheless, the images, and their producers, also point to their limits, and the significance of the senses, whether smell or sound, in the experience of oil.

Finally, it should be noted that our aim in focusing on these images is not simply to extend oil's visibility, as though such extension would necessarily expose the truth of its ubiquitous, uneven, or destructive or illicit presence. We are aware, then, that visibility does not in itself produce consciousness or recognition, though the politics of transparency and truth are often tied to photographic visibility, as Karen Strassler has also noted. And, as scholars of visuality have also recently argued, the proliferation and circulation of images via social media such as Instagram, Facebook or Twitter, where origins and authenticity (the hallmarks of photo-journalism) have been unmoored, also unsettles presumptions about a direct link between images and consciousness or recognition. Rather than staking claims or making presumptions about direct connections between visibility and recognition or visibility and power, the chapters in this volume more specifically detail how images within and beyond the oil industry have been deployed, how they have shaped understandings of the region, the commodity and the industry, and how they have projected particular sensibilities about oil's pasts and futures.

Oil Subjects

The image worlds of the arts and the petroleum industry have rendered oil's presence unevenly visible and detectable but unmistakably dynamic in its relationship to both creators and publics, underlining petroleum's action as mediator, medium and subject of artistic and technical workmanship. The chapters included in Part III of this volume pursue further Middle Eastern oil's often unacknowledged, invisible and ambiguous agency and associations, and challenge its state as a condition of passive being, its fetishisation as an *object* (particularly of geopolitical conflict) and its appropriation and representation as a metonym of war, rentier state, consumerism and authoritarianism. Identifying the four tropes of oil's metonymic quality as the 'resource curse' paradigm, the Malthusian struggle, hyper-consumption, and a visual world that aestheticises pathos, Appel, Mason and Watts call on us to recognise both the power of these discourses and the ways in which they naturalise our understandings of oil experiences.[31] If part of this experience is shaped by oil's objectification and translation, it is also defined by its swift permutations: from mineral to object, political pathology to imaginary, and figure of speech to spectacle. As Timothy Mitchell reminds us, these permutations are not as if 'the powers of carbon were transmitted unchanged', but entail the reconfiguration of oil's outreach and modus operandi, a shift in its power form.[32]

In order to recognise a broader gamut of Middle Eastern oil's transformations, provide a corrective to its narrow and seeming *object* status and complement Part II's discussion of the power of oil images to define historical experience and future expectations, Part III of this volume ('Oil Subjects') focuses on one of petroleum's demiurgic qualities, its ability to 'forge' new subjects through the experience of oil's changing material worlds. We understand subjecthood in two distinct yet interrelated ways. The first section of Part III (Ecologies and Geologies of Petroleum Knowledge) focuses on key areas of the geological, technical and microbiological science of petroleum, and on the relationship between oil as a scientific subject and the natural and human environments in which it breeds. We take knowledge literally as regimes of science producing expertise, but also as culturally and politically mediated 'truth' that generates social and political imaginaries, and public policy and performance. Oil's subject matters analysed in this section contrast and complement the matters of subjecthood discussed in the second section (New Spaces and Mobilities) that relate to sovereignty, citizenry, bondage and socio-cultural belonging. The chapters included in this section examine oil's significance in the transformation of individual and societies through the refashioning of a sense of space, mobility and territory, and the formation of political and social attachment and estrangement.

We argue that the two sections in Part III are intimately connected, as it is often difficult to disentangle actors, practices, and processes of knowledge production from national subject formation, border politics and spatial imaginaries. Rogers' and Biglari's chapters are a case in point. In the early 1990s, after petroleum microbiology fed the food security rhetoric of the Pahlavi regime and OAPEC, oil-for-food research programmes based in Kuwait turned to bioremediation to clean up the pollution of the post-invasion oil spills, metaphorically to heal the traumas of the nation (Rogers). Oil expertise among the young students at Abadan's Technical Institute coalesced into anti-colonial nationalism and functioned as a space of social reproduction of the patriarchal values of oil's global industry (Biglari). Rogers' and Biglari's chapters also remind us that oil technoscience is not omnipotent but often unpredictable, not always successfully performing the task of mastering nature and creating orderly subjects.

This volume's focus on Middle Eastern oil subjecthood brings us back to a Western tradition of political and philosophical thought that considers the constitution of fossil fuel subjectivities as unmistakably bound to that of modern capitalism. As one of fossil fuels' mass industrial enterprises originating in the USA, the Fordist revolution ushered in mass consumption and the privatisation of social space. In Matthew Huber's words, it made segments of the American workforce 'live, think, and feel an individuated sense of *power* over the geographies of everyday practices'.[33] As 'First World' capitalism defined by large-scale mass production and consumption, Fordism had little fortune in the Middle East, unlike in Europe. Yet the boom of the oil industry in Saudi Arabia and Gulf States in the second half of the twentieth century in large measure determined new patterns of regional consumption because of its ripple effect on other Arab countries via the migrant workers who brought money, lifestyles, taste and commodity desires back home. We recognise the importance of the growing material politics of consumption and luxury in the constitution of oil subjecthood, a trend that is particularly visible in artistic critiques of neoliberalism noted in the previous section. Yet in Part III of this volume, we wish to address Huber's experiential shift in the 'geographies of everyday practices' with reference to oil's power to 'move' things and modify spatial and geographical horizons of experience, creating visible and invisible outcomes and flows of people, technical knowledge and ideas.

Like the increased popularisation of petroleum images discussed in Part II of the volume, the development of the oil industry has enhanced circulation of all sorts, a key trope in normative narratives of economic growth and expansion. During the energy crisis of the 1970s, for instance, cultural theorists keen to dig deep into the fragmentation of capitalism started to ponder the spatial valence of speed. Yet while oil can be construed as having accelerated the matrix of modernity (and more recently, globalisation), it can be argued that it has also impeded movement because of the 'con-distancing' effect explained above and its natural state as a viscous substance.[34] In Part III of the volume, we wish to highlight how the binary of oil's velocity/viscosity has played out in the realm of mobility and automobility and in the formation of power differentials. Citino, for instance, understands the Saudi–American petro-encounter in Dhahran in terms of the replicas and resonances of US suburban norms of racial and gender segregation, and as a 'colonial' ordering of indigenous society reminiscent of the colour lines drawn by British and French regimes in Asia and Africa. His account of how the modern experience of speed and masculine norms of freedom and control of physical distance were reproduced among the local population also brings into sharp focus the reversed effect of the principles of individual and collective empowerment and emancipation that characterised the historical expansion of oil's American frontier. As Citino explains, the Saudis' lack of access to cars and airplanes, alongside their unfamiliarity with oil-fuelled travel, stood as a mark of their presumed backwardness. In 1950s Trucial Oman/United Arab Emirates, car travel revolutionised landscapes and soundscapes, making the nation-building experience tangible, visible and accessible to the whole population. Yet MacLean's chapter shows that this newly acquired mobility encroached upon the traditional movement of urban and rural populations, creating conflict and hindering socio-political cohesion. In Oman, the case of the Harasiis tribe discussed by Chatty in this volume is not dissimilar. Their social status and land rights waned as a central government became increasingly manifest following the expansion of oil drilling and the advent of motorised transport.

To say that oil is mobile, ubiquitous and pervasive is not to accept at face value the global reach of the industry and its ability to distance itself from local conditions, prompting, in the words of Vadén and Salminen, the 'death of locality'.[35] Part III's focus on oil subjects offers an opportunity to emphasise the analytical valence of oil as a global assemblage by tackling the question of petroleum subjecthood as Middle East-specific entanglements and liabilities with place. The gaze of Big Oil has tended to obfuscate the scalar intersection of locality with nation-state, region and globe, particularly in a diverse and extraordinarily oil-rich area like the Middle East where fossil fuels have been central to the constitution of particular geographies and temporalities of oil production. This has also been apparent in the contemporary USA. Bret Gustafsson, for instance, has argued that knowledge about fossil fuels produced by the industry, academia and think tanks has created particular geopolitical/territorial configurations giving rise to expressions of regionalism and nationalism.[36]

Perhaps not surprisingly, the space-times of Middle Eastern oil acquire special visibility in the volume through the flexible prism of nation (rather than that of state or petro-state), showing a propensity to make and unmake attachment to and conceptions of boundaries, territory, sovereignty and citizenship. The articulation of the Harasiis' tribal territory by the oil company as *terra nullius* described in Chatty's chapter on Oman precipitated the severance of ancestral ties with their land as the site of communal belonging. MacLean tells us that motor vehicle circulation – as a local, as opposed to a colonial, praxis of mobility – shaped the idea of the British-controlled modern state in Trucial Oman/United Arab Emirates as *al-Dawla*, an Arabic term that hints at transformation through movement. In her account of the mobilisation of Indian workers in the Gulf, Wright critiques the lens of the nation-state but ultimately shows how workers petitioned the young postcolonial Indian government against company abuse by appealing as Indian nationals to liberal notions of citizenship and human rights. In Wright's chapter, the prism of nation serves to highlight oil's transnational connections as she demonstrates how the global managerial practices of the industry in the Gulf contributed to the evolution of Indian nationalism outside independent India.

As in Biglari's account of the AIOC refinery in Abadan, in Wright's chapter the oil industry emerges as a scalar site that produces oil's subjectivities sandwiched in between local labour disputes, the colonial and postcolonial state and the global assemblage. At key historical junctures, oil infrastructure also appears unmistakably site-specific, particularly during violent and ambiguous processes of territorialisation and deterritorialisation of the industry. Al-Nakib in this volume shows that plans made by the Iraqi army to enact the 'maximum destruction' of the oil fields in case of withdrawal and those by the firefighters extinguishing the fires turned Kuwait's oil infrastructure upside down. Defying the industry global standard, firefighting teams, for instance, repurposed pipelines for the transport of water from the coast. In Abadan – as Biglari makes clear in his chapter – AIOC schemes to train Iranian oil experts witnessed a twist of fate on the eve of nationalisation. Supposed to shelter the refinery as a modular/global technological zone from nationalist and labour politics, these schemes ended up shaping it as a hub of Iranian resistance by fashioning notions of nationality, gender and class among students.

Uncovering petroleum subjecthood makes oil individuals, communities and societies visible and nameable, helping to address oil's prevalent objectification. In exposing the effects of the politics of petroleum knowledge and mobility we wish not only to personalise the Middle East's very 'personal' natural and industrial resource but also to grasp and fine-tune the significance of petroleum for past and present societies, and the significance of oil as a life world of political subjectivities, social imaginaries and individual action, and as a world of knowledge that should guide future academic and public intervention.

*****

In bringing together studies of oil's visual cultures and modes of subjecthood we hope that this volume will encourage more scholarly and public efforts to counter the reification of Middle Eastern oil as a dependent of and an appendix to the global, statist and petroleum order that has shaped the twentieth and twenty-first centuries. Without losing sight of the powerful national, transnational and global interests that continue to lubricate the worlds of oil and the many enduring myths surrounding the geopolitics of petroleum (what Robert Vitalis has called 'Oilcraft'[37]), the task at hand for us is to reorient our view away from Big Oil scenarios and embrace petroleum as a set of practices, experiences and representations that need disentangling in order for us to understand how this natural substance has penetrated the Middle East. Ironically, petroleum's in-your-face factuality, materiality, and quality of excess and grandeur – that is, its supposed visibility and knowledgeability – have precluded measuring its effects as a more hidden and subtle force that has moved and orchestrated the grand theatre of oil life.

Notes

1. Hannah Appel, Arthur Mason and Michael Watts (eds), *Subterranean Estates: Life Worlds of Oil and Gas* (Ithaca: Cornell University Press, 2015). On oil's 'critical topography' as defined by its overlapping and elusive circles of reach, see 17.
2. Terry Lynn Karl, *The Paradox of Plenty: Oil Booms and Petro-States* (Berkeley: University of California Press, 1997) and Stephen Haber and Victor Menaldo, 'Do Natural Resources Fuel Authoritarianism? A Reappraisal of the Resource Curse', *The American Political Science Review*, 105, no. 1 (2011), 1–26. The literature on the political economy of Middle Eastern oil is vast. For a reappraisal of oil resources and economic development see Ryan Kennedy and Lydia Tiede, 'Economic Development Assumptions and the Elusive Curse of Oil', *International Studies Quarterly* 57, no. 4 (2013), 760–71.
3. Stephanie LeMenager, *Living Oil: Petroleum Culture in the American Century* (New York: Oxford University Press, 2014), 17.
4. LeMenager, *Living Oil*, particularly Section II 'Writing Oil', 11–19. Imre Szeman, 'Literature and Energy Futures', *PMLA* 126, no. 2 (2011), 323–6. Tere Vadén and Antti Salminen, 'Ethics, Nafthism, and the Fossil Subject', *Relations* 6, no. 1, June 2018, 37.
5. A relatively recent influential study of Middle Eastern oil is Timothy Mitchell, *Carbon Democracy: Political Power in the Age of Oil* (London: Verso, 2011), which has inspired work on technopolitics and infrastructure such as Khatayun Shafiee, *Machineries of Oil: An Infrastructural History of BP in Iran* (Cambridge, MA: MIT Press, 2018) and Mattin Biglari, 'Refining Knowledge: Expertise, Labour and Everyday Life in the Iranian Oil Industry, c.1933–51' (Ph.D. diss., SOAS, University of London, 2020). In the last ten years or so, historians and anthropologists have broached the study of oil experience by focusing on oil cities, societies and natural environments. See Farah Al-Nakib, *Kuwait Transformed: A History of Oil and Urban Life* (Stanford: Stanford University Press, 2016); Nelida Fuccaro (ed.), 'Introduction: Histories of Oil and Urban Modernity in the Middle East', Special Issue in *Comparative Studies in South Asia, Africa and the Middle East* 33, no. 1 (2013), 1–6 and 'Arab Oil Towns as Petro-Histories', in Carola Hein (ed.), *Oil Spaces: Exploring the Global Petroleumscape* (New York: Routledge, 2022), 129–44; Toby Craig Jones, *Desert Kingdom: How Oil and Water Forged Saudi Arabia* (Cambridge, MA: Harvard University Press, 2010); Mandana Limbert, *In the Time of Oil: Piety, Memory, and Social Life in an Omani Town* (Stanford: Stanford University Press, 2010) and 'Reserves,

Secrecy, and the Science of Oil Prognostication in Southern Arabia', in Appel, Mason and Watts, *Subterranean Estates*, 340–53.

6. LeMenager, *Living Oil*, 13.
7. See Mitchell, *Carbon Democracy*, particularly 1–11 and Michael Watts, 'Righteous Oil? Human Rights, the Oil Complex and Corporate Social Responsibility', *Annual Review of Environment and Resources*, 30 (2005), 373–407.
8. See chapters by Fuccaro, Biglari and Wright in this volume. For a discussion of the invisibility of oil labour see Kaveh Ehsani, 'Disappearing the Workers: How Labor in the Oil Complex Has Been Made Invisible', in Touraj Atabaki, Elisabetta Bini, and Kaveh Ehsani (eds), *Working for Oil: Comparative Social Histories of Labor in the Global Oil Industry* (Cham: Palgrave Macmillan, 2018), 11–34.
9. Andrew Lakoff and Stephen J. Collier, 'Ethics and the Anthropology of Modern Reason', *Anthropological Theory* 4, no. 4 (2004), 419–34. Quotation from 427.
10. Douglas Rogers, 'Oil and Anthropology', *Annual Review of Anthropology* 44 (2015), 367–8.
11. Gökçe Günel, *Spaceship in the Desert: Energy, Climate Change, and Urban Design in Abu Dhabi* (London: Duke University Press, 2019).
12. Dubai Municipality, *Future of Dubai Gallery*, https://www.dubaiframe.ae/en/discover/discover-details?id=3 (last accessed 29 January 2022).
13. Limbert, *In the Time of Oil*.
14. David H. Finnie, *Desert Enterprise: The Middle East Oil Industry in Its Local Environment* (Cambridge, MA: Harvard University Press, 1958), 86.
15. LeMenager, *Living Oil*, 6.
16. An excellent ethnography that highlights the diversity of opportunities offered to local communities by encounters with the oil industry is Michael Cepek, *A Future for Amazonia: Randy Borman and Cofán Environmental Politics* (Austin: University of Texas Press, 2012).
17. In this regard oil creates a state of exception in the Schmidtian sense as operating in a vacuum not subject to rules of engagement. Tere Vadén and Antti Salminen, *Energy and Experience: An Essay in Nafthology* (Chicago: MCM, 2015), 27–9. Vadén and Salminen, 'Ethics, Nafthism, and the Fossil Subject', 38.
18. This has been also discussed in Appel, Mason and Watts, *Subterranean Estates*.
19. William Mazzarella, 'Culture, Globalization, Mediation', *Annual Review of Anthropology* 33 (2004), 345–67. For oil mediation see also Appel, Mason and Watts, *Subterranean Estates*.
20. Murtaza Vali, *A Crude History of Modernity* (Dubai: Art Jameel, 2018), 13–21.

21. For different takes on the contemporary oil archive see Andrew Barry, 'The Oil Archives', in Appel, Mason and Watts (eds), *Subterranean Estates*, 95–107, and LeMenager, *Living Oil*, 142–82. On the archaeology of petroleum sites as state-promoted heritage and memory see Rasmus Elling, 'From Palm Groves to Petroleum: The Oil City, Heritage and Memory' (paper presented at the workshop 'The Life Worlds of Middle Eastern Oil', NYU Abu Dhabi, April 2019).
22. Vali, *Crude*, Jameel Arts Centre, Dubai, 11 November 2018–30 March 2019.
23. See for instance Ross Barrett and Daniel Worden (eds), *Oil Culture* (Minneapolis: University of Minnesota Press, 2014).
24. Mitchell, *Carbon Democracy*, 5, and Fernando Coronil, *The Magical State. Nature, Money and Modernity in Venezuela* (Chicago: University of Chicago Press, 1997).
25. Karen Strassler, *Refracted Visions* (Durham, NC: Duke University Press, 2010) and *Demanding Images* (Durham, NC: Duke University Press, 2020).
26. On the intersection of filmmaking and oil extraction in the Arab World and Iran see Mona Damluji, 'Petroleum's Promise: The Neo-Colonial Imaginary of Oil Cities in the Modern Arabian Gulf' (Ph.D. diss., University of California at Berkeley, 2013).
27. Arthur Mason, 'Energy Image: Hydrocarbon Aesthetics of Progress and Form' (paper presented at workshop 'The Life Worlds of Middle Eastern Oil' NYU Abu Dhabi, April 2019).
28. Georgiana Banita, 'From Isfahan to Ingolstadt: Bernardo Bertolucci's "La Via del Petrolio" and the Global Culture of Neorealism', in Barrett and Worden (eds), *Oil Culture*, 145–68.
29. Murtaza Vali, 'A Plastic Petrotopia: Petromodernity and Neoliberalism in the United Arab Emirates' (paper presented at workshop 'The Life Worlds of Middle Eastern Oil', NYU Abu Dhabi, April 2019).
30. Appel, Mason and Watts, *Subterranean Estates*, 9.
31. Ibid., 10–16.
32. Mitchell, *Carbon Democracy*, 7.
33. Matthew T. Huber, *Lifeblood: Oil, Freedom, and the Forces of Capital* (Minneapolis: University of Minnesota Press, 2013), xiv.
34. Peter Hitchcock, 'Velocity and Viscosity', in Appel, Mason and Watts (eds), *Subterranean Estates*, 45–60.
35. Vadén and Salminen, 'Ethics, Nafthism, and the Fossil Subject', 36. Vadén and Salminen use 'death of locality', drawing on the work of Elmar Altaver, the German political theorists who in 2016 coined the term 'Capitalocene' as

opposed to 'Anthropocene' to stress the centrality of capitalism to the changing relationship between men and nature.

36. Bret Gustafsson, 'Fossil Knowledge Networks: Industry Strategy, Public Culture and the Challenge for Critical Research', in Owen J. Logan and John A. McNeish (eds), *Flammable Societies: Studies on the Socio-Economics of Oil and Gas* (London: Pluto Press, 2012), 315–16.
37. Robert Vitalis, *Oilcraft: The Myths of Scarcity and Security That Haunt U.S. Energy Policy* (Stanford: Stanford University Press, 2020).

2

'NO WORDS AROUND TO DESCRIBE': BETWEEN SEEING AND COMPREHENDING KUWAIT'S OIL FIRES

Farah Al-Nakib

On 17 January 1991, US-led coalition forces launched Operation Desert Storm to liberate Kuwait from Iraqi occupation. By mid-February, as a ground offensive became imminent, the Iraqi regime put into effect its 'Plan for Deferred Destruction' (*khittat al-takhrib al-mu'ajal*), which entailed detonating Kuwait's oil wells. When the war ended on 26 February, 735 of around 870 active wells were ablaze and burned 5 million barrels of oil per day for the next nine months. It was initially estimated that it could take five to ten years to extinguish the fires, which, if left alone, could burn for over a hundred years. The first team of American oil firefighters sent by the Kuwaiti government immediately after the liberation to assess the damage got the first glimpse of the enormous task that lay ahead. As Joe Bowden, Sr, founder of Wild Well Control, put it:

> I went back home, I tried to tell them what I'd seen on the trip, what I thought we'd be up against, what they would need to expect. And then I told them, there's no such words that I know that will explain what you're fixing to go into. And I couldn't describe it to them, what I'd seen. I could not. There's no words around to describe what we saw when we came here in March.[1]

One of Bowden's firefighters, George Hill, thought that Joe had 'lost his mind . . . his ability to converse – you know, his vocabulary'.[2]

Figure 2.1 Burning wellheads. Photo by Nicholas Kamm. Courtesy of Getty Images.

The incomprehensibility and indescribability of Kuwait's burning oil fields reflects the 'intellectual vertigo' that characterises the global oil and gas sector. Despite the ubiquity of oil in our everyday lives, 'the inner workings' and 'infrastructural guts' of the industry remain largely invisible to the vast majority of the world's population, and the full 'scale and reach' of the world oil sector is therefore 'impossible to fully grasp'.[3] Like the oil industry itself, the crisis of the oil fires at first glance appeared impossible to comprehend in its totality. Neither the world nor the industry had ever seen a man-made crisis of this magnitude. But while the dizzying scale of the catastrophic destruction seemed incomprehensible, like other oil-related crises the fires paradoxically made oil itself more visible and tangible than ever before. Kuwait's oil infrastructure lay outside the physical boundaries of the city's metropolitan zone in heavily restricted areas. Only those who worked in the petroleum sector had any regular exposure to the technological world of Kuwaiti oil, and few ever really saw the wells themselves before the whole country was exposed to them in 1991. Rather, oil in Kuwait had, since its discovery in 1938, largely existed in public imagination, scholarly discourse and artistic representation in its 'metonymic register': as stand-in for modernity, wealth, welfare, democracy, urbanisation.[4] But after February 1991, oil in Kuwait became hyper-visible 'as itself' – as both technological infrastructure and material substance.[5]

Despite the fact that there seemed 'no words around to describe' the magnitude of Kuwait's oil disaster in March 1991, the Iraqi invasion simplified and reduced Kuwait's oil *industry* (part of the global sector's 'massive, sprawling, high-tech engineering and financial infrastructure, presided over by some of the largest corporate and state-owned enterprises in the world') to some basic, quantifiable, discernible facts, rendering it much more contained.[6] During the occupation, oil production was reduced from 2 million barrels per day to 250,000 barrels, just enough to maintain the country's power and water desalination plants without which nobody, including the occupying army, could survive. The KOC staff were reduced from 5,000 to only fifty individuals who worked through the occupation. Kuwait's oil industry, in other words, was pared down to its bare bones; the vast majority of its wells, pipelines, refineries and ports were taken offline. The KOC headquarters were abandoned, its entire archive (in microfilm) secretly removed from the head office and stashed under the floorboards of KOC engineer Sara Akbar's bedroom closet.[7]

In his analysis of global technological zones – zones of common measurements, connections, infrastructures, standards and practices that cut across specific national borders – Andrew Barry argues that 'although the territorial location of oil appears fixed, the inclusion of the oil industry within various zones of measurement and qualification is certainly not'.[8] Indeed, despite oil's physical location under the ground of a particular territory – the very presence of which catalysed the British Empire's desire in the 1920s to carve up the Gulf region into clearly bounded territories with which Western oil companies could sign concessions – the oil industry has been transnational, global and deterritorialised since its inception. Since the first tanker departed Kuwait's port in 1946, the country has exported more than 70 per cent of its crude oil production. However, with the Iraqi occupation and the reduction of Kuwait's oil production for immediate and local consumption, Kuwait's oil industry took a hiatus from the various technological zones that constitute the global oil assemblage. For over a year between Iraq's invasion in August 1990 and the extinguishing of the last oil well in November 1991, Kuwait's oil industry was highly territorialised. New localised measurements, guidelines and practices were developed first by the Iraqis to destroy the oil fields and then by the international firefighters to extinguish the fires: technologies that applied solely and uniquely to Kuwait's oil infrastructure under siege. Indeed, during the occupation the country's vast, vertiginous, complex oil industry was reduced to its most basic material fact: its (idle) oil wells, the fixed number of sites where oil is extracted from the ground.

After February 1991, oil was also made visible in its own toxicity. To those looking from afar – an airplane window or a television screen on the other side of the world – the fires paradoxically made oil invisibly visible. Though it was now easy to see precisely where in the desert each oil well was located, the black substance itself was burnt off into smoke, and the magnitude of the flames hid the wellhead from all but those who got up close and personal with the inferno. In addition to the firefighters, one such person was the famous Brazilian documentary photographer Sebastião Salgado, who travelled to Kuwait in 1991 to document the 'apocalyptic spectacle' with his camera lens.[9] His photographs capture the reality of the fires (and the process of extinguishing them) in harrowing and intimate detail, with vivid imagery of black oil gushing from burning wellheads, soaking the desert landscape, and covering the bodies of men, birds and horses. The fires also made oil hyper-visible to the people of Kuwait, who were exposed to the dark smoke that covered the skies and to the smell of petroleum in the air. Whereas before 1991 oil oozed through one's life in less explicit forms – asphalt, plastic, drugs – and permeated every aspect of Kuwaitis' everyday lives from the food they ate to the cars they drove,[10] now oil oozed through their lives in the most literal of ways: over their skin, into their nasal passages and lungs, down the walls of their white villas, through their soil in acid rain. Kuwaitis were literally touching their oil, and it was touching them, for the very first time.

But visibility does not automatically render clarity or equate with unambiguous certainty. Despite the unprecedented hyper-visibility of Kuwait's oil from February to November 1991, and even though Kuwait's oil industry was stripped down and simplified while under siege, the dizzying scale of the oil fires themselves rendered them incomprehensible to those who could see them. Just as the smoke and flames obscured the visibility of the wellheads, the magnitude of destruction made it difficult even for those whose task it was to extinguish the fires to fully make sense of what was being seen (as with Bowden). For all their experience fighting oil fires, the renowned international teams like Wild Well Control could only take on the inconceivable challenge of bringing the fires under control by understanding the behaviour of each individual burning wellhead, knowledge that they received on the spot from a handful of local KOC engineers. Filmmakers like Werner Herzog and Monira al-Qadiri, meanwhile, have used their cinematographic talents to provide us with their own alternative ways of not only seeing the oil fires, as Salgado did, but of trying to understand and analyse them: the former as critical commentary on our global obsession with oil and the latter as personal reflection on a national tragedy. Recently, the archive of the Ba'ath Arab Socialist Party of Iraq has emerged as a new way of seeing and interpreting the oil fires; though at first it is difficult for the researcher to identify the fires in such a totally different form as ink on paper, the archive has, somewhat paradoxically, contributed to making Kuwait's oil disaster hyper-comprehensible by reducing the fires to the banal world of bureaucratic efficiency. As this chapter examines, each of these case studies – the firefighters' lived experiences in 1991, films that depict vivid imagery of (and critical commentary on) the oil fires, and the Iraqi archives – reveals a push and pull between visibility and incomprehensibility exemplified by Bowden's experiences of not being able to explain or describe what he clearly saw when he came to Kuwait in March 1991. As these case studies also reveal, despite the fact that the Iraqi occupation of Kuwait served to re-territorialise Kuwait's oil by taking it offline from the transnational global industry, depictions of the fires and the herculean task of extinguishing them tend to deterritorialise the disaster in various ways. For instance, Salgado and Herzog's famous works transform the fires into an extreme and dramatic aesthetic event to be captured on film for international audiences in haunting images devoid of all space–time context, making salient Graeme Macdonald's claim (as quoted by Michael Watts) that 'oil representations, like oil itself, have "significant global routes"'.[11] In such visual imagery, the oil fires become an apocalyptic global phenomenon rather than a distinctly Kuwaiti environmental, economic and public health crisis. The global notoriety and fame of Western experts like Red Adair, who are credited the most – even in Salgado's recent book – with extinguishing the fires in what is always described as an enormous international effort, also erases the critical role played by Kuwaitis themselves in the process. Kuwaiti voices like those of Sara Akbar and Monira al-Qadiri, amplified in this paper, as well as the perspectives of the Iraqi regime itself, as revealed in the archive, serve to re-territorialise the fires back to their specific local context: the oil fields of Kuwait in 1991.

Firefighting on a One-well Basis

Although most Kuwaitis did not 'see' their oil until exposed to the toxicity of the fires in 1991, Sara Akbar was one of few who knew the oil wells intimately before the invasion. In her own words: 'I was born in an oil field. My childhood was around oil fields. And my whole life I spent in oil fields.'[12] Akbar was born in 1958 in Magwa, a village close to the oil town of Ahmadi where her father worked as a KOC driller. Throughout her childhood she watched the burn-off gas flares of the nearby oil fields and snuck under the fence of the KOC refinery to play among the wells. In 1976 Akbar entered Kuwait University's first cohort of the new chemical engineering programme, after which she joined KOC. At age twenty-five she became the petroleum department's first and only female engineer working in the oil fields, tasked with maintaining more than seven hundred of Kuwait's active wells and managing the drilling of new ones. Each well had its own character, and Akbar likens her job to that of a doctor caring for patients: she measured each well's pressure, tested its fluids like blood, took its temperature, gauged its electrical signals. Whenever there was a problem, she devised the treatment plan. To Akbar, Kuwait's oil had always been extremely site-specific and visible, though it remained largely invisible to most of her fellow Kuwaitis before 1991.

Akbar was one of only fifty members of KOC who continued to work through the seven-month occupation, negotiating and co-ordinating with the Iraqi petroleum engineers sent to take over Kuwait's oil operations. The KOC team's primary task was to maintain oil production at one eighth of the country's pre-invasion output 'to ensure the livability in Kuwait' [13] by powering the water and electricity plants; the remaining oil wells had to be safely shut down. The KOC team knew that the Iraqis were developing a plan to destroy Kuwait's oil industry in the event of a forced retreat, and so Akbar and her colleagues also devised a plan to save the oil wells and to rapidly restart production after the liberation. But they never imagined or planned for the fact that Saddam Hussein would detonate practically all of Kuwait's oil wells.

The Iraqi army started exploding the wells in mid-February, a month after the US-led air strikes began and just days before the ground offensive that ultimately drove the occupying forces from Kuwait. When the first wells were detonated, Akbar and her mother drove out to the Magwa oil fields where she had played as a child. She felt a visceral reaction to what she saw: 'These wells were like my friends. I felt horrible. It was like killing my friends.'[14]

The Kuwaiti government exiled in Saudi Arabia also worked on a comprehensive plan for Kuwait's post-occupation reconstruction. They hired OGS Associates of Houston to run the operation to extinguish the oil fires under the leadership of 35-year-old Larry Flak, who enlisted four veteran oil firefighting firms: Red Adair Co., Boots & Coots and Wild Well Control, all from Texas, and Safety Boss of Calgary, Canada. Virtually all of the men of these companies 'grew up near oil fields, worked as roughnecks during high school and graduated (or didn't) to take a job in the Texas oil industry', working their way to 'the top of the heap' of oil firefighting.[15] Twenty-eight teams from around the world joined the massive co-operative effort to extinguish Kuwait's oil wells, with 10,000 people from forty countries (including China, Iran, Romania, Hungary, Britain, Russia, Canada and the USA) and thousands of tons of machinery contributing to the largest non-military mobilisation in history. Only a couple of hundred of those involved were actually specialists in oil field firefighting. Despite the renowned expertise of men like Red Adair (played by John Wayne in the 1968 movie *Hellfighters*), 'Boots' Hansen and 'Coots' Matthews, who between them extinguished one third of Kuwait's fires, much of the firefighters' knowledge had to be learned and improvised on site. For instance, they had to figure out how to dispose of tens of thousands of landmines planted by the Iraqis and undetonated coalition bombs hidden beneath the soft sand around and between the oil wells. Roads had to be constructed over vast oil slicks so the firefighters and their machinery could reach each well, creating a whole new infrastructure in the desert designed specifically for the extinguishing of the fires. Kuwait's existing oil infrastructure was adapted for the same purpose: the pipelines that normally pumped oil from each well to the coast were reversed, pumping water from the sea to the oil wells. The constant stream of water not only helped extinguish the fires but also kept the firefighters from collapsing in the heat and their machinery from melting.

Figure 2.2 Battling a burning wellhead. © Sebastião Salgado.

But even for the most experienced oil firefighter, the magnitude of the crisis in Kuwait was 'still hard to comprehend'. As a firefighter from Wild Well Control said, 'you can't look at the destruction and the devastation and the senseless ruination of the country. You have to look at it on a one-well basis.'[16] Understanding one well at a time and bringing it under control was the only way to make the unprecedented social, economic and environmental crisis comprehensible, and the overwhelming task at hand manageable. Each of the 735 burning wellheads had to be tackled differently, as each well and therefore each 'fire [had] its own personality and [required] its own approach'.[17] Lee Hockstader reported in April 1991 that '[e]very fire, every shattered wellhead is a new problem awaiting a makeshift solution fashioned with the benefit of common sense and engineering'. As Boots Hansen claimed, 'You just got to know what to do and get the job done.'[18] But regardless of the vast understanding these Texas 'roughnecks' – who were earning from $1,000 to $5,000 a day in Kuwait – had from decades of oil field firefighting, their ability to get the job done depended on detailed knowledge of the unique characteristics of each well in order to predict how it would behave during the extinguishing process. For this, they relied on

Figure 2.3 Oil gushing out of a wellhead and soaking the firefighters. © Sebastião Salgado.

now 32-year-old Akbar's intimate and diagnostic knowledge of every single oil well, her 'friends'.

According to Akbar, every evening at six o'clock, she held a meeting with the firefighters to answer their questions about the wells they hoped to cap the following day. With this information the firefighters came up with diverse, and often creative, strategies to extinguish the fires. For instance, one method entailed pumping heavy mud into the throat of the well to choke off the oil supply; another used dynamite to explode the wellhead to cut off oxygen to the fire for a split second, during which water could extinguish the flame. The Hungarian team famously recycled an old Soviet T34 tank for the job; they took off the gun turrets, replaced them with jet engines from a MiG21 fighter plane, injected water into the jet stream, opened the throttles and blew the fire out. Regardless of the method used, once each fire was safely extinguished the well had to be capped to stop the gushing of the oil (which soaked the bodies of the firefighters), and then shut off.[19] After observing the international crews at work, the Kuwaitis assembled their own firefighting team, of which Akbar was second in command. Without any prior firefighting experience, the KOC team successfully extinguished forty-two fires, including the last well in a public celebration on 6 November 1991.

Although Akbar became famous in Kuwait as the only woman to participate in the firefighting operation, and although the Kuwaiti team received much publicity during the final well-capping event, the critical mediating role that Akbar and her colleagues played in the entire process remains largely unknown and unacknowledged. In his book of oil fire photographs published in 2016, Salgado, who drove 'from one burning or gushing well to another' with his cameras in 1991, credits 'some 300 expatriate experts' with the task of not only extinguishing the fires but also rescuing Kuwait's 'battered economy' and helping to stabilise the world oil market.[20] Indeed, the speed with which the job was completed, years ahead of schedule, is most often attributed to the 'intense, if good-natured, competition among the 16 international companies on the scene', while the 'salty Texans' like Red Adair, Boots and Coots and the men of Wild Well Control remain the permanent 'heroes' that the narrative of Kuwait's oil fires 'celebrates most vividly'.[21] Though Salgado mentions that '[the] problem of resuming production was left to other teams working many months later', he makes no mention of who those teams were.[22] It is Akbar's own voice (captured in an oral history interview conducted for this piece) that returns locality and specificity to the narrative of Kuwait's oil fire disaster. 'I don't know if the world actually recognizes the amount of skills in engineering, execution, construction . . . and management, that was put to work in order to [put out the fires and] bring production back' – that is, the skills that the KOC team brought to the entire process of not only fighting the fires but also bringing Kuwait's oil industry back online. By 1993, KOC's petroleum department, of which Akbar was now head, succeeded in bringing Kuwait's production back to 2 million barrels per day.[23]

Through the Cinematic Lens

For the vast majority of people both in Kuwait and around the world, the story of Kuwait's oil fires has largely been told through the mediated lenses of photographers and filmmakers who shot the blazing wells in 1991. In her excellent work on oil company public relations films in Iran and Iraq, Mona Damluji argues that most documentaries on the region's oil industry have tended to 'equate the story of oil with the experience of modernity'.[24] This was certainly true of cinematic representations of Kuwait's oil before the invasion, such as the 1961 Kuwait Oil Company film *Close-Up on Kuwait*, directed by Rodney Giesler, which spends only one minute discussing Kuwait's oil industry against footage of its new pipelines, refineries and tanker ports. The remaining twenty-five minutes of the film focus on the country's advancements in architecture, communications, water desalination, healthcare, education, agriculture, political participation, and so on.[25] This and similar films produced by both KOC and other corporate entities in the decades before the Iraqi invasion narrated Kuwait's story of petroleum by '[leaving] crude oil out of the picture altogether'.[26] When the oil industry does make an appearance in these films, it does so 'as nation-building project . . . [tying] the project of oil to the promise of modernity and social progress'.[27]

The films produced on the oil fires in the early 1990s, by contrast, give petroleum centre stage. The fires generated their own genre of petro-imagery in which oil appears quite jarringly as itself: as burning wellheads, as massive oil lakes, as burned and abandoned infrastructure, as thick black smoke, as oil-slicked wildlife, and as the heroic firefighting men (and one woman) who battled the wells for nine months. This genre includes well-known works such as Sebastião Salgado's iconic photography of the fires and oil-soaked firefighters compiled in his recently published Taschen book *Kuwait: A Desert on Fire*,[28] David Douglas's 1992 Academy Award Nominated IMAX film *Fires of Kuwait*[29] and Werner Herzog's 1995 film *Lessons of Darkness*;[30] less known is Kuwaiti artist Monira al-Qadiri's 2013 film *Behind the Sun*.[31]

As described by film critic Michael Wilmington in his 1993 *Los Angeles Times* review of Douglas's *Fires of Kuwait*, such visual works 'give us something we really *have* never seen before, never will again. They show us a cataclysm, vast and hideous, that beggars most nightmares, reduces most other catastrophes to small change. And a battle against it, equally vast, strange and marvelous.'[32] They show us something else we have rarely seen before: oil, as itself, in such a hyper-visible state. *Fires of Kuwait* is the most context-specific of the aforementioned films in that it provides some background on the war, explains the complex firefighting process, gives voice and relevance to the Kuwaiti team and assesses the lasting impact on Kuwait – all the while capturing stunning cinematographic imagery of the fires, the oil, the machinery and the firefighters. I would like to focus here, however, on Herzog's film and al-Qadiri's response to it, to examine the extent to which his work, while making *oil* visible on film, deals with the question of the *oil fires*' incomprehensibility and incommunicability.

In *Lessons of Darkness*, Herzog eschews all attempts at making the oil fires factual, evidentiary or understandable and embraces instead the poetic, the opaque, the incomprehensible (the film is, indeed, a lesson of darkness over clarity). As described by Imre Szeman, 'Deliberately pushing against the indexical qualities of documentary cinema, [Herzog] has repeatedly intervened in the construction of purportedly "real" films by staging scenes or inventing story elements to enhance dramatic narrative . . . and has made use of documentary and found footage to construct fictions out of documented realities.'[33] Indeed, *Lessons of Darkness* is much more science fiction than informative documentary about the oil fires, one that deterritorialises the oil fires in every way. It is told from the perspective of an alien creature that has landed on Earth, or an earthling that has landed on some other planet. The first creature the narrator encounters 'tried to communicate something to us',[34] and we see a firefighter with his face completely covered gesturing at the camera without speaking. Right away we are met, in fictional style, with the incommunicability of the scene around us – just as we are at the start of *Fires of Kuwait* when Bowden says that 'there's no words around to describe what we saw'.

Oil is hyper-visible (as itself) in this film, with endless aerial shots not only of the fires but also of the oil-stained desert, oil lakes, oil shooting into the air and raining down on firefighters; with visual traces of a rusted and melted oil infrastructure; and, in a chapter titled 'Dinosaur on the Way', with bulldozers evoking the popular myth that dinosaurs put the oil there to begin with. Yet for the most part oil is entirely decontextualised, no longer site-specific; although the film opens with shots of Kuwait City, we never actually know where we are. Rather than saying something about Kuwait's oil crisis, Herzog seems more intent on saying something about our global dependence on oil. In Szeman's reading, 'Herzog takes oil as the name for a complex problem which requires formally innovative methods of exploration if one is to do more than produce an already known object lesson about fuel consumption and the evils of SUVs'.[35] In one of the final scenes oil gushes out from the ground and two firefighters throw a flame and reignite it. The alien voice once again speaks: 'Two figures approach an oil well and set it ablaze again . . . Has life without fire become unbearable for them? Others, seized by madness, follow suit. Now they are content, now there is something to extinguish again.'[36] The message is clear: these creatures are so obsessed with oil that, 'even given its evident destruction of nature and culture, they would not know what to be or how to live' without it.[37] Yet here, oil is represented not as itself, but as fire. The creatures are obsessed with fire; if oil were appearing as itself, setting the well ablaze would destroy the very substance Herzog's creatures supposedly cannot live without. Elsewhere in the film, '[t] he oil is trying to disguise itself as water', as massive lakes.[38] Oil is thus both hyper-visible and incognito: appearing as itself on screen, yet, in Herzog's voice, obscured into something else and rendered into metonym once again (destroyer of nature and culture).

Lessons of Darkness is certainly more science fiction than it is a film about Kuwait's experiences of the oil fires. But, as Szeman also detects, Herzog seems uncertain how to approach his topic and to best make use and sense of the footage he collected in Kuwait. There are moments when Herzog slips out of character as the inquiring and bewildered alien by providing some prior knowledge of what he observes; he knows, for instance, that there was a war. In these moments, science fiction gives way to documentary. Herzog perhaps realises that he needs to provide some context for his viewers to understand and appreciate where he was – the remarkable history and tragedy he captured on film. But in those moments, history is disfigured. In the voiceover during the aerial shots of Kuwait City in the second scene, Herzog, as narrator, says, 'Something is looming over this city, this city that will soon be laid waste by war. Now it is still alive, biding its time; nobody has yet begun to suspect the impending doom.'[39] The scene then cuts to the infamous news imagery of the January 1991 bombardment of Baghdad, with dark skies and bombs lit up in green night-vision, during which he says, 'The war lasted a few hours. After that, everything was different.'[40] We are led to believe that the bombing is of the unsuspecting city we flew over moments before (that is, Kuwait City rather than Baghdad). By means of the statement that the war lasted only a few hours, the history of the seven-month occupation that preceded the actual war to liberate Kuwait (which itself lasted over a month) is erased.

Again, Herzog is known to intervene in the construction of reality in his films by staging scenes and inventing or enhancing the narrative for dramatic effect. He clearly 'betrays an uncertainty about how to explore' the topic of the oil fires in this film: either as critique of our global dependence on oil in which he takes much creative licence, or as documentation of the traumas of Kuwait's war.[41] Twice in the film, Herzog seemingly breaks away from the science fiction theme and enters into the world of documentary. In a chapter titled 'Finds from Torture Chambers', the narrator tells us that a woman who had been forced to watch her sons being tortured to death 'wanted to tell us something' on camera. But then, we find that she cannot speak; her trauma has rendered her unable to communicate her experience. 'But she still tries to tell us what happened', as we painfully watch. The only clear phrase that emerges is 'Salah mako' ('Salah is no more'), which Herzog does not translate. A few scenes later, in a chapter titled 'Childhood', we meet another Kuwaiti woman carrying her young son. She describes how her son's tears, nasal mucus and spit come out black from smoke. Herzog's narration then drowns out her voice as he tells the rest of her story: that a soldier crushed her son's head with his boot and then shot her husband. The boy then allegedly said, 'Mama I don't ever want to learn how to talk', and has never spoken since – another Kuwaiti who has lost the ability to communicate. Herzog's English translation does not match up with what the mother can faintly be heard saying in Arabic, but we are meant to hear (and believe) what the narrator is saying for her.[42]

Herzog may employ this diagetic approach to make his ambivalence over how to approach his subject – his own 'failure of communication' – appear not 'as an aesthetic limit or failure of his approach, but as the very problem which he hopes to foreground in the film'.[43] However, the discomfort here is that the Kuwaitis interviewed in the film are real people, trying to communicate their real trauma. The only voice that comes through, besides Herzog's, is that of a Kuwaiti translator mediating between his fellow citizen and the German filmmaker. The silencing of Kuwaitis by finding voices that are mute or by talking over the voices that speak decontextualises this film about oil. The film also silences the firefighters, who are depicted as aliens who communicate with each other only by hand gestures (which was in fact due to the deafening sound of spewing oil, gushing water and roaring fires). This silencing does not come across as the same incommunicability of the destruction that Bowden depicted; Herzog himself is not silenced by the horror and magnitude of what he sees.

After the scene of the (fictitious) bombardment of the city, the camera flies over a barren desert with circular tracks in the sand, and the narrator says, 'All we could find were traces that human beings actually lived here. Had there ever been a city? The battle had raged so ferociously that afterwards grass would never grow here again.'[44] *Lessons of Darkness* thus evacuates Kuwait of its people and its nature, creating a blank canvas upon which Herzog can address our global problem of oil. Two additional creative approaches further contribute to the film's deterritorialisation of Kuwait's oil fires. One is the narration of passages from the Book of Revelation, relating the scene of the fires to the bottomless pit of hell. The second is Herzog's choice of music: selected scores from Grieg, Mahler, Prokofiev, Wagner and other European composers that are 'cemented into social consciousness' – *Western* social consciousness – 'through their use in popular cinema to gesture to sublime experiences'.[45] In the aerial shots of Kuwait City at the start of the film we hear a call to prayer which, in archaic Orientalist fashion, gives the only indication of where in the world we might be. But as the tension mounts as the narrator discloses the city's impending doom, the *athan* is suddenly drowned out by Edvard Grieg's score to Henrik Ibsen's play *Peer Gynt*, yanking us out of Kuwait and into that other oil state: Norway. As Iranians once complained about AIOC's depiction of their oil industry in early twentieth-century petrofilms, *Lessons of Darkness* presents Kuwait and its people 'as little more than an exotic oriental backdrop' to Herzog's own cinematic genius and (valid) commentary on our global oil obsession. The evacuation of local context 'effectively renders invisible' the deterioration of the Kuwaiti population's health and the destruction of the natural environment caused by the oil fires.[46]

The absence of place–time context in *Lessons of Darkness* can be unsettling for Kuwaitis, for whom both the war and the fires were very real. The Kuwaiti artist Monira al-Qadiri, for instance, describes herself as having a 'love/hate relationship' with the film. When she first watched *Lessons of Darkness* as a young girl, she could not understand why 'this German man [was] making up stories about our war. As if I owned it somehow.' She had assumed it was a regular documentary, and found it 'disturbing, a film full of lies'.[47] With time, however, she appreciated that an artist could transform real political events into something otherworldly and give them new meaning. In 2013, while living in Beirut with war raging in Syria, al-Qadiri was prompted to reconsider her own memories of the Kuwait war, as an adult and as an artist. As a rejoinder to Herzog, she produced a ten-minute film entitled *Behind the Sun*. While just as abstract and otherworldly as, and no more informative or explanatory than, *Lessons of Darkness*, the film re-territorialises (and thus 'reclaims', as al-Qadiri puts it) the oil fires in multiple ways.[48] The footage was shot by Kuwaiti photographer Adel al-Yousifi, who in the months after the liberation recorded amateur home videos of the oil fires. Rather than adopting Herzog's top-down 'God's perspective', captured from a helicopter gliding smoothly overhead, al-Yousifi films from his car and we feel the rugged terrain of the desert rushing beneath us as his camera shakes from the bumps. For audio, al-Qadiri went into the archives of the state-run Kuwait Television (KTV) station and retrieved audio monologues from 1980s Islamic television programmes that overlaid recitations of Sufi poetry in the deep voice of a skilled orator onto hypnotising cinematographic scenes of nature – waterfalls, mountains, wild animals, volcanos – 'geared towards visualizing god through natural miracles'.[49] While mimicking these old programmes by overlaying their familiar narrations onto al-Yousifi's captivating imagery of the oil fires, al-Qadiri here is responding to what she describes as Herzog's 'out of place' passages from the Book of Revelation not just by replacing biblical scriptures with Islamic ones, but by grounding the oil fires in a specific time and place. Anyone who lived in Kuwait in the 1980s will recognise that deep voice slowly reciting verses over the mesmerising imagery of molten lava, arguably more so than they might Herzog's 'triumphant Wagner'.[50]

Seeing in the Archive

Al-Qadiri's artistic intervention constituted an act of archival retrieval. Unlike Douglas, Herzog, Salgado or al-Yousifi, she had not been in the burning oil fields in 1991 to produce her own imagery, and her ability to tell the story of Kuwait's oil fires her way had to be mediated by the documentation of others. Archival materials of the oil fires mostly record the history of the wells after they were detonated: from film and photographic footage to newspaper articles, and from technical documents on the enormous firefighting effort to environmental and medical data gathered on the effects of the fires. Historical knowledge of the life of Kuwait's oil wells after August 1990 but before they were detonated in February 1991 has, until recently, been more opaque. Initially, this knowledge was disseminated by second- (and third-) hand testimony: in an April 1991 *Washington Post* article, for instance, Larry Flak, the Houston engineer co-ordinating the firefighting effort with KOC, recounted what he was able to piece together about the 'method and chronology of the destruction from physical evidence left strewn in the desert, interviews with Kuwaiti resistance members, intelligence reports and reams of documents and notebooks left behind by the Iraqis when they fled in the face of the United States-led advance'. This included notebooks found in an office in Ahmadi used by the Iraqi petroleum engineer in charge of blowing up the fields, now used by Flak. Flak found that the Iraqis began wiring the wells in the first weeks of occupation and described 'with a mixture of disgust and something approaching professional admiration' the intricate command system the occupiers developed 'to make sure their job got done right'.[51]

It is unclear what happened to the documents Flak used to piece together his knowledge of the Iraqis' plans, but the documents described by the *Washington Post* are similar to those published by the Center for Research and Studies on Kuwait (CRSK) in a 1998 book analysing the Iraqi regime's most serious crimes against Kuwait as violations of international law, with evidence 'drawn from the actual Iraqi documents left behind by the Iraqi forces when they fled Kuwait after their defeat by the Coalition Forces in the war to liberate Kuwait'. The CRSK – which the Kuwaiti government established after the liberation with the task of 'filing, classifying and analyzing information and documents on Iraq's aggression on Kuwait' – does not make such documents available in their public archives and only includes descriptions or verbatim reproductions (not scanned copies) of their contents in such publications.[52]

It is only recently that scholars have been able to directly see, read and interpret documents detailing the Iraqi occupation of Kuwait (including the oil fires). In the early 2000s, the US Department of Defense handed over to the Iraqi Memory Foundation 800,000 pages of Iraqi military and political documents gathered by the Coalition forces after the liberation of Kuwait in 1991. Since 2015, these documents have been made accessible to researchers as part of the Ba'ath Arab Socialist Party of Iraq collection housed at Stanford University's Hoover Institution. Referred to as the 'Kuwait Dataset' (KDS), they '[provide] in harrowing detail a view of the treatment of the civilian population as well as the conduct of war'.[53] The documents in the KDS that relate to the Plan for Deferred Destruction confirm what Flak had seen in the documents left behind in Ahmadi as reported by the *Washington Post*, and that the CRSK included in its 1998 publication on Iraqi war crimes. It is possible that all these sources refer to the same set of documents that now reside at Hoover, or that duplicates ended up in different places. Their accessibility in the archive means that historians can now see these documents directly, without their content being filtered through the 'disgust' and 'professional admiration' of men on the spot like Flak whose job it was to undo the destruction planned out in such documents, or newspaper articles from the time reporting this information third-hand largely to Western audiences, or CRSK reports prepared for the purpose of establishing a legal case against Iraqi war crimes. The oil fires can therefore now be seen in a new light.

For instance, we see that after being removed from the complex technological zones that constitute the global oil industry, the infrastructural guts of Kuwait's oil became highly localised and territorialised. New kinds of measurements, guidelines and practices had to be adopted by the Iraqi regime to enact maximum destruction of the oil fields if and when necessary. The KDS contains multiple copies of 'how to' instructions issued by the Iraqi 8th Field Engineering Battalion in November 1990 to all units tasked with arming and eventually exploding the oil wells.[54] The Iraqis set up a three-tiered command structure, at the top of which was the commander responsible for planning and preparing targets, communicating instructions and determining the right time and conditions to detonate. Next was the team leader of the Republican Team of Destruction Guards, who was in charge of all demolition teams and ensured that all target sites were properly destroyed. Finally, the leader of each demolition team was tasked with actually carrying out the detonation order. The demolition team was also assigned a security team that was 'responsible for keeping any enemy forces from interfering with the demolition'. The instructions on how the wells should be set up, charged and detonated were very intricate, and included eighteen 'general recommendations'. For instance, no fewer than thirty pounds of explosives were to be used per well, and two methods of detonation devised. While each well was to be armed with detonation wire, the wells in each group also needed to be connected to each other into one detonation circuit. The instructions required that 'the current wind direction should be monitored. Then the group of wells downwind should be detonated first followed by the second group of wells so that the smoke from the first group does not interfere with the detonation of the second group of wells.'[55] The commanding officers of each unit needed to record the quantities of materials received to arm their group of wells and note any additional surplus, while tests of the circuits and explosives were to be conducted at least twice daily (and the timing of the tests recorded). The officer in charge of the detonation would receive a code-word to carry out the destruction, and would then have immediately to report the results to the authorising commander, including 'the approximate size of the destroyed area, the number of targets destroyed, the size of the destroyed area in the road and runway, and the size of the hole for planting the mines'.[56] All pertinent information – the location of the oil wells to be detonated, the date and time of detonation, the designated code-word, the name, rank and position of the officer in charge of the detonation, the authorising commander – was to be filled out in 'Form 822: Security Level Commands to Demolition Firing Group Co'. The form was to be signed by both the authorising commander and the unit leader, the latter of whom was identified as being 'technically responsible to prepare and destroy the designated target of destruction' and who therefore should 'select a representative from the group who is experienced'.[57] The detonation of Kuwait's oil wells thus required the creation of an entirely new oil infrastructure based on unique forms of engineering, expertise and efficiency which, although as complex and high-tech as the global oil assemblage, were highly localised to meet the very site-specific demands of oil field sabotage.

But the archive is not just a fixed repository of factual knowledge – of documents detailing precisely how the Iraqi regime carried out its Plan for Deferred Destruction. As with any other storehouse of cultural memory, the archive repeatedly opens up opportunities to reinterpret the past in novel ways. Indeed, the KDS reveals the story of the oil fires from a vantage point that we are entirely unaccustomed to: that of the Iraqi regime. As Joseph Sassoon and Alissa Walter argue, '[q]uestions about how the Iraqi regime viewed the people and property of Kuwait during the occupation could not be thoroughly examined' before the availability of this archive.[58] For instance, the detonation of Kuwait's oil wells and electric power stations by the retreating Iraqi army is most commonly described as a scorched earth strategy: destroying everything that must be left behind that might be useful to the enemy. (Indeed, an image of Kuwait's oil fires appears on the Wikipedia page for the concept of 'scorched earth'.) But what we see in the archive is that, aside from simply destroying the enemy, the Iraqi regime believed that 'deferred destruction' – defined in the documents as the destruction of targets 'at the last moment in front of the enemy' – would '[play] an important role in the uplifting of morale' among its own forces.[59] To see the enemy approaching would indicate failure of Saddam Hussein's intended 'merger' (*indimaj*) of Iraq and Kuwait, a demoralising blow to the forces whose job it had been over the previous seven months 'to hold onto Kuwait indefinitely'.[60] In this context, then, destroying Kuwait's oil wells just 'as the enemy approaches' would perhaps assure the Iraqi forces that all their efforts over the previous months had not been entirely in vain.[61] This last-ditch morale boost was perhaps a face-saving gesture for Saddam Hussein aimed at his own people, as much it was a scorched earth strategy.

But despite the opportunity to understand and interpret the oil fires in a new light by such direct and unfiltered access to the Iraqi archive, various aspects of archival practice itself mediate the relationship between the historian and the documentary record, particularly in relation to an institution's accessioning policies when creating archival collections. The extent to which provenance (the recorded chronology of ownership, custody and location of collections) is revealed, and 'original order' or 'respect des fonds' (the practice of collecting and preserving records as they were created and/or received) is maintained in the accession of documents, influences how a historian relates to the historical evidence. With regard to the latter, the documents of the KDS were collected, digitised and remain archived in batches as they were found by the coalition forces in the wreckage of liberated Kuwait (though this does not necessarily reflect how they were actually created by the Iraqi regime). The minimal re-organisation of the documents actually makes the KDS quite cumbersome and time-consuming to navigate. A single PDF file (which corresponds to a single 'box' of collected documents) can contain hundreds of pages of documents dealing with a wide range of topics that are not always in chronological or coherent thematic order. The KDS folder on the computer station at the Hoover Institution (the only place these documents can be viewed) contains a list of sixteen main topics, each leading to a long list of sub-topics, each of which links to a PDF file that contains documents related to that sub-topic. For instance, the topic 'Sabotage' leads to sixty sub-topics, nineteen of which relate to the destruction of the oil fields. But many of these sub-topics (such as 'Kuwait oil field destruction plans', which also appears under the topics 'Military directive' and 'Treatment of population under occupation') links to a PDF file (in this case, 274 pages) that includes not only documents pertaining to the oil fields but also all the other documents found and scanned in the same batch, many of which have nothing to do with sabotage or oil. The 'respect des fonds' method of scanning all the documents as they were found without separating them into truly separate topic files (e.g. on the oil fires), though minimising the level of archival mediation between the historian and the source, unwittingly amplifies the incomprehension often associated with the oil fires. Researching the fires requires reading through reams and reams of scanned pages on a computer screen before finding something relevant. In this process, oil is once again rendered temporarily invisible in the archive, something to be located, mined and extracted from the depths of more than 800,000 handwritten and typed pages.

But once this material is found, the archive provides an entirely new way of *seeing* the oil fires. For instance, some documents contain either requests for, or notices of receipt of, materials with which to arm the wells: hundreds of kilometres of electric wire, hundreds of 'capsules' of explosives, and so on. These items are often listed in neat, handwritten tables, carefully drawn with a ruler. It is difficult to 'see' the oil fires this way: as text and numbers, straight lines and tables, without any visual imagery except for the occasional hand-drawn sketch of how to arm and detonate an oil well. That is, it is much harder to immediately identify the oil fires for what they are in ink on paper than it is when looking at a photograph or film footage.But once you do so, the archive makes the 'painstaking' and 'enormous undertaking' that was Iraq's Plan for Deferred Destruction – with all the 'extensive bureaucracy and deft engineering' with which it was carried out – intellectually accessible.[62] Whereas images of the oil fires as captured on film tend to make them appear otherworldly or incomprehensible, on paper in the archive the fires become legible in the most banal, simplified and quantifiable way. The KDS contains detailed lists and documents planning out what was known as operation 'Babylon Lion' (*asad babel*): the 'fiery plan' (*al-khittah al-nariyyah*) for the destruction (*takhrib*) of Kuwait's oil wells. Unlike the instructions distributed among the detonation teams mentioned above, these documents reveal how the Iraqis actually organised and planned out the operation. For instance, there are numerous tables showing Kuwait's oil fields divided into a number of groups, with each numbered group in a particular area containing a certain number of wells: thirty-eight wells in Group 15 in Raudhatain, for example. The tables identify twenty-six groups and a total of approximately 735 wells. By clearly identifying the geographic coordinates of each cluster as well as the name of the officer responsible for each detonation, these documents re-territorialise the oil fires by situating them explicitly within Kuwait's physical borders, and back into their historic context of early 1991.

Conclusion

In 2017, KOC opened the new Ahmed al-Jaber Oil and Gas Exhibition, the country's first national museum, which, according to its mission, 'creates a learning experience for visitors based on the story of oil'[63] – that is, the story of oil as itself rather than as a stand-in for something else. The exhibition features a short film about the 1991 firefighting mission, during which real columns of fire unexpectedly shoot upwards in the small space between the screen and the audience, the intense heat and bright flashes exposing visitors to the fires in the most literal way, and in a way that most had never experienced first-hand. I myself first witnessed the oil fires in June 1991 from an airplane window, on one of the flights bringing Kuwaiti families like mine, that had spent the occupation in exile, back home after nearly a year away. As we flew over Kuwait's southern airspace and the airplane lights were switched off for landing, everyone took turns looking out the windows at the bright bursts of fire punctuating the darkness below in complete silence; nobody could speak. Nearly three decades later I was in the Hoover archive perusing the KDS for a project on the Kuwaiti resistance. When I first came upon documents detailing the Iraqis' Plan for Deferred Destruction, it took me several minutes to realise that the neat and simple charts, diagrams, forms and lists I saw on the computer screen before me had led to the unspeakable scene I had witnessed outside that airplane window. Seeing the oil fires this way in the archive, as a Kuwaiti, was even more jarring than when I felt the unexpected burst of flames in the KOC exhibit. And it was the archive that eventually triggered my own interest in revisiting the story of Kuwait's oil fires as a historian – to 'reclaim' it, to quote al-Qadiri again, 'as if I owned it somehow'.[64]

Notes

1. *Fires of Kuwait*, directed by David Douglas (IMAX, 1992).
2. Lee Hockstader, 'Fighting the Great Balls of Fire', *Washington Post*, 1 April 1991.
3. Hannah Appel, Arthur Mason and Michael Watts, 'Introduction: Oil Talk', in Appel, Mason and Watts, *Subterranean Estates: Life Worlds of Oil and Gas* (Ithaca: Cornell University Press, 2015), 5.
4. Appel, Mason and Watts, 'Introduction', 10.
5. Ibid., 18.
6. Ibid., 6.
7. Sara Akbar, interviewed by Farah al-Nakib, 19 August 2018, Rumaithiya, Kuwait. Oral History and Documentation Project, American University of Kuwait, https://oralhistory.auk.edu.kw/items/show/7 (last accessed 17 September 2022).
8. Andrew Barry, 'Technological Zones', *European Journal of Social Theory* 9, no. 2 (2006), 246.
9. Sebastião Salgado, *Kuwait: A Desert on Fire* (Cologne: Taschen, 2016), 14.
10. Appel, Mason and Watts, 'Introduction', 13.
11. Michael J. Watts, 'Specters of Oil: An Introduction to the Photographs of Ed Kashi', in Appel, Maon and Watts, *Subterranean Estates*, 169.
12. Sara Akbar, interviewed by Farah al-Nakib, 9 August 2018, Salhiya, Kuwait. Oral History and Documentation Project, American University of Kuwait, https://oralhistory.auk.edu.kw/items/show/7 (last accessed 17 September 2022).
13. Akbar interview, 19 August 2018.
14. Ibid.
15. Hockstader, 'Fighting'.
16. *Fires of Kuwait.*
17. Ibid.

18. Hockstader, 'Fighting'.
19. *Fires of Kuwait.*
20. Salgado, *Kuwait*, 12, 7.
21. Michael Wilmington, 'Beauty, Horror Emerge from IMAX's "Fires of Kuwait"', *Los Angeles Times*, 11 June 1993.
22. Salgado, *Kuwait*, 12.
23. Akbar interview, 19 August 2018.
24. Mona Damluji, 'The Image World of Middle Eastern Oil', in Appel, Mason and Watts, *Subterranean Estates*, 147.
25. *Close-Up on Kuwait*, directed by Rodney Giesler (Kuwait Oil Company, 1961).
26. Damluji, 'Image World', 147.
27. Ibid., 161.
28. See note 9.
29. See note 1.
30. *Lessons of Darkness*, directed by Werner Herzog (New Video Group, 1995).
31. *Behind the Sun*, directed by Monira al-Qadiri (2014).
32. Wilmington, 'Beauty, Horror'.
33. Imre Szeman, 'The Cultural Politics of Oil: On *Lessons of Darkness* and *Black Sea Files*', *Polygraph* 22 (2010), 37.
34. *Lessons of Darkness.*
35. Szeman, 'Cultural Politics of Oil', 41.
36. *Lessons of Darkness.*
37. Szeman, 'Cultural Politics of Oil', 38.
38. *Lessons of Darkness.*
39. Ibid.
40. Ibid.
41. Szeman, 'Cultural Politics of Oil', 41.
42. *Lessons of Darkness.*
43. Szeman, 'Cultural Politics of Oil'.
44. *Lessons of Darkness.*
45. Szeman, 'Cultural Politics of Oil', 40–1.
46. Damluji, 'Image World', 158.
47. Monira al-Qadiri, 'Apocalyptic Aspirations', talk given at the American University of Kuwait, 30 September 2015.
48. Al-Qadiri, 'Apocalyptic Aspirations'.
49. Mona al-Qadiri website, https://www.moniraalqadiri.com/videos (last accessed 14 July 2022).
50. Al-Qadiri, 'Apocalyptic Aspirations'.

51. Hockstader, 'Fighting'.
52. Hussain 'Isa Malallah, *The Iraqi War Criminals and Their Crimes during the Iraqi Occupation of Kuwait: A Legal Reading in the Documents of the Iraqi War Crimes against Kuwait and Its People* (Kuwait: Center for Research and Studies of Kuwait, 1998).
53. Iraq Memory Foundation Website, http://www.iraqmemory.com/en/projects/documentation (last accessed 14 July 2022).
54. See for instance, 8th Engineer Regiment, 'Instructions on how to destroy oil fields and pipelines', n.d., KDS box 00586, pp. 16–19 (English translation) and pp. 20–2 (Arabic original).
55. 8th Engineer Regiment, 'Instructions', p. 18.
56. Form No. 822: 'Security Level Commands to Demolition Firing Group Co', n.d., KDS box 00586, pp. 23–4 (Arabic original) and pp. 26–8 (English translation).
57. Ibid.
58. Joseph Sassoon and Alissa Walter, 'The Iraqi Occupation of Kuwait: New Historical Perspectives', *The Middle East Journal* 71, no. 4 (2017), 607.
59. Summary translated orders for the destruction of the Al Wafra oil fields, and 'how to' instructions from the 8th engineering regiment on the destruction of oil fields, n.d., KDS box 00586, pp. 4–5.
60. Sassoon and Walter, 'Iraqi Occupation', 612.
61. 8th Engineer Regiment, 'Instructions', p. 16.
62. Hockstader, 'Fighting'.
63. Kuwait Oil Company Ahmed al-Jaber Oil and Gas Exhibition website, https://www.kocexhibit.com/Default (last accessed 14 July 2022).
64. Al-Qadiri, 'Apocalyptic Aspirations'.

PART II

IMAGE WORLDS

VISUALISING PETROLEUM PASTS

3

PHOTOGRAPHING CRUDE IN THE DESERT: SIGHT AND SENSE AMONG OIL MEN

Laura Hindelang

Setting the Scene – From Europe to the Persian Gulf

It was early spring 1924 when Swiss petro-geologist Arnold Heim (1882–1965) departed Zurich by train for the Persian Gulf with the purpose of investigating at first-hand oil deposits in the region. Earlier that year, the Eastern and General Syndicate Limited, a London-based company, had contacted Heim requesting that he carry out a 'geological examination and exploration' in Eastern Arabia of various concessions the company held or hoped to obtain, including the Hasa Concession, the Neutral Zone Concession, the Koweit Concession and the Bahrain Concession, 'in respect of their mineral and especially oil possibilities'.[1] For a period of five months (until August 1924), Heim's task was to assess the availability of exploitable resources in the region, especially but not exclusively petroleum. According to early twentieth-century methods of preliminary geographical surveying, Heim did so by immersing himself into the landscape of Eastern Arabia, effectively sensing crude in the wild. In the 1920s, petroleum had not yet been found in substantial quantities in the region. It was still unknown if such fossil deposits even existed. Also, the extent to which the black gold would mould life into a global petro-culture was probably far from imaginable.

From today's perspective, neither the worldwide importance of the discovery of petroleum in the Gulf in the first half of the twentieth century nor the effects of oil industrialisation and petro-modernity on local communities can be overestimated. Yet, the naturalization of fossil energy usage in the region and the world at large has mainly worked to abstract petroleum-the-raw-material in the process. Today, oil gives way to a complex regime of (in)visibility because it is somehow everywhere and in everything, but its synthetisation redirects our experience and knowledge of petroleum via other materials, forms of energy, infrastructure and images. Our encounter with crude oil is negotiated through the car's speed, the view from an airplane, a trainer's plastic material, or oily skin care. Sensing petroleum-the-raw-material, as smell, fire or sticky substance, in contrast, has become emblematic of calamities such as the Gulf War in 1990/1 and Deepwater Horizon. But what about the traces of crude in its natural habitat prior to the pipeline–refinery complex? Pre-industrial sighting and sensing of crude oil in the Arabian Peninsula desert have so far not received much scholarly attention.

To this end, the first half of the chapter explores the life of Middle Eastern oil through the archive of the Swiss petro-geologist Arnold Heim relating to his 1924 Eastern Arabia travel. Reading through his diary entries, field notes, reports, maps and photographs provides for a strategy of seeing through Heim's eyes how a European effectively used not only his sight but also his other senses to ascertain and locate potential deposits of crude. It reveals the fundamental uncertainty of finding oil, therefore opening up new insights into the life world of Middle Eastern oil shortly before its discovery, its industrialisation, and the concomitant socio-economic, cultural and ecological rupture that Abdelrahman Munif has so vividly and violently conjured in *Cities of Salt*. Although this lens – the Heim lens – is tainted with Orientalism, this 'pre-oil petroleum archive' offers a rare occasion to encounter one of the earliest (photographic) portraits of wild crude from which to engage with questions such as: what is the 'ideal', the most 'authentic' or 'natural' form of oil?; what happens when petroleum is idle and not (yet) industrially used?; how can one ascertain that a crude oil finding will allow for industrial exploitation in quantity and quality?

In juxtaposition, the second part of the chapter then investigates the life of Middle Eastern oil during the early heydays of oil industrialisation and petro-modernity. Notably, by the mid-twentieth century, the certainty of abundant petroleum deposits loomed large. I discuss the illustrated travel report

Im Auto nach Koweit (*To Kuwait by Car*) (1953) by the Austrian automobilist Max Reisch and the ways in which crude oil underlaid and thus materialised (in) his Middle Eastern road trip. Thirty years after Heim, Reisch set out to 'explore' and witness the Arabian Peninsula's oil boom, a phenomenon which, in contrast to Heim's experience, already implied not seeing one drop of crude. Rather, urbanisation in concrete forms, and, especially, asphalted roads providing for the perfect driving experience, came to stand in place materially, visually and sensuously for the crude encounter. To Reisch, too, photography was important as a means of evidencing the fossil-fuelled experience of tracking oil pipelines across the Middle East, although resulting in a totally different set of motifs. Both Heim and Reisch embodied, each in his own way, the life world of Middle Eastern oil of their periods. Both engaged with the (un)certainty and the secretive visuality of crude oil through sensing, sighting and image-making, for which I interpret their textual and photographic documentations not only in terms of their social configurations but also in terms of their modes of visual as well as sensory evidence.

Aiming to discern petroleum's visual materialities or material visualities, in the words of geographers Gillian Rose and Divya P. Tolia-Kelly, implies asking: 'What is made visible? (And what is rendered invisible?) How is it made visible, exactly – what technologies are used, and how, and what are the specific qualities of the visual objects thus enacted? And what are the effects of those visualised materialities and materialised visualities, particularly for the people caught up in those practices, as researchers, and as those researched?'[2] Overall, the chapter provides a typology of sights and senses related to oil exploration in the 1920s and the 1950s that were concomitant with particular ways of travelling and photographing. This typology of oil-related experiences moves from early uses of petroleum as illumination and heating material, to the geologist's sensory search for oil in the desert, and finally to the automobilist's asphalted and fossil-fuelled driving experience.

Incorporating the Arabian Peninsula into the Encyclopedia of the West through/as Images

As an art historian sifting through these petroleum-related materials of the 1920s and 1950s, I situate Heim's and Reisch's activities in the Arabian Peninsula within the larger hegemonic and Orientalist project of incorporating

'Arabia' into the encyclopedia of the West, here through both the scientific and the popular exploration of resources. Since the late nineteenth century, Western interest in the Arabian Peninsula and the Persian Gulf had predominantly been of a political-economic nature, and evolved around finding a final destination for the Baghdad railway, securing access to India and, gradually, prospecting for petroleum. Consequently, such diverse personnel as geologists, missionaries, engineers, political officers, ethnographers and travel writers author textual and visual documents on the Eastern part of the Arabian Peninsula. I argue that Heim's photographs may be taken as early modern visual representations of politically-economically stimulated encounters with the region that continue, in a different form, with travel writers such as Reisch.

The historical contexts of Heim's photographs substantiate Ali Behdad's claim that Orientalist photography was seldom the result of an individual's 'manic obsession', but instead relied on certain circles, networks and contacts to receive the 'technical knowledge and logistical support' necessary for a European voyaging the Middle East.[3] Given that so few Europeans travelled through Arabia, these networks were even tighter and more transient across different professions and disciplines. Edward Said, who first meticulously dissected the historical origins and mechanisms of Orientalism, saw Orientalism as characterised by 'its increasing scope, not its greater selectiveness'.[4] Being an Orientalist could imply being a philologist of Chinese languages, an archaeologist specialised in the pharaonic period, or a scholar of fourteenth-century Ottoman literature, or, in this case, a petro-geologist working on Eastern Arabia.[5] Said's initial focus on the Orientalist legacy of textual images, clichés and imaginaries has subsequently been widened to acknowledge the power of the Orientalist visual imagery.[6] The heterogeneous, interdisciplinary Orientalist zeal to document/know/subjugate everything that Said identified is therefore evident in the visual sources that document, construct, imagine and conjure the image of Middle Eastern oil. In an attempt to explore sources across disciplinary boundaries, this chapter analyses the visual history of the Arabian Peninsula through the (camera) lens (and sensual apparatus) of European petro-geologists and travellers between the 1920s and the 1950s, demonstrating the historical relevance of disparate visual sources in the process of visually creating and representing the Gulf region and the experience of petroleum in the twentieth century.

Geology, Petroleum and Photography

The fact that a Swiss petro-geologist produced what is probably the first photographic portrait of crude oil in Eastern Arabia is less surprising than we might think. When the London company Eastern and General Syndicate Limited had first inquired whether Arnold Heim would be interested in and qualified for the job, he self-confidently responded that he was 'known as one of the most experienced and skilled petroleum geologists'.[7] In fact, the 'Swiss excellence of field work' that was said to have started with the Genevan geologist Horace Bénédict de Saussure (1740–99) roughly 150 years earlier 'was responsible for the employment of a disproportionate number of Swiss geologists' worldwide in the early twentieth century.[8] Correspondingly, 42-year-old Heim had already worked for Royal Dutch in East India, Oklahoma and California as well as for Royal Dutch Shell in Australia and the South Sea Islands by the time Eastern and General intended to employ him.[9]

As a geographer, Heim operated in a discipline that is argued to be extremely visual.[10] As an experienced traveller, Heim paid special attention to his equipment, which included not only maps and geological instruments, but also bags of dried fruit, a head-to-toe tropical suit, shoe nails and insect powder. But of greatest importance to him for the Arabia expedition was the expensive photographic equipment of which Heim noted in his travel diary with pride:

> I have four devices, plates and films with me and the latest have partly been sent directly to Bahrein, Kuwait and Baghdad. The old 9 × 12 tropical [plate camera] Nettel, which, although it is heavy, has always been reliable; the old Alpin as a backup; the 6 × 9 roll film [camera] Icarette and a 'contemporary' Sept cine camera model. What kinds of result I will be able to achieve, remains to be seen, although it seems to me I would need to learn something new . . . The photo equipment alone has cost around 2000 [Swiss] francs, not to speak of the expenses for transport, development and copies, diapositives, coloring, and the work. And then people find 150 francs for a slide lecture exaggerated, 100 francs still expensive![11]

Good photographs were essential to Heim's geological research and represented a substantial yield of this expedition given that he also intended to turn his travel experiences and photographs into commercial slide lectures.

Curious to see early photographic results already under way, Heim had the first batch of photographs developed in Baghdad after two weeks of travelling before he had even reached the Gulf. The corresponding diary entry reads: 'Film developed at Kerim [studio] – Ilford – perfect emulsion. Most of them great. But now the humid Gulf region is next.'[12] In 1920s Baghdad, photography as a commercial business and a visual technology was evidently well-established. The Gulf, in contrast, was not a well-known photographic environment, at least to Heim, who assumed that photographing would be demanding under the climate conditions awaiting him. Despite the environmental challenges, Heim had spared no expenses regarding his equipment, which attests to the importance the geologist attached to photography.

Photography and petroleum share deeply entangled histories in the Gulf and elsewhere. Petroleum is not only represented in visual media as a motif as we will explore shortly, but also constitutes visual media in its own right.[13] In the 1820s, even before the invention of the daguerreotype, modern photography's preceding technology, French Nicéphore Niépce (1765–1833) experimented with so-called 'bitumen of Judea' for a new process of light-sensitive photo-engraving and successfully produced first pictorial results from bitumen-coated bases that became non-soluble after being exposed to light. Later, with the invention of the modern roll film, much of the light-sensitive film was made from celluloid and polyester that often consisted of petroleum-based synthetic materials.[14]

In case of the Arabian Peninsula, the entanglement between petroleum and photography also played out historiographically. In *Kuwait by the First Photographers*, the first and still the only work on photography taken in the first half of the twentieth century in Kuwait, William Facey and Gillian Grant define the 'early photography of Kuwait' (1900–50) not just in terms of early technical advances and by the early history of the medium itself but as something that comes to an end with the first export of crude in 1946, 'when the first professional photographers begin to appear with the oil company, and images start to proliferate'.[15] Although they rightly highlight the liaison between photography and petroleum, their periodisation blurs the fact that the two coincided much earlier than the first Kuwaiti oil export (itself a prominently photographed event), as the archive of Arnold Heim demonstrates, although the period from the mid-twentieth century onwards indeed marks a much more drastic intersection between petroleum and photography in the region.[16]

Building on Mimi Sheller's claim that 'energy transitions have histories' – visual, material and sensory histories, I would add – an analytical reading of the particular sensory and visual histories that accompanied the shift in material culture and energy usage in early twentieth-century Kuwait towards petro-culture is crucial here.[17] Arnold Heim's travel documentation substantiates that petro-culture already began manifesting in the 1920s in the ways in which bodies moved through space, and in how the world became perceivable (fossil-fuelled heat, illumination) and visualised in a permanent carbon form, the photograph. While Heim's exploration of petroleum deposits materialised as a fossil-fuel-driven experience only occasionally (he mostly travelled by camel, for example), his photographing, however, must be understood as an intrinsically fossil-fuelled technology, especially in the context of the Gulf.

Travelling through Eastern Arabia on the Brink of Petrolisation

At that time, traversing the distance between Europe and the Arabian Peninsula involved numerous stop-overs, and Heim's voyage was no exception, as his trip included connections at Genoa, Naples, Alexandria, Cairo, Palestine, Beirut, Damascus and Baghdad and involved switching between train, ship and car to arrive at his destination. After Heim had stopped in Baghdad, he took the train to Basra and finally met with the British-New Zealander mining engineer and geologist Major Frank Holmes (1874–1947), a fellow oilman and driving force behind the establishment of Eastern and General Syndicate Limited. Holmes was in charge of preparing Heim's expedition on site. Together they took a motorboat, which the Eastern and General Syndicate had apparently gifted to Sheikh Ahmad al-Jaber Al Sabah of Kuwait (1885–1950), to cross over from Basra to Kuwait the next day.[18]

The walled-off coastal town made a positive impression on Heim (Figure 3.1):

> All of Kuwait looks good and massive, like a fortress in contrast to Baghdad. Subrecent shell bed (Pliocene) as bedrock = building material. I have clean water, an impeccable American cot, the floor is covered with woven reeds . . . The houses look mostly clean and are plastered with lime that is produced locally.[19]

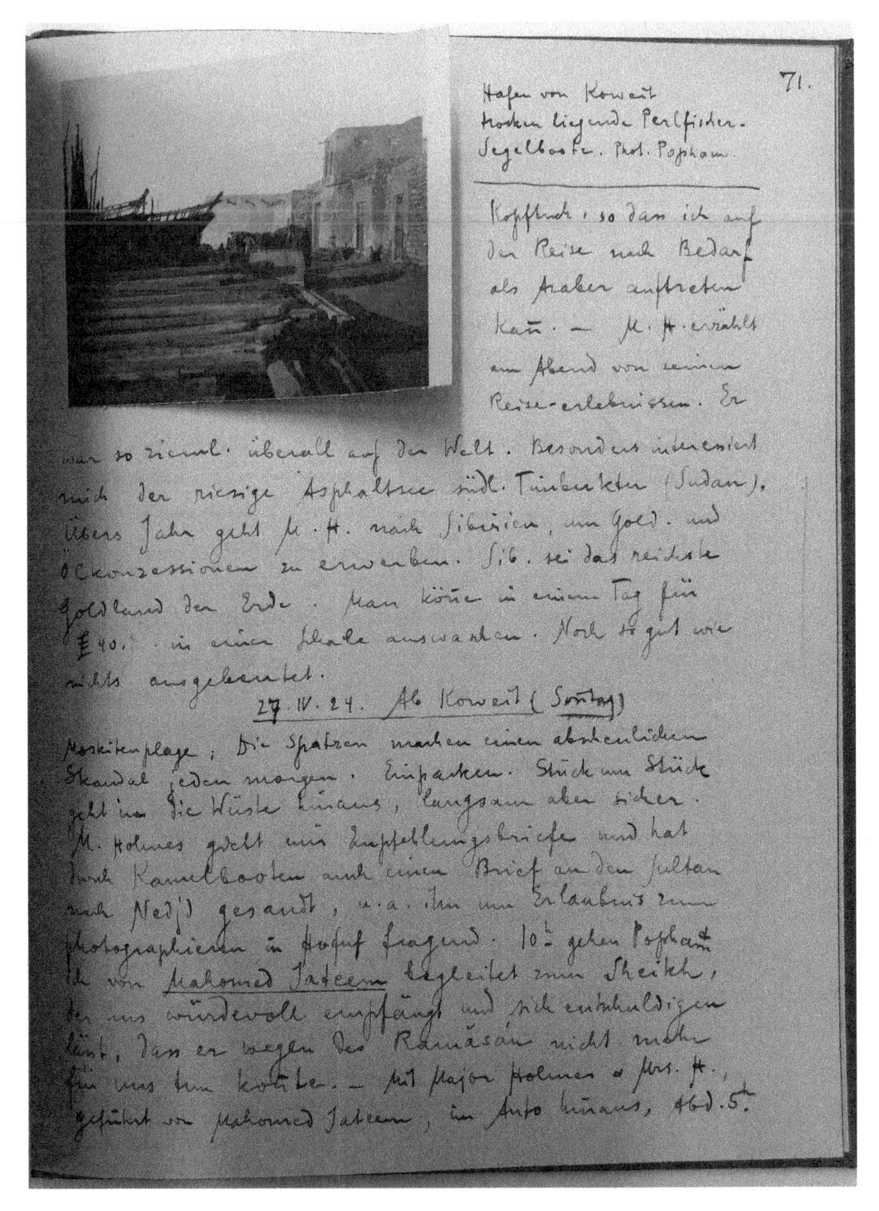

71.

Hafen von Kowait
trocken liegende Perlfischer-
Segelboote. Phot. Popham.

Kopftuch, so dass ich auf
der Reise nach Bedarf
als Araber auftreten
kann. — M. H. erzählt
am Abend von seinen
Reise-erlebnissen. Er
war so ziemlich überall auf der Welt. Besonders interessiert
mich der riesige Asphaltsee südl. Timbuktu (Sudan).
Übers Jahr geht M. H. nach Sibirien, um Gold- und
Ölkonzessionen zu erwerben. Sib. sei das reichste
Goldland der Erde. Man könne in einem Tag für
£ 40.– in einer Schale auswaschen. Noch so gut wie
nichts ausgebeutet.

27. IV. 24. Ab Kowait (Sonntag)

Moskitenplage; Die Spatzen machen einen abscheulichen
Skandal jeden Morgen. Einpacken. Stück um Stück
geht in die Wüste hinaus, langsam aber sicher.
M. Holmes gibt mir Empfehlungsbriefe und hat
durch Kamelboten auch einen Brief an den Sultan
nach Nedjd gesandt, u. a. ihn um Erlaubnis zum
Photographieren in Hofuf fragend. 10h gehen Popham
& ich von Mahomed Jateem begleitet zum Sheikh,
der uns würdevoll empfängt und sich entschuldigen
lässt, dass er wegen des Ramasan nicht mehr
für uns tun könne. — Mit Major Holmes & Mrs. H.,
geführt von Mahomed Jateem, im Auto hinaus, 46d. 5t.

Figure 3.1 Entry in Arnold Heim's Eastern Arabia travel diary no. 1 with a small-format photograph of Kuwait's harbour, taken in late April 1924. ETH Zurich University Archives, Arnold Heim Collection, Hs 494: 240.29. Courtesy of Dore Heim.

In addition, Heim noted that 'with exception of the missionaries and the [British] consul, the Holmes and us geologists are the only white people in Kuwait'.[20]

Indeed, Kuwait was not a popular destination for Western travellers at the time, and the town therefore does not feature prominently in nineteenth- and early twentieth-century English- and German-speaking travel writing on the Arab world.[21] Early photographs of Kuwait are similarly rare. Although Kuwait was part of the Arabian Peninsula, as a coastal settlement it was not representative of the endless desert and of Bedouin life, of the 'mysterious' interior lands that highly individualised Western travellers usually aspired to explore while crossing the Arabian Peninsula overland. Rather, Kuwait was a commercial hub that connected the sea and the desert and allowed for transition from boat to camel or foot and vice versa. For example, Harry St. John Philby (1885–1960), an Arabist and British colonial intelligence officer, travelled the Peninsula in the company of Bedouins in 1917/18 on a mission to meet with Ibn Saud and reached Kuwait as his point of exit at the end of the journey in October 1918. Like Philby, travellers mostly visited Kuwait as a short stop-over on their way to somewhere else, catching Kuwait fleetingly. This changed radically once petroleum made its appearance and geologists, oil company emissaries and engineers arrived in flocks. In 1924, however, this was not yet the case.

Still, the rare arrival of individual geologists and travellers interested in petroleum, such as Heim and Holmes, indicated a growing shift in modes of locomotion and slow but radical changes in indigenous and travellers' lifestyles, as Philby foresaw. In the preface to his travel account *Arabia of the Wahhabis* (1928), Philby would emphasise that his was the last desert voyage prior to motorised (read: fossil-fuel-powered) modes of travelling:[22] modes that would dramatically alter the level of speed and accessibility of this secluded environment, of social interaction between travellers and sedentary and tribal communities that Philby and other travellers considered characteristic of the Arabia voyage.[23] For Philby, this development equaled a 'desecration' that destroyed the 'air of mystery', which he considered the lure for most of the Western travellers interested in 'Arabia'.[24] Literature scholar Helen Carr argues that this critique, which many travel writers shared at the time, expressed 'the perturbing recognition that the lines of demarcation between Europe and the other are becoming disturbingly blurred . . . For others, modernity, in the shape of tourists if not colonialists, is about to sweep away the picturesque customs they have come to seek.'[25] As a matter of fact, part of this growing 'blur' was that fossil-fuelled modes of transportation and other petroleum-derived products slowly made their way into the Gulf, altering not only landscapes but also lifestyles and modes of perception.

Traces of Early Petro-culture

Heim's travel diaries and field notes provide invaluable first-hand documentation of the ways in which fossil-fuelled products and mobility affected the life worlds of both Western travellers and the indigenous people of the Peninsula in the 1920s. The mode of travelling was first of all a pragmatic decision. The impressive caravan comprising around sixty people and sixty camels that Frank Holmes had organised for Heim undertook the planned overland journey from Kuwait to Hofuf and back by foot and on camel. This choice was most likely determined by the limited number of off-road cars available, although a few cars already existed in the region despite a lack of service infrastructures. For instance, Sheikh Ahmad of Kuwait owned a car as well as a motorboat.[26] Also, Mr and Mrs Holmes had a car at their disposal at Kuwait and undertook some trips with Heim. Regarding fossil fuel energy, the geologist Heim reported on a combustion engine that produced electrical light for the American Mission hospital in Kuwait Town.[27] Otherwise, electricity did not exist in Kuwait, and throughout the desert journey the geologist wrote his notes under the stars in the company of a petroleum lantern.[28] Generally, households used charcoal imported from Karachi for cooking and not gas.[29] Petroleum also marked everyday life by its consumed absence, in its negative form so to say, as empty petroleum tins were repurposed as containers for all kinds of products, such as water.[30]

Also, while staying in Baghdad, Heim sketched a 'pissoir system' that he observed in hotels and apartments (Figure 3.2). It consisted of a commercially available oil tin with a storage capacity of four gallons that was modified with a funnel to collect the urine. These instances demonstrate that refined oil products were known and partly used in the region, although not in abundance. Shapes of petroleum in absence foreshadowed a horizon of expectation for possible industrialisation. Where were these products coming from?

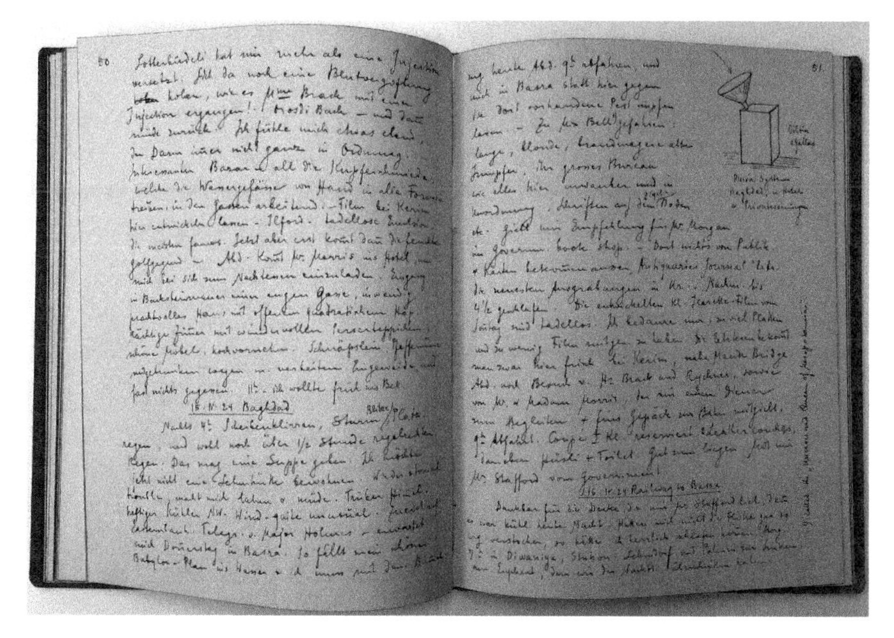

Figure 3.2 Entry in Arnold Heim's Eastern Arabia travel diary no. 1 with a drawing of the oil tin 'pissoir system', entry Baghdad 15 April 1924. ETH Zurich University Archives, Arnold Heim Collection, Hs 494: 240.29. Courtesy of Dore Heim.

Elsewhere, Heim mentioned that 'the oil residues from the Persian fields are relatively cheap' and therefore easily available.[31] This suggests that the (for the time typical) oil-based products such as bitumen, paraffin and gasoline that were already used sporadically on the Western shore of the Persian Gulf were most likely imported from across the Gulf, possibly from the Abadan refinery (completed in 1912). Overall, these observations Heim recorded in 1924 indicate that petro-culture was slowly but gradually manifesting in Eastern Arabia, although the region was not partaking in petroleum production yet. Prospecting for locally occurring crude in the 1920s, in order to determine the availability of industrially exploitable oil deposits, proved to be a surprisingly hands-on, multi-sensory and embodied engagement.

Prospecting for Petroleum as an Embodied Experience

To discover natural reservoirs of the fossil matter with the methodologies available at the time required a particular geological skills set. On the one hand, Heim had to rely on local hearsay, as well as on the little geographic literature available on the region that he had consulted in advance, to find hints on naturally occurring crude in the vast desert along the Eastern shore of the Arabian Peninsula. On the other hand, in order to determine whether fossil matter existed at a particular site, Heim focused on discerning petroleum by its sensory qualities, by its smell, taste, feel and looks.

For example, on the island of Bahrain, Heim observed the asphalt deposit at Ain el Kar, which had already been described in Heim's most important preparatory literature *The Geology of the Persian Gulf and the Adjoining Portions of Persia and Arabia* (1908) by Guy Ellcock Pilgrim. On site, several pits contained tar to various extents. Heim attached importance to tasting the water of the nearby water well, concluding that 'it was not bad for drinking, and had no taste of oil'.[32] Similarly, at a location named Ainain, a warm water spring around six kilometres from the town of Al-Jubail, Heim reported that 'it was said that the water of Ainain in winter time sometimes has some smell of oil. At the writer's visit (May 12th) no such trace was found.'[33] Heim also took obscure information into account as possible evidence of fossil matter, such as the story that 'two weeks ago again two men have died of toxic gas while digging water holes. Apparently, this happens every year.'[34] Could the mortal vapours be indicative of natural gas deposits, Heim probably wondered when recording it. Obviously, prior to a complete modern cartography and technological innovations in oil prospecting such as seismic and aerial surveying, finding oil was a multi-sensory undertaking that relied on the versatility of the raw material's diverse characteristics. Searching for petroleum clearly required one to be embedded, to fully integrate oneself into geography, topography, and to embody its sensory strata in order to detect it.

For the Swiss geologist, Kuwait appeared the most promising locality in which to find substantial amounts of petroleum. Several oral and written sources reported not only on tar deposits but also on oil seepages inside and outside the town of Kuwait. Outside the town's jurisdiction, the area of Burgan appeared as a promising location 'where oil sites were reported to be'.[35] But how difficult it was to actually find the exact spots where oil was actively leaking to the surface and to assess them is demonstrated in Heim's report to his contractor: 'The place [Burgan] was considered as of the greatest importance. Amongst the Arabs it is generally believed that an oil seepage is existing, and there was even talk of a dozen of such seepages. But notwithstanding special attention, we were unable to find any trace of oil.'[36] In a later addendum to the report's Burgan entry, Heim wrote, surely with a bitter taste, '[Burgan] became later the largest oil field, on the surface nothing to be seen.'[37]

In addition to Burgan, a place on the north-western side of Kuwait Bay where, according to hearsay, some people regularly collected tar and sold it at the market in Kuwait appeared even more important.[38] Due to circumstances, Heim and his crew only focused on this locality once they had returned from Hofuf, but they experienced great difficulties in finding this oil seepage, too.[39] In order to travel, Heim required Sheik Ahmad's approval, who finally agreed but cautioned to cross the sea by night in order not to alert the British, who had been refused an oil concession a year earlier.[40] Next, Heim and colleagues had to find the man who was known to bring oily tar to the market of Kuwait, where it was said 'to be used for sail boats'.[41] The man was crucial because, as Heim was told, 'only two to three men exist that know the oil location'.[42] Finally, the man, Mesfer of Sabiya, was located and the same night a boat was arranged, and the party left the town to land on the swamp coastline north of Kuwait Bay. At the shore, Mesfer showed them the next batch of tar he had prepared awaiting transportation. Heim summarised that 'this tar is gathered by Mesfer every six months with greased hands, and brought to Kuwait. The output is 10 oil tins of 4 gallons each for about every six months. This would make 1 litre daily, or 0.6 cubic cm per minute.'[43] Petroleum extraction at the time materialised as oily hands. Departing from the shoreline, the party continued by foot and donkey, traversing wet swamps, dried mud fields with salty crusts, then plains with shrubs, and finally a tableland of red sandstone. At some point, they crossed the telegraph line between Kuwait and Basra and from here another 1.5 kilometres south there finally lay the oil seepage of 'El Bohara', as indicated on a geographical map later prepared by Heim (Figure 3.3). The first photographic portrait of sighting oil in this area subsequently evidenced the bituminous encounter.

The First (Photographic) Sighting of Oil

Two black-and-white photographs taken by Arnold Heim with the large Nettel camera encapsulate the moment of finding and exploring the oil seepage of El Bohara in participatory observation.

In the first picture, the sandy soil-plain expands up until the horizon

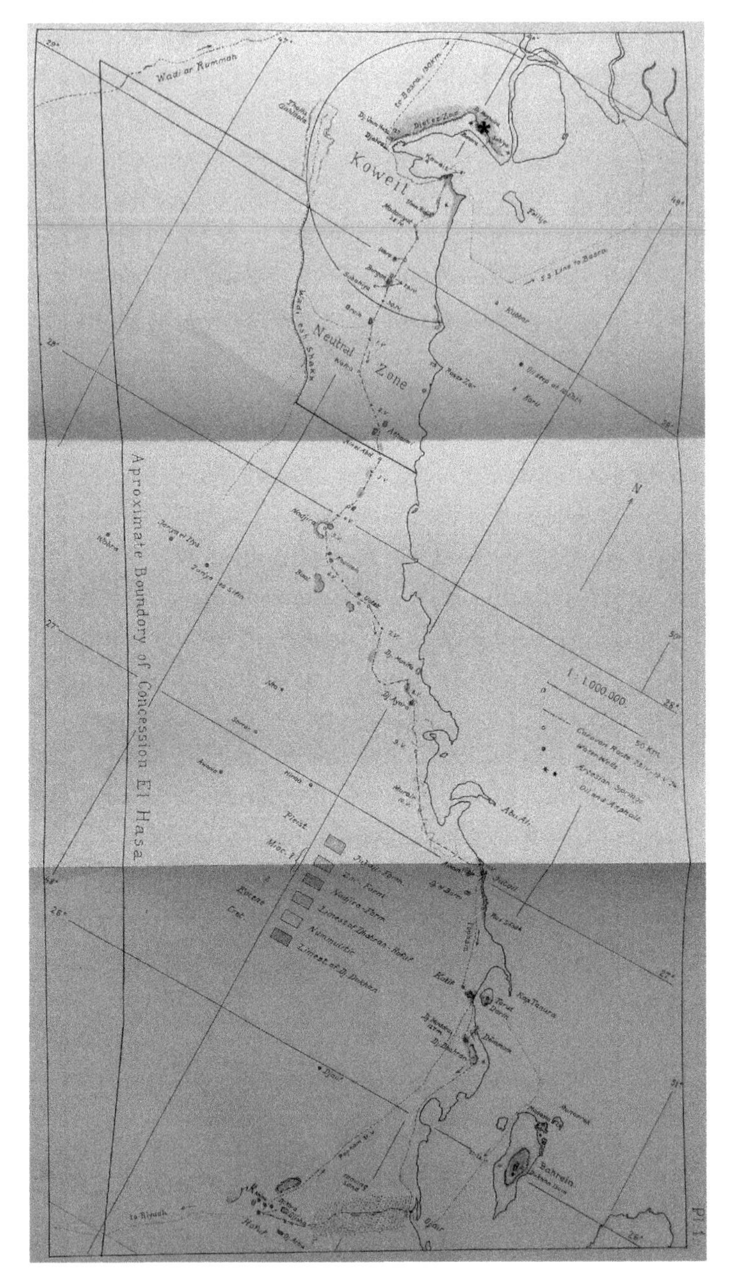

Figure 3.3 Detail of the map drawn by Arnold Heim of his journey through Eastern Arabia showing the location of the oil seepage 'El Bohara' (see asterisk). ETH Zurich University Archives, Arnold Heim Collection, Hs 494: 70. Courtesy of Dore Heim.

(here the edge of the photographic plate) (Figure 3.4). At the centre, among small bushes and shrubs, a man dressed in a white dishdasha with head cover and a firearm appears prominently. The photograph probably re-enacted the moment when Mesfer, who had spotted the site where he collected tar, called for the party to draw closer. Shown as 'Rückenfigur' (a figure from behind), Mesfer appears as if in motion of turning and pointing forward. He therefore serves as the point of visual entry into the picture in his role as key figure of the moment of discovery. Just ahead of Mesfer, a site of heaped earth emerges, a distinct rupture of the otherwise seemingly homogeneous and rather flat landscape. The site and its surroundings give no hint of the presence of crude. No tar sands or oil spills, just plain sand and shrubs pave the way towards the oil seepage referenced in the photograph's title and anticipated by the key figure Mesfer. There is an astonishing discreetness to petroleum in the wild

Figure 3.4 Arnold Heim, *Seepage Bohara, Norden von Koweit (Kuwait)*, 27 June 1924, black-and-white glass diapositive, 8.5 × 10 cm. ETH-Bibliothek Zürich, Bildarchiv / Fotograf: Heim, Arnold / Hs_0494b-0019-282-A-F / CC BY-SA 4.0. http://doi.org/10.3932/ethz-a-000052836

Figure 3.5 Arnold Heim, *Ölsee Bohara, Norden von Koweit (Kuwait)*, 27 June 1924, black-and-white glass diapositive, 8.5 × 10 cm. ETH-Bibliothek Zürich, Bildarchiv / Fotograf: Heim, Arnold / Hs_0494b-0019-283-F / CC BY-SA 4.0. http://doi.org/10.3932/ethz-a-000052839

prior to industrialisation, a silent secretiveness that undoubtedly triggered petro-geologists' sense of 'exploration'.

The second photograph captures the seepage of El Bohara from the edge of the crater with the camera looking over and into the greyish-black pond (Figure 3.5). Heim measured that 'the main seepage consists of a roughly oval hole of about 8 by 5 metres, dug out of the sandy muddy ground to a depth of some 2½ metres below the surface, the salt water being at 1 metre below it'.[44] Opposite the camera and close to the edge of the seepage, we see Mesfer standing and looking down with an unknown man next to him kneeling and facing the camera. The crater with its dark furrowed edges emerges distinctly from the surrounding flat bush-pierced steppe. In his writings Heim comments that Mesfer had dug this crater over the years in order to extract the tar. Therefore, even in this initial stage of wild crude, human interest had forced the dark matter further towards the surface. Yet, besides this meddling, the tar pit had been left to itself, displaying the best possible photographic portrait of a 'natural' formation of oil on the eastern side of the Arabian Peninsula.

Looking at this photograph today, we see that the camera's angle triggers a sense of diving into a miniature landscape with a large lake at its centre whose content is hard to describe. A light-grey layer has settled on the otherwise pitch-black surface of the seepage, like crystallising ice floes that have not yet covered the entire surface, apparently crystalline formations of salt crusts, as Heim noted.[45] The geologist furthermore wrote, in describing the dynamic at play, that

> from the ground, oil is rising at more or less regular intervals in the form of black drops which rapidly spread on the surface, and become surrounded by thin skins of iris [iridescent] colours made of the lighter parts of the oil. They chiefly emanate from the SW part of the hole, and drive away to the SE corner, where the lighter parts entirely evaporate and black tar is accumulated.[46]

El Bohara, we learn here, consisted of heavy and light crude oil, but mostly bitumen, the latter a material whose 'natural state is liminal, being not quite liquid and not quite solid', as Hannah Tollefson and Darin Barney have pondered elsewhere.[47] Clearly, there is something inhomogeneous about the substance gathered in the seepage. What the photo (in my interpretation) and Heim's description reveal is that 'petroleum' in the wild embodies and performs a state of liminality, uncertainty. Petroleum exists as a mélange ranging from liquid to viscous content, and in cohabitation with (salt) water. In its material ambiguity the seepage does not disclose whether it will be economically viable to drill and explore it further. Like an inverted salt-crystal-covered tip of the iceberg, its dimensions and future economic and energetic potential for being liquidated remain hidden from view. Moreover, crude in the wild offered a multi-sensory experience of smell, sound, movement and sight.

Commissioned to assess the availability of exploitable resources in the region, Heim was confronted with this landlocked image of bitumen and compromised insofar as he captured petroleum through a historically tar-tainted medium. It appears that he took some material samples to be analysed in Switzerland, but overall it is remarkable that he made his assessment on site, not later in a laboratory. His final reports sent to London are fair copies of his notes and drawings included in the field diaries. The early history of oil exploration in Eastern Arabia rested on being on site, being embedded in the (potential oil) field, and embodying petroleum by tasting, touching, smelling and observing it.

When evaluating El Bohara in his final geological report to the London-based company, Heim argued that the seepage seemed not liquid enough, therefore diminishing its estimated industrial potential to be exploited through pipeline infrastructures. 'Koweit is a country of some possibility but not of high promise', he tentatively concluded, as a prospective second trip to the Arabian Peninsula for Eastern and General Syndicate Limited never did materialise.[48] Still, to Heim, El Bohara was 'the only important seepage encountered on the writer's voyage'.[49]

Notably, the photographic album that resulted from Heim's Eastern Arabia trip contains only two photographs of oil sites, which are the two of El Bohara.[50] Besides those, Heim photographed the town of Kuwait and other settlements, the caravan, people and camels in the desert, but no other oil findings. As regards Heim's expectation of producing a slide lecture, other crude sites apparently did not qualify as picturesque or significant enough to be captured, or else circumstances might have made it impossible to photograph them. His multi-sensory sensing and sighting of petroleum effectively boiled down to two images. We see here what Bruno Latour described vividly in *Pandora's Hope* about a field trip and soil sampling in the Amazonas: 'Scientists may dominate the world, but only to the extent that the world meets them in the form of two-dimensional, superimposable, and combinable inscriptions . . . They construct artificial representations that seem to be increasingly distant from the world and yet bring them closer.'[51] Latour further argued that the two-dimensional result-yielding image (diagram, photograph) only produced meaning insofar as it maintained the reference, as a chain of stages, to the actual site, coining the term 'circulating reference'. How close do these two historical photographs bring us to petroleum's 'nature' (or ontology) in re-visiting them? What is lost in following the trace forward and in reverse, in the sense of Latour's 'circulating reference'? Clearly, we need to re-situate the photographs in the historical moment of sensing and sighting petroleum, as I have attempted.

Heim's multi-sensory embodied experience of petroleum was hardly photographically conservable and hence demonstrable. The common practice in geology (and art history) today of using slides (or nowadays, PowerPoint) as 'faithful signs of what was photographed' and thereby of what was found has its ontological limits, as the geologist Gilian Rose has stressed: 'Slides always have a certain flatness; they can't convey taste or smell, and they're also always still.'[52] The multi-sensory experience which crude in the wild offered in 1924 and which it effectively characterised became subsequently reduced to a photographic argument that was also hard to decipher without Heim's additional descriptions. Crude's first portrait nevertheless encapsulated a visual representation of petroleum's uncertainty, a highly sensory material encounter that was unsteady, disintegrated and dynamic, but locally fixed, deep and unfathomable.

Petroleum Professed

Twelve years later, in 1936, the newly found Kuwait Oil Company, a joint venture between the Anglo-Iranian Oil Company and Gulf Oil, discovered petroleum in abundance at the Burgan Oil Field. Due to already productive fields in Iran and to resource scarcity during World War II, the industrialisation of Kuwaiti oil, however, paused to (re)start on an unprecedented scale in 1946 when the first oil shipment left Kuwait for Britain. With rapidly rising post-war energy demand and mass consumption, fossil-fuelled technologies and oil-based practices also entered private households and everyday life around the globe, spurring the industrial extraction of petroleum. In this atmosphere, Kuwaiti petroleum was on everyone's lips. However, although petroleum had finally been struck in Eastern Arabia, for a general public on site and in everyday life, it remained effectively hidden from view as a raw material. The efficient and 'clean' control over this dark matter became vital for the multinational oil company in justifying its extraction, making the resource's visual absence in the operational landscape of extraction a constant that has only been overruled at times of new oil fields (in the form of the gusher) being struck, occasional oil fires, and devastating oil spills such as during the Gulf War. In retrospect, Heim's sensing and sighting of crude in the wild, his embeddedness in the site and his embodiment of petroleum gave way to a material-centred, multi-sensory performance of future uncertainty that remains unique to this day.

With petro-culture taking root as the common way of living from the mid-twentieth century onwards, petroleum as a natural matter dissolved into abstraction. Yet, as fossil energy and chemically modified resource it manifested in such diverse and quickly omnipresent forms as cement mixers, asphalt stripes and neon signs, that also came to crowd Kuwait City's urban landscape as part of the consumerist oil boom that swept the country. Urban sprawl covered the sandy surface of the Arabian Peninsula, that relied on fossil-fuelled technologies for air-conditioning, the generation of electricity and desalinated water, and lavishly greened front yards, unleashing a new reality in terms of the life world of Middle Eastern oil. The experience of oil encounter as multi-sensory sighting in the 1920s became a mediated, representational encounter of motorised driving and of photographing the urban transformation in the 1950s, as contemporary travelogues communicated especially vividly.

The Pipeliners' Road Trip on the Tracks of Middle Eastern Oil

One of these documentations is the travel account *Im Auto nach Koweit* (*To Kuwait by Car*), in which the Austrian travel writer Max Reisch (1912–85) describes a road trip from Beirut to Kuwait undertaken in 1952 together with a colleague.[53] Calling themselves 'true pipeliners', Reisch and his partner followed the route of the Trans-Arabian Pipeline (Tapline) and, on their way back, of the recently completed Kirkuk–Baniyas Pipeline.

Loosely tracing the course of petroleum in reverse – as visualised in an illustrated map inserted into the book (Figure 3.6) – they drove from Lebanon, where the oil harbour at Sidon serves as the Tapline's final destination, via Syria, where the pipeline crosses the Golan Heights, and then from Jordan onwards alongside the Tapline into Saudi Arabia to the initial in-take station at Qaisumah. Subsequently, their route diverted to Kuwait. In contrast to Heim's experience thirty years prior, motorisation was now common and finding petrol stations along the way not a problem. Emblematic of the 'desecration' Harry St. John Philby had warned of, the desert road trip had clearly superseded the camel-paced desert voyage and subsequently gave way to a new experience of petroleum as a comfortable, but also disembodied, environmentally disconnected way of travelling. Crude, however, remained a constant.

Kuwait took centre stage again, and figured not only as the final destination of Reisch's petroleum parkour, but also as the climax of petro-pleasures. As he finally arrived in Kuwait City, Reisch experienced a 'huge adobe village

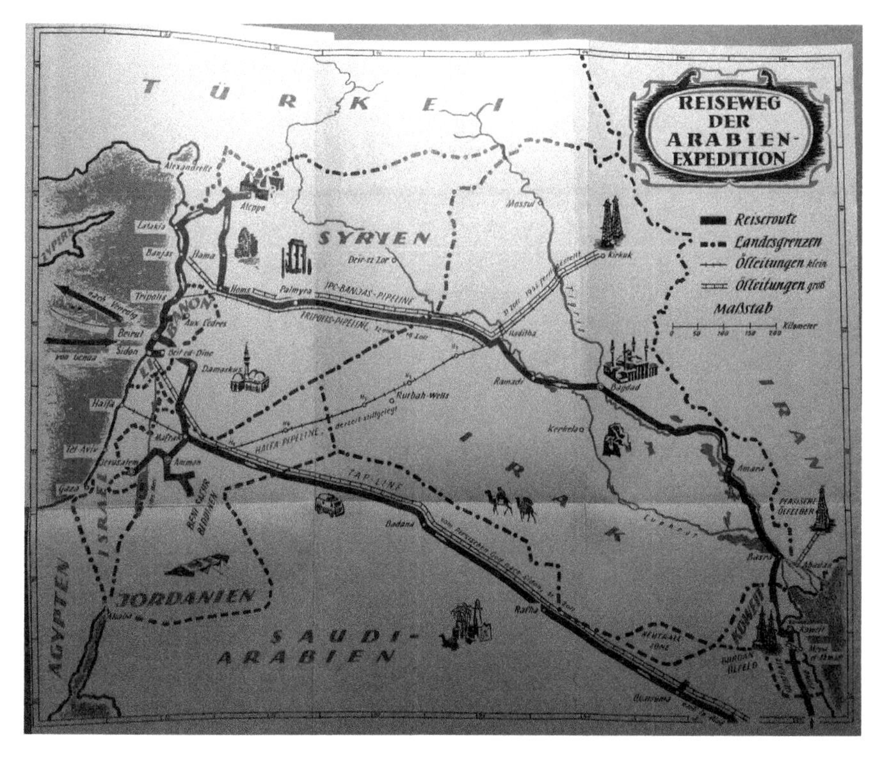

Figure 3.6 Illustrated map of Max Reisch's road trip from Beirut to Kuwait and back in 1952 as published in Im Auto nach Koweit. Courtesy of Peter Reisch.

from which a bank, a department store, and a garage of reinforced concrete have sprung up here and there'.[54] To Reisch, the Shuwaikh Secondary School (later the university of Kuwait) was emblematic of this dynamic (Figure 3.7).[55] In Reisch's colour photograph, we see the school under construction located on a vast sandy site outside the town as part of the 1950s urban expansion. It seems to rise like a *fata Morgana* out of nowhere, set against the flatness, vastness and apparent emptiness of the desert. The accelerated growth of Kuwait into a modern metropolis in an area that had been characterised by a lack of the most basic natural resources became symbolic of petro-modernity.[56] Representatives of this change were the vertically growing, solidified concrete forms of the metropolis and the desert's urbanisation. These visible changes allowed for the industrial infrastructure of derricks, pipelines, tank farms and loading piers to fall into the (back)ground, effectively disconnecting petroleum extraction and the pleasures of petro-modernity even in producing

Figure 3.7 Colour photograph taken by Max Reisch in 1952 of Shuwaikh Secondary School (later Kuwait's University) under construction. Reproduction of Kodachrom color diapositive as published in *Im Auto nach Koweit*. Courtesy of Peter Reisch.

countries like Kuwait. The mirror image of petroleum extraction became the vertical growth of Kuwait City as a simulacrum of the petro-fuelled city of comfort and prosperity, an economy of flows from bitumen underground to asphalted roads and mesmerising heights high above ground.

The most emblematic moment occurs when Reisch first entered Kuwaiti territory after a particularly strenuous ride through Saudi Arabia. Driving across the Kuwaiti desert near to the Burgan Oil Field, he experienced a fossil-fuelled epiphany:

> And suddenly the miracle happened: within the empty desert the road turned unexpectedly into a broad and asphalted road. As if conjured out of nowhere, suddenly this blue shimmering symbol of civilization was under the wheels of the 'Sadigi' . . . To my left and right there still was deadly solitude, but underneath me, in the depth, a vast treasure was lying. I was in the Burgan Field, the most productive oil field in the world. Engineers of the Kuwait Oil Company had laid asphalted roads cross and horizontal through the desert. On one of these I stormed north.[57]

Reaching the final destination of this road trip culminated in driving across a landscape that had been moulded into the smoothest driving experience for the Austrian automobilist, subsequently triggering Reisch to interpret the asphalted roads of Kuwait as a symbol of (Western) civilisation stamping desert sand. Experiencing the effects of Kuwait's oil-boom unfolded also for Reisch as a multi-sensory experience, but one ultimately mediated by petro-culture: the smell of petroleum was that of asphalted roads and exhaust gases, its touch was that transmitted through the car's tyres rolling over – the car as Reisch's bodily extension. And gliding over vast oil reservoirs yet to be extracted provided for the additional thrill. Thirty years after Arnold Heim's visit, Kuwait had become a land paved with black gold in the eyes of the passionate European motorist, who experienced the desert, 'the wild', mitigated by petroleum on multiple levels.

Travelling by car with the intoxicating sense of endless resources (while simultaneously exhausting fossil energy) provided the reason for travelling, simply because one could. *Im Auto nach Koweit* is clearly a declaration of love to motorised fossil-fuelled travelling, where automobility becomes stylised as oil consumption without the negative side-effects, therefore symbolising prosperity and individual freedom. Reisch's frictionless movement sharply contrasts with Heim's strenuous journey and the labour of Mesfer, who could only collect as much from the seepage as his oily hands could scratch and the oil tins would carry. Instead, Reisch thrived on speed and comfort, under

Figure 3.8 The 'Sadigi' parked on a square in Kuwait, a small van of the model Gutbrod Atlas 800 converted into a modern prototype camping car. Photograph taken by Max Reisch in 1952, reproduction of Kodachrome color diapositive as published in *Im Auto nach Koweit*. Courtesy of Peter Reisch.

constant illumination by his car's shining headlights; a car that he referred to as 'Sadigi' (Arabic for 'My Friend') taking great interest in the personified car's well-being. In fact, automobility in form of a customised 'house on wheels' attracted so much attention that crowds of people circled the car at every stop (Figure 3.8).

Heim, by contrast, had been confronted with breakdowns and interruptions just when travelling (and not even residing) in one of the few cars available, often getting stuck along the un-asphalted sandy tracks. In the mid-twentieth century, gasoline-fuelled automobiles formed alliances with asphalted avenues, parking lots and service points to provide for the ultimately smooth and pleasure-giving experience of petro-modernity encapsulated in Reisch's account.[58] The experience of freedom resulted in particular from asphalt having been transformed from liquified bitumen into a solid road track on which the car flowed freely.

From Bitumen Underground to Asphalt on the Road and Back

The geographer and the automobilist, despite all these differences, were both looking for petroleum and encountered it in form of bitumen and asphalt 'in the wild' of some sorts. For Heim, the Eastern Arabia expedition's highlight was the bitumen seepage of El Bohara. For Reisch, gliding over Kuwait's smooth asphalted roads 'out of nowhere' in the Burgan Oil Field became the ultimate thrill. Historically and technically, the term 'asphalt' denotes bitumen, mineral pitch, and tar or tarmac interchangeably. Heim and Reisch therefore experienced the same material insofar as bitumen describes both natural and industrially processed, both liquid and solidified asphalt-bitumen complexes.[59] In a diachronic reading, artist Robert Smithson's video work *Asphalt Rundown* (1969) performatively encapsulated petroleum's multifaceted bituminous appearance. The video shows a truck pouring liquid bitumen into an abandoned gravel pit. A black stream runs from the loading deck of the truck at the edge of the pit down to its centre. It dramatically performs not only the natural viscous flow of bituminous petroleum but also an aesthetic rendering of the generic plastering of the earth's surface with asphalted roads. In our particular context, it is also the utopian re-filling of El Bohara with the re-melted asphalt the Sadigi glided over.

Asphalt's contribution to petro-modernity is to allow for unlimited speed, to pacify the ground, and to gain distance from everything unruly that is going on underneath it.[60] As sedimentations of yet another layer of the human past, these bituminous traces of asphalted roads will become, next to plastic detritus, the stratigraphic findings in future archaeological excavations of the Anthropocene.[61]

The historical accounts of a geological field trip to the Persian Gulf in 1924, as well as a road trip across the Middle East to Kuwait in 1952, gave evidence that prospecting for petroleum and exploring the oil boom produced a visual history of Eastern Arabia, in this case of Kuwait, and of petroleum.

In Max Reisch's photographs of Kuwait, modern architecture and asphalted roads emerge 'out of nowhere' in the middle of the desert in a neo-Orientalist perspective onto Eastern Arabia. As solidified carbon form, the new architectural landscape can be read as a vertical extension of El Bohara's wild crude that the geologist Arnold Heim had captured three decades earlier. To answer the questions as to what the 'ideal' or the most 'natural' form of oil is remains a difficult task. It is nevertheless clear that crude in the wild embodies and performs a state of uncertainty. In its material ambiguity the seepage of El Bohara in Heim's photography and descriptions does not disclose whether it will be economically viable to drill and explore it further. At the time of Reisch's visit, petroleum's abundance in Eastern Arabia had been proven. But what crude oil looks like, and what its future will bring to the region, remained uncertain, too.

Bitumen or asphalt, as a specific formation of petroleum, works as an extremely meaningful continuity between naturally occurring and industrially produced fossil matter in which photography has played a crucial part. The photographic imagination of the extractive petroleum landscape of the Arabian Peninsula has been rolling on the same matter it has been seeking to encounter since its inception.

Notes

1. Memorandum of Agreement between Dr Arnold Heim and Eastern and General Syndicate Limited, 24 March 1924, Aktendossier Ostarabien, ETH Zurich University Archives, Arnold Heim Collection, Hs 494a: 2.
2. Gillian Rose and Divya P. Tolia-Kelly, 'Introduction: Visuality/Materiality. Introducing a Manifesto for Practice', in Gillian Rose and Divya P. Tolia-Kelly (eds), *Visuality/Materiality: Images, Objects and Practices* (Farnham: Ashgate, 2012), 9.
3. Ali Behdad, 'The Orientalist Photograph', in Ali Behdad and Luke Gartlan (eds), *Photography's Orientalism: New Essays on Colonial Representation* (Los Angeles: Getty Research Institute, 2013), 14–16.
4. Edward W. Said, *Orientalism* (New York: Vintage, 1979 [1978]), 50.
5. Ibid., 50–2.
6. See for example Jill Beaulieu and Mary Roberts (eds), *Orientalism's Interlocutors: Painting, Architecture, Photography* (Durham, NC: Duke University Press, 2003); Jocelyn Hackforth-Jones and Mary Roberts (eds), *Edges of Empire:*

Orientalism and Visual Culture, New Interventions in Art History (Oxford: Blackwell, 2005).

7. Letter to M. Francke, Selection Trust, 15 February 1924, Aktendossier Ostarabien.
8. Edgar Wesley Owen, *Trek of the Oil Finders: A History of Exploration for Petroleum* (Tulsa: American Association of Petroleum Geologists, 1975), 18.
9. Letter to M. Francke, Selection Trust, 15 February 1924, Aktendossier Ostarabien. On Heim as petro-geologist, see Hans-Jürgen Philipp, 'Arnold Heims erfolglose Erdölsuche und erfolgreiche Wassersuche 1924 im nordöstlichen Arabien', *Vierteljahrsschrift der Naturforschenden Gesellschaft in Zürich* 128, no. 1 (1983); Monika Gisler, *'Swiss Gang': Pioniere der Erdölexploration*, Schweizer Pioniere der Wirtschaft und Technik (Zurich: Verein für wirtschaftshistorische Studien, 2014).
10. See for example Denis Cosgrove's historical work *Apollo's Eye: A Cartographic Genealogy of the Earth in the Western Imagination* (Baltimore: Johns Hopkins University Press, 2001).
11. Reisetagebuch Arabien 1924, No. 1, ETH Zurich University Archives, Arnold Heim Collection, Hs 494: 240.29, 5–6 (author's translation).
12. Ibid., 51 (author's translation). Ilford provided photoplates of a 9 × 12 format that Heim used with the Nettel camera.
13. See also Ross Barrett and Daniel Worden, 'Oil Culture: Guest Editors' Introduction', *Journal of American Studies* 46, no. 2 (2012).
14. See also Mona Damluji's work on documentary film and oil (companies), Mona Damluji, 'Petroleum's Promise: The Neo-Colonial Imaginary of Oil Cities in the Modern Arabian Gulf' (Ph.D. diss., University of California at Berkeley, 2013).
15. William Facey and Gillian Grant, *Kuwait by the First Photographers* (London: I. B. Tauris, 1999), 28.
16. Laura Hindelang, *Iridescent Kuwait. Petro-Modernity and Urban Visual Culture since the Mid-Twentieth Century* (Berlin: De Gruyter, 2022), Chapter 4: 'The Kuwait Oil Company, (Color) Photography, and Iridescent Impressions'.
17. Mimi Sheller, 'The Origins of Global Carbon Form', *Log*, no. 47 (2019), 59.
18. Reisetagebuch Arabien 1924, No. 1, 58.
19. Ibid., 64, 70.
20. Ibid., 70.
21. Noteworthy exceptions are, for example, the German cartographer and explorer Carsten Niebuhr (1733–1815) and the German scholar and traveller Hermann Burchardt (1857–1909).

22. Harry St John Bridger Philby, *Arabia of the Wahhabis* (London: Constable, 1928).
23. The concept of the pioneering desert journey derived from nineteenth-century (Victorian) traditions of exploring unknown territory as part of a scientific quest. On the one hand, the desert voyage was known to be a risky adventure infused with the notion of man taming nature. On the other, the desert landscape and Bedouin lifestyles stimulated projections of an alternative way of living that offered spiritual redemption from Western industrialisation and modernisation. See Eva J. Holmberg, 'The Middle East', in Carl Thompson (ed.), *The Routledge Companion to Travel Writing* (London, New York: Routledge, 2016), 377–9; Billie Melman, 'The Middle East/Arabia: "The Cradle of Islam"', in Peter Hulme and Tim Youngs (eds), *The Cambridge Companion to Travel Writing* (Cambridge: Cambridge University Press, 2002), 112–19.
24. Philby, *Arabia of the Wahhabis*, vii.
25. Helen Carr, 'Modernism and Travel (1880–1940)', in Hulme and Youngs (eds), *The Cambridge Companion to Travel Writing*, 81.
26. Reisetagebuch Arabien 1924, No. 1, 76.
27. Ibid., 168.
28. Ibid., 74.
29. Ibid., 186.
30. This also connects to Gisa Weszkalnys' observation of empty oil barrels in the backyards of people in Sao Tomé and Principe as a signifier of a future oil industrialisation that is yet to be materialised, as discussed during the workshop 'The Life Worlds of Middle Eastern Oil', NYU Abu Dhabi, April 2019.
31. Arnold Heim, Geological Report No. 1, ETH Zurich University Archives, Arnold Heim Collection, Hs 494: 70, 22.
32. Ibid., 20–1.
33. Ibid., 13.
34. Geologisches Feldbuch, 1924: Arabien, ETH Zurich University Archives, Arnold Heim Collection, Hs 494: 158.49, 9 (author's translation).
35. Reisetagebuch Arabien 1924, No. 1, 79.
36. Heim, Geological Report No. 1, 8–9.
37. Reisetagebuch Arabien 1924, No. 1, 78.
38. Ibid., 69–70.
39. Reisetagebuch Arabien 1924, No. 2, ETH Zurich University Archives, Arnold Heim Collection, Hs 494: 241.30, 167.
40. Reisetagebuch Arabien 1924, No. 2, 170. Arnold Wilson, a British colonial

officer who had overseen the initial development of the oil fields at Masjid-i-Sulaiman, today Iran, and who in 1923 became manager of the Anglo-Persian Oil Company (APOC) in Abadan, visited Kuwait to deliver the APOC's oil concession proposal to Sheikh Ahmad, but was rejected.

41. Heim, Geological Report No. 1, 5.
42. Reisetagebuch Arabien 1924, No. 2, 174 (author's translation).
43. Heim, Geological Report No. 1, 6–7.
44. Ibid., 6.
45. Ibid.
46. Ibid., 6–7.
47. Hannah Tollefson and Darin Barney, 'More Liquid than Liquid: Solid-Phase Bitumen and Its Forms', *Grey Room* 77 (2019), 40.
48. Heim, Geological Report No. 1, 26.
49. Ibid., 24.
50. Fotoalben Arabien 1924, ETH Zurich University Archives, Arnold Heim Collection, Hs 494b: 19.
51. Bruno Latour, 'Zirkulierende Referenz', *ARCH+*, no. 238 (2020), 20 (author's translation); for the full-length argument see Bruno Latour, *Pandora's Hope: Essays on the Reality of Science Studies* (Cambridge, MA: Harvard University Press, 1999), Chapter 2.
52. Gillian Rose, 'On the Need to Ask How, Exactly, Is Geography "Visual"?', *Antipode* 35, no. 2 (2003), 215.
53. Attesting to its popularity, the book was published in two editions in 1953. In 2011, it was translated into Arabic and published by the Center for Research and Studies on Kuwait. For a biography on Max Reisch, see Horst Christoph, *Max Reisch: Über alle Straßen hinaus* (Innsbruck: Tyrolia-Verlag, 2012).
54. Max Reisch, *Im Auto nach Koweit* (Vienna: Ullstein, 1953), 2nd edn, 162 (author's translation).
55. Ibid., 171–2.
56. Nelida Fuccaro has frequently emphasised that 'cities and urban environments [have] constituted the primary setting where oil modernity unfolded'. Nelida Fuccaro, 'Introduction: Histories of Oil and Modernity in the Middle East', Special Issue in *Comparative Studies of South Asia, Africa and the Middle East* 33, no. 1 (2013), 1. On the importance of overcoming the lack of drinking water in Kuwait and petroleum's role in it, see Laura Hindelang, 'Precious Property: Oil and Water in Twentieth-Century Kuwait', in Carola Hein (ed.), *Oil Spaces: Exploring the Global Petroleumscape* (New York: Routledge, 2021), 159–75.

57. Reisch, *Im Auto nach Koweit*, 157 (author's translation).
58. See also architect Mirko Zardini's poetic and thought-provoking essay on asphalt's cultural history, Mirko Zardini, 'Homage to Asphalt', *Log*, no. 15 (2009).
59. Arthur Danby, *Natural Rock: Asphalts and Bitumens, Their Geology, History, Properties and Industrial Application* (London: Constable, 1913).
60. Laurence Lumely, 'The Invisible Bituminous Desert', *Log*, no. 47 (2019), 29.
61. The installation *The Period of Great Contaminations* by Gaetano Pesce in the Museum of Modern Art, New York, in 1972 morphed human and non-human forms into one rampant subterranean plastic mass. See Ingrid Halland, 'Being Plastic', *Log*, no. 47 (2019), 35–44.

4

THE OIL COMPANY'S FIELDS OF VISION: PUBLIC RELATIONS AND LABOUR IMAGES IN THE ARAB WORLD

Nelida Fuccaro

Sargon Joseph Hallaby, an Iraqi Christian born in the oil city of Kirkuk in 1934, joined the Basra Petroleum Company (BPC) in the early 1950s as a communication apprentice and was sent to Lebanon and England as a company trainee. In 1971 he left Iraq and migrated to Australia where he started working as a telecommunications technician for the Sydney Postal Services. Interviewed in the early 1990s by the Migration Heritage Centre of New South Wales for a project that collected memories of homeland among immigrants, Sargon chose to be photographed holding a copy of *The Basra Auto Radio System* – one of BPC's publications – with a photo of himself as a young apprentice in one of the company's audio equipment rooms (Figure 4.1). He recalled: 'One day in 1954, a photographer came to the company and photographed us for the book. In England, in 1967 one of the engineers said to me "You know there is a journal about your company?" and he arranged to give me a copy. It is a very good memory of those days in Iraq.'

The copy of the *Basra Audio System* and Sargon's photo as a young man taken by an unknown BPC photographer are part of a large corpus of public relations materials produced by foreign oil companies in the Arab world since the early 1950s: photographs of employees, public projects, oil infrastructure and modern townships; magazines, technical reports, brochures and posters; and videos and films used to train, educate and entertain the workforce.

Figure 4.1 Sargon Joseph Hallaby, c. 1990. Courtesy of Migration Heritage Centre and the Fairfield City Museum (Australia) 'Belonging project 2008'. Photographer: Shirley McLeod.

Much of this public relations archive is elusive. It is 'slippery' like the crude hidden underground, but also piecemeal, reflecting the painful incompleteness of and lack of access to many archives on and in the Middle East, an area where document absence and displacement are often the work of authoritarian regimes, political upheaval and, in some cases, simple carelessness. Oil company materials related to the period before the nationalisation of the industry are often scattered across several libraries and private collections in the West and Middle East, not having been made available to the public in bulk by the British and American consortia that controlled the petroleum industry.[1] As David Nye has argued, the control exercised by corporations over their publicity materials represents a metaphor for the hegemony they strived to achieve in their self-serving ideological production of capitalism and Taylorism. In an important respect, the pre- and post-nationalisation Middle Eastern oil archive is one of what Omnia El Shakry has called the 'vexed' archives of de-colonisation, a trace of an earlier 'colonial' encounter (in this case masqueraded as a corporate enterprise) that has often forced historians to write histories 'without documents'.[2]

As if the long-defunct BPC were still able to exercise the power of 'obfuscation' over its public relations images, the photo of young Sargon published on the Heritage Centre website is small and fuzzy. The page of the *Basra Audio System* he is showing to the camera includes several photographs, a confusing ensemble that precludes the viewer from recognising Sargon's facial features in BPC's audio room. While Sargon's life as a company trainee remains concealed, his photo brings to light a story of affect. By choosing to sit in front of the camera holding on his lap black and white photos of his young self, Sargon evokes his past life in Iraq. He also uses BPC's photography as an object of truth in the aftermath of the catastrophe of the First Gulf War, as material evidence of the prosperous and forward-looking Iraq of his youth. For many company employees like Sargon who toiled in oil fields, industrial workshops, refineries and oil stations, the publications and photos that featured them fashioned their affective reality and continued to evoke pride and nostalgia long after these images had been produced by company public relations offices.

We can read 1950s and 1960s labour photography through the prism of selfhood, memory and longing, yet images of workers also enable us to explore the fields of vision of the oil company as a corporation. These images disclose the combination of technology and human agency that underpinned their creation and circulation. They show how cameras and lenses interacted not only with photographers and workers as subject matter, but also with a new public relations world of oil publicists, magazine editors and writers. Moreover, labour photography provides evidence of the company's values and concepts of work and social relations, and of its power to create imaginary sociologies and geographies of petroleum production in order to entertain, instruct and ultimately subdue the workforce. Like thousands of other images, that of young Sargon served to articulate a model for the new corporate lives of oil that made acceptable the introduction of modern material and professional cultures, the transformation of living spaces and natural environments and, ultimately, the exploitative and often discriminatory corporate practice of foreign oil companies. The petroleum imagery produced by oil's public relations enlisted the company's own workers in the project of corporate extraction and efficiency and wrote them into a narrative of modernisation. Discussing the power of petroleum imagery, Pendakis and Wilson remind us that sighting oil necessitates a 'triple passage through vision, space and discourse'.[3] In catching sight of photographs as instruments of social construction and persuasion this chapter performs this triple passage by also analysing images in conjunction with stories of oil development published in the company press which were set in specific spaces of corporate life, particularly the oil township and the suburban villa.

Company Public Relations and its Mediums

In Iraq, Bahrain, Kuwait and Saudi Arabia – countries that had already developed or were in a process of developing large-scale oil extraction – British and American companies started to produce photographs and films for local audiences in the early 1950s. These still and moving images were the brainchild of newly established public relations offices that aimed to influence and, in some cases, engineer public opinion in an age of national development and militant decolonisation. The local audiences they targeted were diverse: Arab and Western company employees, politicians, governments and ordinary people alongside regional publics, particularly in downstream countries such as Lebanon which hosted refineries and pipelines that processed Arab crude for international consumption.

Across the region, the single most important factor that prompted companies to develop new strategies of communications with host countries was the negotiation of the 50/50 royalty agreements with local governments, a process spearheaded by the Arabian American Oil Company (Aramco) in Saudi Arabia in 1950.[4] The objective of these new 'prestige publicity' campaigns – as they were called by the head of the Public Relations office of the Kuwait Oil Company (KOC) – was to present the companies as the ideal partners of governments in nation-building projects, popularising their role as benevolent modernisers and dispensers of welfare, education, and material and spiritual improvement. In other words, through their public relations exercise companies sought to create an informed public opinion that advanced their corporate interests while feeding into the developmental plans of local regimes. It is no coincidence, for instance, that the Iraq Petroleum Company (IPC) opened its first Public Relations office in 1951 soon after the establishment of the Iraq Development Board, a government body in charge of overseeing public projects funded by oil revenue. In Aramco's case metropolitan audiences were also important, as American oil companies operating abroad sought to counter demands by domestic oil producers concerned about petroleum imports.[5]

Public relations also gained momentum after the Abadan crisis of 1951 that led to the nationalisation of the oil industry in Iran, and after 1956 when violent anti-British and anti-imperialist agitations triggered by the Suez crisis shook the Arab world. As labour unrest under the umbrella of anti-colonial resistance rocked oil-producing countries, the general managers of petroleum corporations in the oil metropoles of the Arab World – London and New York in particular – started to view their young public relations offices overseas as the saviours of the industry. By the end of the 1950s, the outreach activities of these offices had acquired such visibility as to attract bitter contestation, as attested by this poignant and very personalised criticism of KOC that appeared in a progressive Arabic newspaper published in Kuwait:

> To you [Kuwait Oil Company], spender of millions of rupees on cheap propaganda which you publish in yellow newspapers in praise of your great services and generosity. To you . . . who ignoring your purely commercial status are interfering in the country's affairs . . . and in the affairs of people, whose destiny compelled them to join your employment to secure an honest living . . .[6]

Written by one of the many journalists who were engaging in a fierce battle against KOC, these words speak volumes about the resentment felt by young politicos against foreign-owned companies inspired by a new wave of Pan-Arabism, Arab nationalism and left-wing activism. Such resentment was also shared by students. In a newspaper article published in 1957, the Federation of Kuwaiti Students in Egypt accused the company of oppressing its Arab workers, condemning its actions as 'thieving and embezzlement'.[7] Elsewhere in the Arab World a group of technocrats and public intellectuals also started to mobilise against foreign companies, focusing their attention on the pernicious nature of their public relations efforts. Omar Haliq, a Saudi diplomat of Palestinian origin and radical inclinations based in New York, accused public relations men of inflating the companies' 'technological missionary' rhetoric by posing as 'compassionate' promoters of local needs.[8]

The many critics of the oil companies often accused them of disseminating propaganda (*diʿayiah*), a term which after World War II was associated with the Nazi regime. Aware of the necessity for distancing themselves from this damning connection, corporations in the USA had developed public relations into a form of 'ethical persuasion', a practice that fed into a global market-place of petroleum publicity initiated by Standard Oil at the beginning of the twentieth century.[9] Like their parent companies in the USA and Europe, KOC, IPC, the Bahrain Petroleum Company (Bapco) and Aramco started to promote 'prestige publicity' that essentially aimed at building up loyalty to the company brand. KOC and Bapco employed prominent advertising agencies like Mather & Crowther and Hill & Knowlton to design company adverts for the Arabic press, and exhibition stands, logs and publicity posters for the Arab Petroleum Congresses organised by the Arab League.[10]

While advertising their brand, companies started to pay increasing attention to local customs, attitudes and policies so as to tailor their messages to intended audiences. As public relations offices extended their gaze over Arab societies, their work resembled old-style colonial information gathering, particularly in British companies where public relations was much less professionalised than in their American counterparts. Young political activists often accused KOC and BPC of harbouring spies as their public relations offices were perceived to be an instrument of informal empire, the *longa manus* of the Foreign Office in London.[11] In Kuwait, many identified the official name of KOC's Town Office – 'Arab Relations Bureau'– with the language of empire. Established in 1955 in Kuwait City as the headquarters of the company's public relations team, the Town Office supervised the company's 'friendly relations' with the ruling family and the powerful merchant class.[12] In Basra, BPC had a relatively small public relations apparatus with an Orientalist outlook. Since 1953 the head of the office had lived in an Arab House in town and held weekly majlises with the locals, like the British political officers stationed along the Arab coast.[13]

The advancement of the companies' 'prestige publicity' did not generally rely on the new media of the age, radio and television. By the late 1950s, radio stations were either state monopolies or exposed to the infiltration of what companies perceived to be 'subversive' political ideas. In 1958, public relations executives of large oil corporations with bases of operations in the Middle East extolled the virtues of radio communication, encouraging companies to contribute to national radio with programmes of great entertainment value so listeners could be distracted from Radio Cairo's 'inflammatory propaganda'.[14] Yet locally based public relations strategists viewed radio as unreliable, particularly in Iraq, Bahrain and Kuwait where it was susceptible to the Pan-Arabist message from Nasserist Egypt. Companies in the Arab world faced a further dilemma, as their participation in national radio broadcasting could be construed as direct interference in the country's affairs, exposing them to unwanted criticism.[15]

To reach their publics inside Iraq, Kuwait, Bahrain and Saudi Arabia companies relied extensively on their own press and cinema. Before the 1950s, the propaganda of companies like the Anglo Iranian Oil Company (AIOC) relied on films that were produced in London, primarily for British and international audiences. The nationalisation of Iranian oil in 1951 forced AIOC's sister companies in the Arab world – IPC and KOC – to initiate the production of in-house cinematography and magazines to prop up their legitimacy.[16] Typically, companies started their publications in English for their British and American white-collar employees living in oil towns like Awali and Ahmadi, followed by Arabic magazines that targeted local blue-collar workers. As in the case of the radio, the local press was not considered a reliable medium, since it was perceived as being too politicised.

IPC started to publish the monthly magazines *Iraq Petroleum* and *Ahl al-Naft* (Oil People) in 1951. The company's Film Unit – also established in 1951 – had an annual production of approximately ten documentaries that were screened in Iraqi cinemas and schools and, after 1957, occasionally broadcast on the young national television. Bapco and KOC started a number of publications after the nationalist agitations during the Suez crisis: *The Islander* in 1956 and *Al-Najmah al-'Usbu'iyyah* (The Weekly Star) in 1957; in Kuwait, *The Kuwaiti* in 1948 and *Risalat al-Naft* (The Oil Newsletter) in 1957, the latter replaced by *al-Kuwayti* in 1960. In 1958 Bapco and KOC also established Film Units that produced short features and newsreel cinematography, also with entertainment and educational advertising. Aramco in Saudi Arabia had the largest and most professional Public Relations Department in the Middle East and was the only company that sponsored its own television station, established in 1957. From 1945, Aramco published *Arabian Sun and Flare* in English for its American employees, followed by *Aramco World* in 1949, and in 1953 it started *Qafilah al-Zayt* (Oil Caravan), the first illustrated magazine in Arabic published in Saudi Arabia.[17]

Labour, Public Relations Photography and Cameras

> Photography is an apparatus of power that cannot be reduced to any of its components: a camera, a photographer, a photographed environment, object, person, or spectator. 'Photography' is a term that designates an ensemble of diverse actions that contain the production, distribution and exchange and consumption of the photographic image.[18]

By the late 1950s, magazines and films sponsored by the companies had a relatively large distribution, particularly in countries like Iraq and Bahrain.[19] The circulation of still and moving images through company publications, reports, brochures, films, documentaries and newsreels created new public relations markets across the region. Like cars, electric fans and fridges, magazines and films became sought-after items of consumption to be read, exchanged and watched, part and parcel of the new material worlds produced by the oil industry. As oil objects, images generated new chains of consumption and formed personal experience, as exemplified by Sargon's story related at the beginning of this chapter. They also fostered intimacy between the local population and the oil industry, as, in particular, the photographs included in Arabic publications depicted workers in workshops, oil fields and company housing.[20] Yet at the same time, these images distorted realities of development by creating a picture of the industry in the image of the corporation, making statements about the world that reflected managerial, rather than workers', perspectives.

Before the 1950s photographs had a very limited public relations value in areas of oil production. For instance, up until World War II the annual reports that IPC and Bapco prepared for the Iraqi and Bahraini governments included very few images and were poorly designed. After the war, as the oil industry started to expand in Kuwait, KOC employed freelance professional photographers like Adolf Morath, a renowned German-born but London-based industrial photographer, to produce portfolios of infrastructure and public projects.[21] With the printing boom of the 1950s, the services of commercial as opposed to art photographers like Morath became very much in demand as magazines, newsletters, pamphlets, educational posters and official reports necessitated images that reflected the achievement of the company and its drive to contribute to local development, while conveying to the outside world the effectiveness of its communication strategies.

In April 1957, the Director of the Printing and Publishing Department of KOC urged the general manager of the company to take into serious consideration the employment of a professional photographer in charge of public relations advertising. He lamented the poor quality of the photos included in the company's annual report to the ruler of Kuwait and noted that they were unable to compete with the high-standard images included in the publicity materials of Bapco and Aramco, the two American companies operating in Bahrain and Saudi Arabia respectively. Bapco fared quite well in this propaganda race. Two years earlier the company had won an award at a prestigious printing exhibition in New York for the quality and design of its annual report to the government of Bahrain.[22] The anti-imperialist agitation that engulfed the Arab World after Suez was, however, the single most important factor that forced oil publicists working for British-controlled companies to focus on local employees as the human face of the petroleum industry. A few months after the publication of the vitriolic article attacking KOC propaganda mentioned in the previous section, the company headquarters in London instructed Kuwait's Town Office to use local personnel – as opposed to expatriates – in publicity photographs with a view to showing an 'appreciation' of local labour.[23]

The iconography of labour emerging from the public relations photography of the 1950s and 1960s is similar to that widespread in Western industrial capitalism. The Arab oil worker is clean, effortless and dressed up in tidy and orderly clothes, even when performing manual labour in workshops and oil fields, in opposition to his or her sweaty Socialist counterpart toiling to fulfil oil's working-class mission. Given the highly mechanised nature of the oil industry, workers were portrayed individually or in groups defined by technical/manual skills. Group photos usually convey the importance of the job rather than that of the person performing it, while portraits of individuals emphasise dedication and ability. Labour photography mirrored the neo-colonial hierarchies that dominated the Middle East oil industry shaped by stark racial divisions that were coterminous with pay grades. Images published in company magazines in particular evoke not only the distance between white British and American management and the local workforce but also that between national labour (Iraqis, Saudis, Bahrainis and Kuwaitis) and Arab and Indian expatriate communities. When nationals meet expatriates, they always occupy a position of subordination, as for instance Egyptian teachers giving language classes to Saudis, and Palestinian foremen supervising Kuwaitis in KOC workshops.

Like public relations, the corporate photography of labour was an American invention, as corporations like General Electric established photographic divisions in the 1890s.[24] The Depression years guided the development of a documentary style of photography that depicted the poor and influenced the visual counter-offensive launched by corporations to protect the interests of capitalist enterprises. After World War II, professional photo-journalism made inroads into the nascent Middle East oil industry keen to publicise it to Americans as the new and friendly face of US industrial capitalism abroad. In June 1945, *Life* magazine published a large photo essay by renowned photographer Dmitri Kessel entitled *Middle East Oil* claiming to be 'the first complete look at this fabulous and troublesome part of the world'.[25] The photo album Kessel shot for the essay includes portraits of Saudi, Iraqi, Bahraini and Iranian workers alongside images of the towering petroleum infrastructure that was shaping the region's deserts in the image of corporate America (Figure 4.2). Kessel's portraiture is intimate and realistic yet staged and purged of any political and social allusion. As will be explained below, Aramco Public Relations Department adopted this photo-journalistic style as one of its key communication strategies in magazines like *Qafilah al-Zayt*.[26]

Like Kessel's photos, the few available public relations photo albums of KOC and IPC are a flawed construction of industrial realities as they hide the precarious working and living conditions of large segments of the workforce. Images of Saudis, Kuwaitis, Iraqis and Bahrainis at work in oil fields, industrial installations, workshops and company towns have often an eerie and

Figure 4.2 Three Bahraini oil workers, Dmitri Kessel, 1945. Courtesy of Shutterstock.

unnatural quality. Workers appear as accessories of and an extension to the machines they operate, an integral part of sites of technology that made up the landscapes of oil extraction and production. To historians, these photo albums constitute an often unsystematic kaleidoscope of public relations thinking and action. Rather than creating a chronological history, they form a sporadic documentation of people, workplaces and living spaces that are filtered through the technical and technological gaze of the company. It is no coincidence that labour was not indexed in KOC and IPC's public relations photo archive but included at random under the categories 'Infrastructure', 'Development Projects', 'Housing Welfare' and 'Training'.[27]

Some images establish a man–machine visual system that directs the attention of the viewer towards the wonders of oil machinery by cutting off the surrounding working environment. In a series of striking photos of wellheads, the focus of the lens is on their intricate mechanical architecture, into which their human operators blend with ease (Figures 4.3, 4.4). Rather than documenting the realism of the workplace, these images give precedence to aesthetics over function, sidelining the plight of the workers. In fact, their sharp focus and factual style of presentation points to an aesthetics of realism that is typical of industrial photography.

Figure 4.3 KOC wellhead operators, 1959, KOCA GT 7532. Courtesy of Kuwait Oil Company.

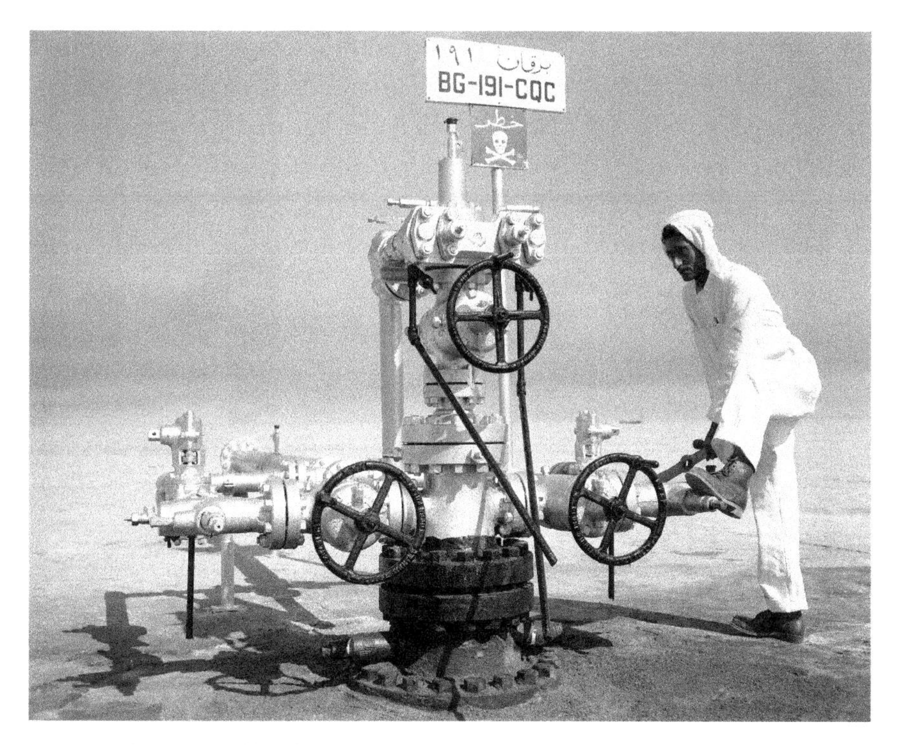

Figure 4.4 KOC wellhead operators, 1961, KOCA GT 9813. Courtesy of Kuwait Oil Company.[28]

Human figures were often used to highlight different landscapes of production linked to oil extraction, particularly company towns and housing developments where the company manufactured modern ways of collective life as the appendixes of petroleum's industrial sites.[29] KOC's large photographic portfolio featuring Ahmadi, Kuwait's oil town, exudes the company's confidence in oil-powered technology and infrastructural development. Traffic policemen, petrol station attendants and firefighters – all Kuwaitis on the company's payroll – are portrayed engaging with cars, the icons of petroleum's new urban age.

In contrast with the alienating images of wellhead operators, the human and machine components of these pictures are more balanced, hinting at the social implications at stake in technological advance. Integrated into landscapes of production, workers and machines are combined to convey an 'ergonomic' performance, in terms of both technical efficiency and human safety.

The traffic policeman looks sternly but confidently straight into the camera lens. He is at the centre of the picture's frame, with the front of an incoming car entering it from his right-hand side. In the background is Ahmadi's 'Arab village', the new housing development built by the company which the traffic warden is sheltering from dangerous driving (Figure 4.5). Another set of images feature bus drivers lining up in front of their buses ready to start their shift, a firefighter operating the communication system of a fire engine from its front seat, and another being trained to use a water hose.[30]

While presenting the relationship between workers and technical equipment as naturalised into the suburban setting of Ahmadi, these images speak to the alienation embedded in this relationship through the portrayal of Kuwaitis as motionless and depersonalised human subjects. The same staged quality of stillness infuses images of young Kuwaitis engaging with state-of-the-art services offered by the company for personal use like public telephones

Figure 4.5 Traffic police in Ahmadi Town, 1953. KOCA, AFI 461. Courtesy of Kuwait Oil Company.

and post-boxes.[31] Establishing the axis workers–technology–modern infrastructure, the Ahmadi photo album encapsulates KOC's self-image as supporter of a benevolent technical order that treated workers as a cog in the industrial machine of oil production. In this respect, photographers continued to hide the human factor of the industry by subjecting workers to a corporate ideology that celebrated the power of machines at the expense of that of workers as the driving force of an increasingly mechanised society.

Reproducing a similar multi-layered visual regime that blended image-technology and image-labour, the illustrations published in workers' magazines provide a further example of how the ensemble of photographers, cameras and lenses entered the management strategies of companies as human and technical corporate meaning-makers. One of the main purposes of labour photography – and indeed one of the main incentives for its development in the 1950s – was to illustrate stories about workers, company achievement and oil development that appeared in company publications. As such, images of workers became one of the pillars of the company's mission to persuade, instruct and entertain the labour force, that also served to operationalise its welfare paternalism.

Aramco led the way with *Qafilah al-Zayt*, a monthly magazine that began publication in 1953. Up until the early 1960s – when its readership remained largely confined to Aramco employees – images of young and old Saudi workers feature prominently on front and back covers and in feature articles. *Qafilah*'s labour photography oozes the confidence of the company's professional and sleek public relations machine and stands out among that of its contemporaries (particularly KOC's *Risalat al-Naft* and Bapco's *al-Najmah al-'Usbu'iyyah*) for its realistic style and bright colour palette. Photos had the key advantage of attracting the attention of workers irrespective of their level of literacy, which in a country like Saudi Arabia was still very low. Despite the magazine's strong visual profile, the name of Aramco's photographers does not appear in print, occasionally only as anonymised initials, unlike those of the public relations officers, intellectuals and poets who wrote feature articles, weekly columns and editorials. Moreover, when browsing the magazine, the presence of the camera is palpable, although we are seldom made aware of it explicitly. When this happens, it is a fleeting moment. The reader, for instance, is casually reminded that the camera lens is 'with us' in a feature article published in 1957 on Rahimah Town, one of Aramco's new housing projects.[32]

While anonymity is an indication of the scant professional and artistic visibility of public relations photographers as Aramco employees, oil publicists were keen to showcase the art of image making as part of the company's technical wizardry. In 1959, *Qafilah* published a long feature on the history of photography alongside a photo essay that explained various aspects of its use by Aramco. Three pictures in particular offer evidence of the company's presentation of photography as a technical and bureaucratic tool in the hands of the company. The first two are images of oil fires which are contrasted to show how photos serve different purposes: the first to understand technical failure, the second for news reporting. The third image explains the use of photography as an instrument of instant 'recognition' that served Aramco's scientific management of the workforce. In the photo, a Saudi expert in personal records (*khabir sajjalat athbat al-shakhsiyyah*) is searching a filing cabinet to add a passport-size photo to an employee dossier.[33]

In the same way as cars, cameras were presented as machines essential to the new oil world brought by the company to Eastern Arabia. Yet unlike cars, they entered the sphere of corporate management also as objects of leisure. The theme of workers' leisure had dominated employee magazines published by US corporations since the early twentieth century, promoted to increase productivity and distract workers from strikes and insubordination. As David Nye has suggested in his study of General Electric's magazine *Work News*, a key aspect of company ideology was 'endless substitutions for and translations of work into other realms'.[34] In the Arab World, companies were also keen to present the after-work life of labour as an extension of company employment. Bapco's magazine *al-Najmah al-'Usbu'yiyyah* prominently featured images of workers listening to the radio, swimming and playing table games as a component of orderly and educational leisure pursuits that served to discipline Bahrain's tumultuous labour force after the agitations of the mid-1950s.[35] *Qafilah al-Zayt* went a step further by including a column on hobbies (*Huwayyat*) that related first-hand personal stories of leisure pursuits by Saudis on the payroll of Aramco.

One of the first articles featured Yusuf ibn 'Abdallah al-Da'iji – a technician in one of the air conditioners' workshops – as an amateur photographer. The large image illustrating the column portrays Yusuf hesitantly holding a state-of-the-art Rolleiflex camera at his chest, looking away from the camera's

Figure 4.6 Yusuf ibn ʿAbdallah al-Daʾiji, an amateur photographer. *Qafilah al-Zayt*, Year 7, vol. 1, July 1959, p. 36. Courtesy of Saudi Aramco.

upper viewing lens (Figure 4.6). Like many images included in *Qafilah*, that of Yusuf can be interpreted as presenting an idealised and humble figure of the archetypal Aramco skilled worker. The caption reads: 'Photography is a skill that requires patience and precision, it can be a remunerating profession, but for Yusuf it is a hobby which he says helps him to overcome boredom.'[36] Yusuf's avoidance of the gaze of two cameras, the one he is holding and that of the Aramco photographer taking the picture, could be interpreted as an act of both submissiveness and defiance. Yet it is difficult to imagine Yusuf making an independent statement. His posing in front of the camera displays an artificial symmetry accentuated by the branches of the oleander tree that frame the picture.

After the establishment of the Aramco television station in 1957, photography was also used to introduce the new medium of cinematography to *Qafilah*'s readers. A feature on the new Aramco television station shows a picture of two Arab cameramen (probably Saudis) shooting outdoors from the roof of a mobile unit. The caption informs the reader that the state-of-the-art equipment they are using can record images and sound simultaneously. In the same feature, the camera enters the recording studio during the filming of the popular programme *Learn Arabic on TV*, an extension of Aramco's literacy programmes for Saudis.[37] The superimposition of photography and cinematography as public relations tools is exemplified by an image published in 1961 as part of an illustrated tour of Aramco's oil exhibition held in Jeddah, an event to publicise the facts and figures of the global oil industry, and Aramco operations and achievements in the Eastern province of Saudi Arabia.[38] The photographer captures one of Aramco's cameraman while filming a group of Saudi visitors standing in front of a large map showing global petroleum consumption (Figure 4.7). Their images are broadcast live on a television screen that hangs from the ceiling above the map. The visitors are photographed from the back with only the upper part of their bodies visible in the frame, standing between the cameraman and the photographer taking the picture. The attention of the audience (and of the viewer) is directed towards the television camera, the cameraman

Figure 4.7 Watching Jeddah's oil exhibition. *Qafilah al-Zayt*, February/March 1961, p. 19. Courtesy of Saudi Aramco.

behind it, and the screen. The large map is distant and out of focus, used as compositional background.

The double gaze of the photographer and the cameraman, and the audience watching themselves on screen while looking at the map, makes this image a powerful montage. This is a double-layered public relations exercise that brings together technology and representation of oil publics, the latter as both subjects and audiences of still and moving images.

Exterior and Interior Ecologies of Oil Lives

As emblematic 'experts', company photographers worked alongside television crews, writers, journalists and public relations specialists to make the workforce increasingly legible and visible to scrutiny and corporate management. In the new public relations age, photography and workers magazines became a powerful instrument for the reproduction of labour alongside the technical training and educational programmes offered in schools, workshops and oil fields.[39] *Qafilah* offers plenty of examples of the use of images as a pedagogical device to develop the general knowledge and key skills of its readers in line with the tests developed by Aramco for the selection of its Saudi personnel.[40] For instance, in the column 'Testing Your Knowledge' (*Imtihan Ma'lumatik*) pictures of oil infrastructure, towns, and Saudis at work are used in multiple-choice drills to hone readers' basic visual and numeracy skills, to familiarise them with ideas of 'traditional' and 'modern', and to develop their general knowledge about Aramco's operations.[41] Photo competitions also tested attention to detail, visual memory and writing skills. In the November 1957 issue, for instance, readers were asked to recognise the jobs performed by six Saudi workers and to write the answers in the slots under each image. Instructions printed at the bottom of the page asked participants to write their answers clearly and legibly.[42]

With workers as target audience, magazines also used photography in feature articles that presented but also represented the lives of labour. Alongside text, images created intertwined rhetorical and visual fields concerning workers, their families, and their professional and private behaviour. For the company this was in part a self-referential exercise that ultimately conveyed very little about the real conditions of the workforce but made statements about the lives the company wanted its employees to lead. Photography was one of the principal means through which companies came to 'see' their employees through the lens of their public relations efforts. This gaze was instrumental in enforcing regimes of social and personal discipline, labour bureaucratisation and corporate indoctrination. In this respect, photography inside magazines offers further insights into how companies performed the reproduction of labour by means other than entertainment and education. Leading through example, the stories they published instructed workers in the ways of modern living so as to direct their thoughts and behaviour towards corporate loyalty while showing an active participation in the modernisation and development plans of local governments.

For maximum effect among a workforce whose level of literacy was low, oil publicists narrated stories of industrial productivity and corporate sociability in the photo-journalistic style widespread in the US media and corporate world. Particularly in Aramco's *Qafilah* but also in IPC's *Iraq Petroleum* and *Ahl al-Naft* – all magazines with high-quality colour images – photographs often led worker-related stories set in industrial sites, company offices and modern townships, and shaped narrative styles and aesthetic conventions, crafting very distinctive yet flawed portrayals of company life. Unlike the photographic tour (*al-Jawlah al-Musawwirah*) that took readers on journeys of technical discovery of company infrastructure, the illustrated story (*al-Qissah al-Musawwirah*) unveiled aspects of the nitty-gritty of oil life by sketching the stylised profiles of young and upwardly mobile Saudi, Iraqi, Bahraini and Kuwaiti employees. The adoption of the 'telling stories through images' formula in the press mirrored the increasing popularity of company-sponsored cinema. For this reason, stories about workers were also often integrated into news items, as for instance in the Illustrated News section of *Qafilah* (*al-Anba' al-Musawwirah*), that attracted readers' attention by using as headings drawings of cinema newsreels.[43]

Company photographers entered offices, worksites and the corporate home. An article published in 1955 follows step by step Saudi jobseekers inside Aramco Employment Office to show registration, medical inspection and aptitude testing.[44] One of the features in the 'Hobbies' column captures 'Abdallah Hasan at home, revealing how the company viewed his after-work pastime (mending watches and alarm clocks) as an extension of company employment. The title of the article plays on the double meaning of the

Arabic word *sa'at* (watches or hours). *Taslih al-Sa'at* hints not only at watch repairing but also at the 'fixing' – in the sense of 'filling up' – of 'Aballah's empty hours of leisure with a technical pursuit. 'Abdallah's photo occupies a large portion of the page and is shot to convey the importance of the task he is performing, and his dedication, confidence and dexterity. Sitting at a desk in his living room, he is repairing an alarm clock using precision tools and a magnifying lens (Figure 4.8).

This picture is also part of a large corpus of labour photography that manufactured workers' corporate domesticities in company literature. 'Abdallah's body is framed by curtains in the background and by a large radio set and an alarm clock to the side, the archetypal items of the modern company

Figure 4.8 'Abdallah repairing an alarm clock at home. *Qafilah al-Zayt*, Year 6, vol. 7, March/April 1959, p. 27. Courtesy of Saudi Aramco.

housing, alongside desks, chairs and Western furniture.[45] The illustrated story also followed daily work and leisure routines across several company spaces: from townships to the workplace; from personnel offices in charge of arranging education for children, distributing company literature and supplying refrigerators and electric fans, to clubs and schools where the company organised recreation and education. The collage arrangement of the illustrative materials of these genre articles often conveyed the sequencing of an assembly line (Figure 4.9).

Alongside the workplace, magazines portrayed housing built or subsidised by companies – the centrepiece of company welfare paternalism – as the hubs of new oil communities. In fact, they became the archetypal accessory that defined the model up-and-coming employee, driven by individualism, ambition and self-worth. New housing developments and oil towns feature abundantly in company public relations literature and photography of the 1950s and 1960s, as also exemplified by KOC's portfolio on Ahmadi discussed in the previous section. Although companies extensively used oil's built environments to make statements about the benefits of modern life, the photos featuring them establish an ambiguous and contrasting visual regime. On the one hand, aerial photography was used to celebrate housing developments as large-scale public projects like refineries and pipelines. Images from above also served to hide or dismiss the 'messy' details of the everyday life of the workforce, and convey a sense of social order, progress, discipline and harmony.[46] On the other hand, this lifeless large-scale spectacle of oil progress and prosperity was countered by close-up encounters with the company suburban villa. The reader expects these encounters to anticipate the company's intrusion into the home as the core of the biological – as opposed to corporate – family.

Yet many of these close-ups effaced domesticity. Oil publicists projected a stylised, and what now appears a dystopian, image of the oil suburban villa, emptied of human subjects, and spatially, culturally and socially alienated from the traditional towns and villages located in areas of petroleum extraction that were becoming integrated into large oil conurbations. The image-building strategy of companies focused on showing villas from a distance, from above or in miniature. Magazines prominently featured photographs of exteriors of unoccupied company villas of different sizes and design, or of villa prototypes

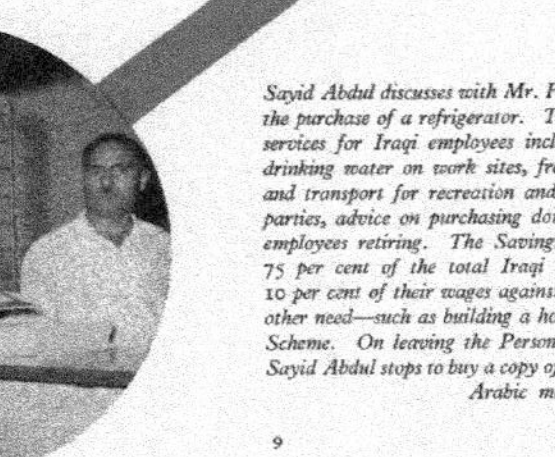

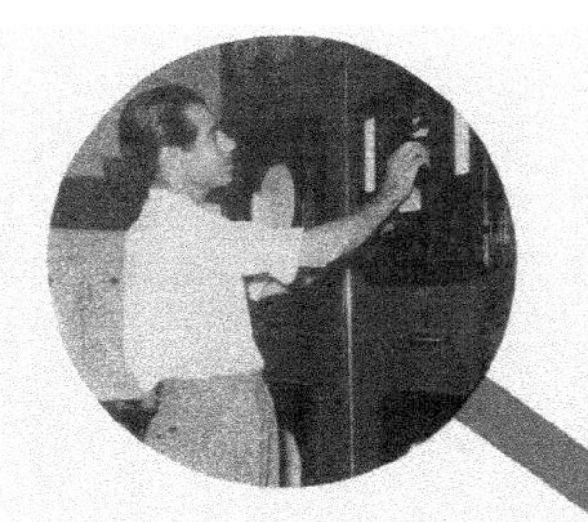

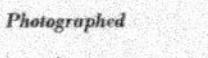

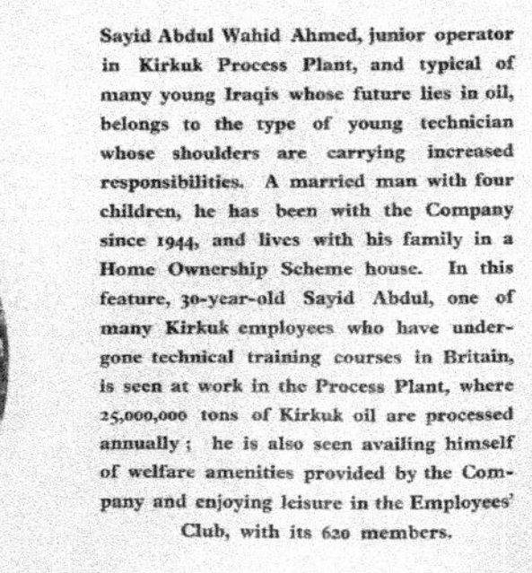

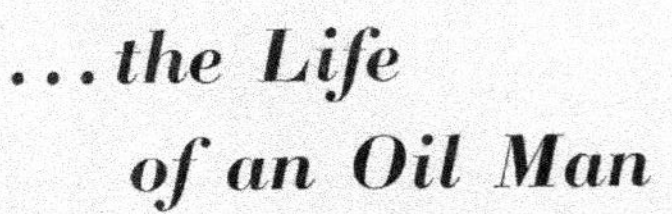

Sayid Abdul Wahid leaves his home in Kirkuk for the railway station where a train will take him to his work in the heart of the oilfield. His home was built under the Company's Home Ownership Scheme. Under this scheme, employees deposit 10 per cent. of the cost of the land and house, and the balance, lent to them by the Company, is repaid in monthly instalments. Sayid Abdul chose his land and style of house, and the services provided by the Company often amount to 20 per cent of the value of the house. When Sayid Abdul has finished paying instalments the house will belong to him entirely: in the event of his death, the property will be handed over to his wife without any further cost, even though he may have just started repayments.

Photographed
by
DAVID KITTAR
Kirkuk Photographic Section

A Day in . . . . . . the Life of an Oil Man

Sayid Abdul Wahid Ahmed, junior operator in Kirkuk Process Plant, and typical of many young Iraqis whose future lies in oil, belongs to the type of young technician whose shoulders are carrying increased responsibilities. A married man with four children, he has been with the Company since 1944, and lives with his family in a Home Ownership Scheme house. In this feature, 30-year-old Sayid Abdul, one of many Kirkuk employees who have undergone technical training courses in Britain, is seen at work in the Process Plant, where 25,000,000 tons of Kirkuk oil are processed annually; he is also seen availing himself of welfare amenities provided by the Company and enjoying leisure in the Employees' Club, with its 620 members.

8

As a junior operator, Sayid Abdul is responsible for the smooth running of from one to three stabilisation columns in the Process Plant at Kirkuk. Inside these columns, gases from heated crude oil rise in vapour form while the stabilised crude falls to the base. The object of this process—known as distillation—is the separation of certain constituents of the crude oil. Sayid Abdul Wahid ensures that the furnaces which heat the crude oil are kept burning, and that levels and temperatures in the columns are kept constant. In cases of emergency, he is expected to call his Supervisor—but he is also trained to use his own initiative. He is seen (opposite, below) supervising the running-up of a heat exchanger with Mr. R. Morrison, Shift Control Operator, standing below; and (left) checking the inlet and outlet temperatures of oil in a still tube.

Sayid Abdul discusses with Mr. Frank Jeune, Personnel Services, the purchase of a refrigerator. This Department's wide range of services for Iraqi employees includes the provision of ice, cold drinking water on work sites, free transport for school children and transport for recreation and shopping, catering for private parties, advice on purchasing domestic articles, and guidance to employees retiring. The Savings Group Scheme, supported by 75 per cent of the total Iraqi employees who contribute over 10 per cent of their wages against retirement, emergency or some other need—such as building a house under the Home Ownership Scheme. On leaving the Personnel Services Department (left), Sayid Abdul stops to buy a copy of "Ahlun Naft," the Company's Arabic magazine.

9

Figure 4.9 'A Day … in the Life of an Oil Man', *Iraq Petroleum*, Year 5, vol. 8, March 1956, pp. 8–10. © BP Archive.

on display in personnel offices for prospective owners/tenants. Villas were also publicised as suburban skylines dissociated from services and amenities, as suggested by several striking images of houses built by IPC on the outskirts of Kirkuk. Published in *Iraq Petroleum*, these images show an empty and desolate suburbia which the caption describes as 'imaginative town-planning'.[47]

The presentation of suburban living spaces in company magazines also impressed upon readers the idea that the straight lines of their modernist architecture and interior design were essential technical ingredients of orderly, privileged, hygienic and aesthetically pleasing lives. Features that publicised home ownership schemes often included floor plans and images of furnished interiors. In December 1959 *Iraq Petroleum* published a set of striking colour images of the model house of the BPC's Home Ownership Scheme with house plans and pictures of a modern kitchen and dining room with linoleum floors and built-in cupboards (Figure 4.10).

Corporate photography often juxtaposed the modern exterior design of company housing with images of family life that represented the household as a nuclear family. Documenting 'moving-in' days often served this purpose, as in the images of Kuwaiti and Iraqi (male) employees taking possession with their children of company-sponsored villas in Ahmadi and Basra published in *The Kuwaiti, Ahl al-Naft* and *Iraq Petroleum* (Figure 4.11).[48]

Yet the camera seldom entered these new homes as domestic environments. Magazines did not publish panoramic photographic tours of villas like those that took readers inside workshops, offices, canteens and leisure centres. Images depicting interiors with their residents are remarkably few and far between. They are stylised and follow compositional clichés, which suggests that they were commissioned as promotional materials to make company housing schemes palatable to the labour force. When the camera entered villas, its gaze did not reach beyond the semi-public reception area of the majlis/living room, capturing the secluded nature of the traditional Arab household. Here, workers were portrayed surrounded by a mixture of 'Oriental' items like carpets, cushions and narghiles, Western furniture like table, desks, chairs and curtains, and corporate objects like radios and clocks. Photographers also captured the patriarchal nature of households as no wives, daughters or mothers feature in these images, only male guests and children.[49]

Female-less interiors displaying an eclectic and bare house décor like that

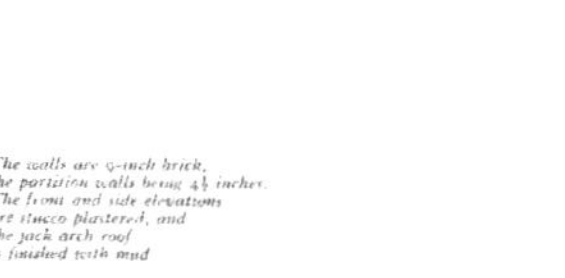

The walls are 9-inch brick, the partition walls being 4½ inches. The front and side elevations are stucco plastered, and the jack arch roof is finished with mud

The internal decoration is complete. There is terrazzo tiling in the sitting-room, dining-room and hall, the floors of the kitchen, living-room and study being covered with linoleum

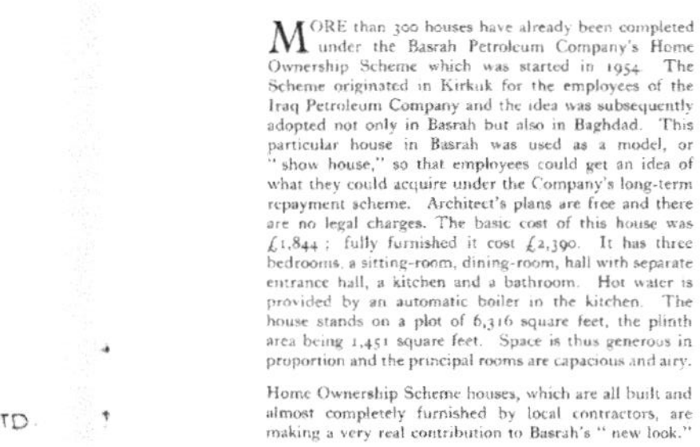
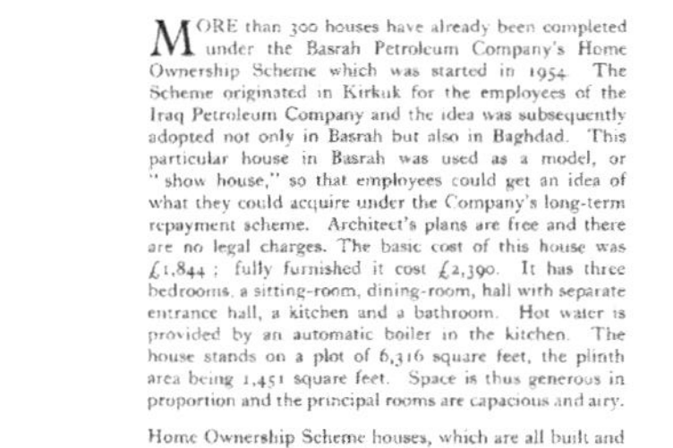
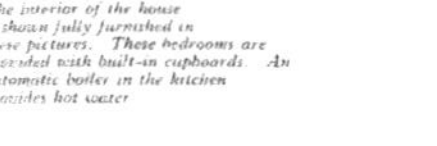

The interior of the house is shown fully furnished in these pictures. These bedrooms are provided with built-in cupboards. An automatic boiler in the kitchen provides hot water

MORE than 300 houses have already been completed under the Basrah Petroleum Company's Home Ownership Scheme which was started in 1954. The Scheme originated in Kirkuk for the employees of the Iraq Petroleum Company and the idea was subsequently adopted not only in Basrah but also in Baghdad. This particular house in Basrah was used as a model, or "show house," so that employees could get an idea of what they could acquire under the Company's long-term repayment scheme. Architect's plans are free and there are no legal charges. The basic cost of this house was £1,844; fully furnished it cost £2,390. It has three bedrooms, a sitting-room, dining-room, hall with separate entrance hall, a kitchen and a bathroom. Hot water is provided by an automatic boiler in the kitchen. The house stands on a plot of 6,316 square feet, the plinth area being 1,451 square feet. Space is thus generous in proportion and the principal rooms are capacious and airy.

Home Ownership Scheme houses, which are all built and almost completely furnished by local contractors, are making a very real contribution to Basrah's "new look."

This "show house" has been bought, through the Scheme, by Sayid Nasir Dawood, an equipment operator in Mechanical Engineering Department, who has been with the Basrah Petroleum Company for seven years.

14 15

Figure 4.10 'The "Show House" of the Basra Petroleum Company's Home Ownership Scheme, *Iraq Petroleum*, Year 8, vol. 9, November/December 1959, pp. 15–16. © BP Archive.

Figure 4.11 Moving in company housing. Composite image. From right to left: KOCA PR 2973 and PR 2990 published in 'First Day Payroll Families Move into Company Accommodation', *The Kuwaiti*, 6 May 1954, n. 318; 'Home Ownership in Basrah', *Iraq Petroleum*, Year 7, vol. 9, April 1958. Courtesy of Kuwait Oil Company and © BP Archive.

in Ahmadi's Arab Village (Figure 4.12) contrast with the many articles that addressed spouses of Saudi employees as housewives, instructing them in the art of arranging modern, beautiful, functional and happy new homes. By the early 1960s company magazines included features and columns that linked the interiors and fittings of the modern oil villa to new models of female domesticity, often with photos of state-of-the-art Western-style furnished kitchens and living rooms. As discussed by Reem Alissa, in Kuwait KOC publications created the image of the Ahmadi housewife which had a profound influence on the lifestyle of young women in the 1960s and 1970s.[50] Yet despite the influence of home features on social behaviour, the photos that illustrated them have an unmistakably *un*-placed quality. They are manicured

Figure 4.12 Interior of worker's house in the Arab Village, Ahmadi. Courtesy of Kuwait Oil Company.

showroom shots of furnished interiors that hide the subjectivity and the cultural specificities of family life. For instance, in 1957–8 IPC's Arabic magazine *Ahl al-Naft* published a series of articles on the sleek interiors of the modernist country home of the British and American upper classes, addressing the increasingly sophisticated and westernised taste of Iraq's bourgeoisie.

The arty and lavishly illustrated house interiors of *Ahl Al-Naft* contrast with the sober and image-less household that emerges from *Qafilah*'s column 'Rukn al-Manzil' ('The Home Corner'), partly reflecting the more traditional social milieu of the Eastern province of Saudi Arabia. The pronounced didactic tone of this column also shows the vigorous efforts by Aramco to enforce the model of the nuclear family in order to lure married employees to live in company-sponsored housing.[51] The column instructs the lady of the house in matters of home economics, house decoration, gardening and family life: how to store perishable food, which flowers and herbs to plant in the garden (and in which season), and how to choose and arrange furniture, curtains and carpets in matching colour schemes. In the late 1950s articles started to focus less on how to beautify house interiors and exteriors, and more on nuclear family behaviour, particularly child rearing and marital relationships.[52] The absence of illustrations and of female voices in the column, and the enforcement of unabashedly Western models of female domesticity, makes the wives of Aramco's Saudi employees an accessory to the model house, the hidden and invisible subjects of company discipline. Though increasingly perceived by the company as important to labour's social reproduction and the functioning of oil operations, local women continued to be ignored in company literature.

Conclusion

Using Sargon Hallaby's photo of a photo as an entry point into the corporate photography of oil in the Arab World, this chapter has traced aspects of and actors in the multi-layered visual universe created by the petroleum industry in the 1950s and 1960s: not only workers as photographic subjects and public relations offices as hubs of company communication, but also photographers, cameras and workers' magazines. The labour-related images they combined to produce and circulate bring into sharp focus the company's dominant gaze, inevitably highlighting absences and exclusions: not only oil in its crude and processed forms and the spouses, daughters and mothers of Saudi employees, but also unskilled workers on daily contracts and non-Western expatriate labour like Palestinians, Egyptians, Indians and Pakistanis. In the 1950s, the newly established partnership between foreign companies and local governments resulting from 50/50 royalty agreements made this omission inevitable as companies became increasingly committed to the 'nationalisation' of the workforce. Sighting labour photography also highlights how the different scales of the company's fields of vision – from the aerial shot of the oil township and the photo of the oil villa's exterior to the close-ups of interiors, individual workers and the machinery they operated – served to engineer new oil communities, and with them the first corporate/industrial landscapes of the Arab world. Far from conveying realism and expressing confidence in oil futures, these images confront us with a distorted and anxious corporate present shaped by companies' attempts to impress, persuade and manipulate their audience in order to make acceptable the dramatic change around them, driven by the capitalist logic that underpinned oil production throughout the region.

Notes

This chapter owes a considerable debt to many colleagues who have offered insightful comments on earlier versions presented as a series of conference papers. In particular, I wish to thank Mandana Limbert, Gabriele Vom Bruck, Nathan Citino, Kaveh Ehsani, Norma Moruzzi, Mona Damluji, and Ulrike Freitag and the research group 'Representations of the Past' at the Zentrum Moderner Orient (Berlin).

1. The exception is the British Petroleum archive held at the University of Warwick (UK), which has been used for this chapter. At the time of writing, the archive has re-opened after a period of closure, but with limited public access. Even in this extensive collection public relations materials are often organised randomly.
2. David E. Nye, *Image Worlds: Corporate Identities at General Electric, 1890–1930* (Cambridge, MA: MIT Press, 1985), 1–2; Omnia El Shakry, '"History without Documents": The Vexed Archives of Decolonization in the Middle East', *The American Historical Review* 120, no. 3 (2015), 920–34.
3. Andrew Pendakis and Sheena Wilson, 'Sight, Site, Cite: Oil in the Field of Vision', *Imaginations: Journal of Cross-Cultural Image Studies* 3, no. 2 (2012), 4–5. Quotation from 4.
4. For a critical discussion on the received wisdom about the 50/50 oil agreement see Robert Vitalis, *America's Kingdom. Mythmaking on the Saudi Oil Frontier* (New York: Verso, 2009), 138–40.
5. British Petroleum Archive (hereafter BP) ARC162027: Memorandum 'The Future of Public Relations within the IPC Group of Companies', 25 March 1966. On the Iraq Development Board see Arthur Salter, *The Development of Iraq: A Plan of Action* (London: Caxton Press and Iraq Development Board, 1955). I wish to thank Nathan Citino for having drawn my attention to the metropolitan dimension of public relations campaigns by American oil majors.
6. BP ARC162027: *al-Sha'ab*, 14 August 1958.
7. BP ARC106863: 'The Kuwait Oil Company Persecutes Arab workers', *al-Ittihad*, 16 February 1957.
8. BP ARC163338: Omar Haliq, 'A Statement of Arab Oil Objectives', 5.
9. Mona Damluji, 'Petroleum's Promise: The Neo-Colonial Imaginary of Oil Cities in the Modern Arabian Gulf' (Ph.D. diss., University of California, 2013), 5–6.
10. BP ARC106975: 'Town Office Monthly Report – February 1958', Item 8 'Prestige Publicity'; 'Publicity in the Arabic Press', Town Office Press Relations

Officer to Head of Town Office, 26 February 1958 and 3 April 1957. Karen S. Miller, *The Voice of Business: Hill & Knowlton and Postwar Public Relations* (Chapel Hill: University of North Carolina Press, 1999), 165–6.

11. For instance, an article in the Kuwaiti newspaper *al-Sha'ab* published on 11 September 1958 accused the Town Office of espionage. BP ARC107975: Memo Head of Town Office to Head of KOC, 30 December 1957.
12. BP ARC108134: file 'Arab Relations: Town Office, 1953–64' and BP 106932: file 'Public Relations Adviser, 1956'.
13. David H. Finnie, *Desert Enterprise: The Middle East Oil Industry in Its Local Environment* (Cambridge, MA: Harvard University Press, 1958), 185–6.
14. BP ARC107965: Report by KOC Press Relations Officer on Public Relations Conference held at Harvard University, 15–19 April 1958.
15. BP ARC58685: 'Report on Broadcasting (Sound and Vision) in Iraq' by Marjory and Edward Ward, 1958 c.a., 55 pp.
16. Damulji, 'Petroleum's Promise', 30–1, 37. The extensive use of the press replicated the public relations turn that had transformed the American oil industry in the early twentieth century. Stephen R. Leccese, 'John D. Rockefeller, Standard Oil, and the Rise of Corporate Public Relations in Progressive America, 1902–1908', *Journal of the Gilded Age and Progressive Era* 16 (2017), 247–8 and 251–5. I wish to thank Nathan Citino for suggesting this article.
17. Nelida Fuccaro, 'Shaping the Urban Life of Oil in Bahrain: Consumerism, Leisure, and Public Communication in Manama and in the Oil Camps, 1932–1960s', *Comparative Studies of South Asia, Africa and the Middle East* 33, no. 1 (2013), 72–4. Finnie, *Desert Enterprise*, 185–7. The first Bapco public relations officer appointed in 1955 was a New York journalist.
18. Ariella Azoulay, *The Civil Contract of Photography* (New York: Zone Books, 2008), 85–6.
19. For instance, in Iraq up until the 1958 Revolution *Iraq Petroleum* and *Ahl al-Naft* (Oil People) had a circulation of 20,000 and 48,000 copies respectively. Given that IPC had approximately 15,000 employees we can assume that both publications reached beyond those on the payroll of the company. BP ARC162027: Memo 'The Future of Public Relations within the IPC Group of Companies', 25 March 1966.
20. This sense of intimacy is less pronounced in English publications such as Aramco's *Aramco World* and IPC's *Iraq Petroleum*.
21. BP ARC106975: 'Town Office Monthly Report – March 1958'. For KOC's Adolf Morath's photography see Laura Hindelang, *Iridescent Kuwait.*

Petro-Modernity and Urban Visual Culture since the Mid-Twentieth Century (Berlin: De Gruyter, 2022), 131–54.

22. BP ARC106863: Town Office Kuwait to Director Public Relations London, 11 April 1957 and press extract from *Najmah al-Bahrayn*, 13 March 1957.
23. BP ARC106899: Public Relations London to Town Office Kuwait, 27 November 1958. For a history of South Ahmadi see Reem Alissa, 'The Oil Town of Ahmadi since 1946: From Colonial Town to Nostalgic City', *Comparative Studies of South Asia, Africa and the Middle East* 33, no. 1 (2013), 49.
24. Nye, *Image Worlds*, 9–30.
25. 'Middle East Oil', *LIFE Magazine*, 11 June 1945, 25–37.
26. Since the beginning of the company's public relations campaign targeting Saudis, the company also produced publicity posters called 'Foto Stories', each including a dozen photographs illustrating aspects of national development and company welfare. Robert Vitalis, *America's Kingdom*, 124.
27. See for instance BP ARC113269: 'Green Album Kuwait Oil Company'.
28. The dates printed on these two images and in Figure 4.5 that follows are not reliable, reflecting the random cataloguing and organisation of visual materials in the KOC archive. It is possible that the image 'Wellhead Operator – 1961' is by Adolf Morath, as it is typical of his stylised portraiture of human figures at work. Yet the photograph would originally have been in colour. I wish to thank Laura Hindelang for this suggestion.
29. For an example of cinematic propaganda manufacturing scenarios of economic and social production that integrated locals with natural landscapes see Todd Reisz, 'Landscapes of Production: Filming Dubai and the Trucial States', *Journal of Urban History* 44, no. 2 (2018), 298–317.
30. Kuwait Oil Company Archive, Ahmadi (hereafter KOCA): GT 6230 (1958), GT 4979 (1958) and P 644 (1960).
31. KOCA: GT 12331 n.d.
32. 'Come with Me to Rahimah Town', *Qafilah al-Zayt* 5, no. 5 (December 1957), 9–12.
33. 'Photography', *Qafilah al-Zayt* 6, no. 2 (September 1958), 23–6.
34. Nye, *Image Worlds*, 83–4. Quotation from 91.
35. Fuccaro, 'The Urban Life of Oil', 74.
36. 'Photography: Yusuf's Hobby' *Qafilah al-Zayt* 7, no. 1 (July 1959), 36.
37. 'Dhahran's Television Station', *Qafilah al-Zayt* 8, no. 12 (May/June 1961), 13.
38. 'Exhibition of the Oil Industry in Jeddah', *Qafilah al-Zayt* 8, no. 9 (February/March 1961), 17–20.

39. See also Biglari in this volume for the role played by technical training in the management of the Iranian labour force by AIOC in the Abadan Refinery.
40. Particularly the General Classification Test developed by the company in 1953. See Finnie, *Desert Enterprise*, 91.
41. For instance, '"What Do You See in These Pictures?" "What Do You Recognise About Your Country?" "Do You Know?"', *Qafilah al-Zayt* 1, no. 3 (October and December 1953); *Qafilah al-Zayt* 1, no. 4 (January 1953); *Qafilah al-Zayt* 1, no. 6 (March 1954); 'Find the Appropriate Job', *Qafilah al-Zayt* 2, no. 6 (March 1955).
42. '*Qafilah* Photo Competition', *Qafilah al-Zayt* 5, no. 4 (November 1957), 37–8.
43. See for instance *Qafilah al-Zayt* 7, no. 7 (January 1960), 21.
44. 'Find the Suitable Job', *Qafilah al-Zayt* 2, no. 6 (March 1955), 6–7.
45. 'Fixing Time', *Qafilah al-Zayt* 6, no. 7 (March/April 1959), 27.
46. This is also true of visual representations of oil urbanism in Iran. See Mona Damluji, 'The Oil City in Focus: The Cinematic Spaces of Abadan in the Anglo-Iranian Oil Company's *Persian Story*', *Comparative Studies in South Asia, Africa and The Middle East*, 33, no. 1 (2013), 75–88 and Pamela Karimi, 'Building the Iranian Oil Space: Architecture Culture in the Early Petroleum Era' (paper presented at the conference 'The Global Petroleumscape', TU Delft, May 2017). For the development of oil company towns in the Arab World see Nelida Fuccaro, 'Arab Oil Towns as Petro-Histories', in Carola Hein (ed.), *Oil Spaces: Exploring the Global Petroleumscape* (New York: Routledge, 2022), 129–44.
47. 'Skyline – Kirkuk City', *Iraq Petroleum* 7, no. 12 (July 1958), 28–9.
48. KOCA: PR 2973 and PR 2990 published in 'First Day Payroll Families Move into Company Accommodation', *The Kuwaiti*, 6 May 1954, n. 318. 'Home Ownership in Basrah', *Iraq Petroleum* 7, no. 9 (April 1958); *Ahl al-Naft* 77 (1958), 49. Despite the fanfare with which they were advertised, Home Ownership Schemes were increasingly used by companies as substitutes for expensive housing developments to accommodate local labour. On Saudi Arabia see Vitalis, *America's Kingdom*, 208.
49. The portrayal of females was considered socially acceptable only in Iraq and with Christian families, particularly in the magazines that covered Kirkuk's oil fields, where Christians like Sargon Hallaby constituted the largest portion of IPC's skilled workforce. See for instance the photo of a nuclear family in the kitchen in BP ARC65577: *Iraq Oil 1951*.

50. Alissa, 'The Oil Town of Ahmadi', 52; and Fuccaro, 'Shaping the Urban Life of Oil', 70–1. On Iran see Pamela Karimi, *Domesticity and Consumer Culture in Iran: Interior Revolutions of the Modern Era* (New York: Routledge, 2013).
51. For a discussion of gender and domesticity in Saudi Arabia as promoted by Aramco see Citino in this volume.
52. See for instance 'The Home Corner', *Qafilah al-Zayt* 6, no. 7 (March/April 1959); *Qafilah al-Zayt* 6, no. 8 (April/May 1959); *Qafilah al-Zayt* 7 no. 7 (January 1960); *Qafilah al-Zayt* 7, no. 9 (March 1960).

5

AL-BAHITHUN: SOUNDS THAT CALL TO THE (OIL) FIELDS

Ala Younis

The Searchers

Al-Bahithun (The (Re)searchers) is an art project that highlights the link between nationalised knowledge and oil wealth. The nationalisation of the Iraqi oil industry in 1972 allowed the Iraqi state to utilise unprecedented revenues from the sale of oil. My work explores how this wealth was dependent on, and tied to, the empowerment of local expertise and how a good portion of oil revenues went into training and educational programmes as well as supporting the welfare of scientists, researchers and university graduates. My project's name is borrowed from an Iraqi film made in 1978 entitled *Al-Bahithun* (The Searchers).[1]

I play on the double meaning of the Arabic term *al-bahithun*: 'the searchers', and 'the researchers'. While the *al-bahithun* featured in the film are primarily oil explorers, assembled to search for oil in the marshes of Southern Iraq (*al-Ahwar*), in my work I explore what has been made possible by the revenue generated by these (oil) discoveries: the development of a national research culture that helped secure the autonomy of oil as an Iraqi national industry as well as the structures of power and politics underlying it. This chapter describes how the objects I featured in my art project *Al-Bahithun* reproduce the condition of (re)searchers who oscillated between the projects of nationalisation, knowledge, art, architecture and war as precipitated by oil. Finally, I aim to highlight how these oil movements are, as in the film, accompanied by attempts to hear the sounds that call(ed) the (re)searchers to the oil fields.

The core team featured in the film *The Searchers* represent a constituency of Iraqi society in the late 1970s – an engineer, a revenge-seeking peasant, a geologist and a treasure-seeking explorer. These men live and work on the water and conduct their explorations with a strange-looking tractor (Figures 5.1 and 5.2), a floating workstation equipped with telecommunication devices. When a phone call alerts the team that they are floating on a sea of oil, they explode with joy, except two of the men, who split from the group to embark on a secret search for a legendary land, a 'lost paradise'. After a long, tedious cruise, guided by science-fiction sounds and lights torching the sky pointing towards a mysterious destination, the two separated men finally

Figure 5.1 This drawing shows some elements featured in the installation *Al-Bahithun*, 2018, by Ala Younis, which the artist discusses in this chapter. This installation was commissioned for Crude, the inaugural exhibition of the Jameel Arts Centre in Dubai held between 11 November 2018 and 30 March 2019. Pictured left to right is an image of oil explorers or searchers on a water tractor from the poster of the Iraqi film *Al-Bahithun* (1978); a redrawing of a stamp issued in 1976 showing Saddam Hussein embracing Ahmad Hassan Al-Bakr, then President of Iraq; a drawing of the film's water tractor merged with one of the Dutch buildings on Haifa Street in Baghdad; the film's two protagonists as they survey the oil field; and the 1966 mural by Nuha al-Radi commissioned by the Iraq Petroleum Company. By Ala Younis.

Figure 5.2 Left: Plastic model of a building and a car juxtaposed over the body of the water tractor that appears in the film *Al-Bahithun* (1978). Right: Special edition stamp issued in 1976 celebrating the fourth anniversary of the nationalisation of Iraqi oil [*Al-Bahithun* (2018) by Ala Younis].

find themselves at the foot of fuming oil towers in the Rumaila oil fields, the largest in Iraq, which had been nationalised a few years earlier. One man cries knowing his lost paradise to be a delusion, while the other is enchanted by the revelation that the oil fields are just this paradise.

The sounds calling the men in *The Searchers* are wavering sounds similar to those popular in the science-fiction films of the era. Only one of the two men can hear them, and he becomes obsessed with the messages he thinks he is supposed to decipher. As he realises that the lost paradise he is looking for is an oil field he collapses to his knees, trembling, weeping and unable to speak. Next to him is his companion: a scientist (an archaeologist or geologist), who did not hear the sounds but believes he can understand the drive of his companion in pursuing them. His encounter with the oil fields, a pleasant surprise, is rechannelled into his scientific reasoning, instinctive optimism and national pride. For him, this is the real paradise, as he utters to the other searcher: 'Real signs for a physical Eden. A paradise that is capable of creating a coherent mix of myths and reality. Symbol and logic. I think this is what you were looking for, isn't it?'[2]

As the two explorers lose contact with their team they are left in 'paradise' under the sun without help or guidance. They can no longer hear the leading sounds. They sit thinking about a solution to their return dilemma, while their colleagues are already on a mission to find them. They are finally located with the help of a helicopter that can hover over the marshes and see them from a distance. As the helicopter lands nearby, the two men shout joyously 'They found us!' and run to the helicopter. The group finally re-unites, while the pilot, in dark sunglasses with a thick moustache, looks at the lost men from a distance. The team return to the workstation in the middle of the river and resume their work riding the same strange water tractor. The last scene of the film shows the men's return from the outdoors (the fields) to indoors (their station).

I am interested in the surfaces, sounds, objects and processes that we encounter when moving between the stories that relate to the production of oil. There is a parallel liquidity between riding a water tractor in search of oil and finding it as another liquid lying under the water. The unknown sounds guiding the searchers on their oil sub-journey are less scientifically understandable. The burning fumes mark the evaporation of doubts (or hopes) at the sight of the oil fields. There is an exchange of gazes in the act of rescue coming in the guise of a helicopter from the sky. And the sun de-colours the time of the final scene of the film as the two men return to the floating station to continue their mission.

The Presidential Embrace Stamp

A special edition stamp, a redrawing of which I have also incorporated in my installation, was issued on 1 June 1976 marking the fourth anniversary of the Republic of Iraq's nationalisation of oil (Figure 5.2). It depicts two men in embrace: the then President Ahmed Hassan Al-Bakr (1914–82) and Saddam Hussein (1937–2006). Saddam Hussein was appointed deputy to the President in 1968, the year the two men and the Baʿath party came to power following a series of military coups in the decade-old republic. The stamp is a map of power relations within a state and provides a window into an intimate moment that portends the gradual transfer of power in Iraq's first office while celebrating the transfer of control of the country's most precious resource. Saddam Hussein led the nationalisation negotiations with the foreign oil companies that controlled the industry. He asked for an adequate share of revenues as well as for the regulation of barrel production and minimum exports, all of which increased Iraq's annual budget. He also entered into a fifteen-year agreement with the Soviet Union which sent a technical team to Iraq to advise on negotiations, to drill wells in the Rumaila oil fields, to train Iraq National Oil Company (INOC) staff, to provide pipelines and equipment, and to acquire a portion of the crude extracted from Rumaila in return for these services.[3] On 1 June 1972, Al-Bakr arrived in a Mercedes sedan with Saddam Hussein at the National Radio and Television building in Baghdad. Al-Bakr was filmed entering the building and reading aloud the decree that nationalised the assets of the British-controlled Iraq Petroleum Company (IPC). This announcement was made into a promotional clip that staged Iraqis of diverse backgrounds, particularly workers, listening to al-Bakr's declaration. The film *The Searchers* was one of several cinematic features that were produced in the wake of Iraq's nationalisation.

Between 1973 and 1975, and then up until nationalisation was completed in February 1979, the state continued to oust one European shareholder after the other, all of whom were associated with the foreign oil companies that had extracted Iraqi oil. The government, however, continued to maintain special relations with France, which entered into a deal with Iraq to provide nuclear reactors, equipment and know-how. In this period, Iraq's Ba'ath party also continued to invest massively in national education. The percentage of technocrats among senior government officials increased and professional administrators were replaced by technical experts and academics.[4] Saddam Hussein assumed his new role as President in July 1979.

In many ways, this stamp can be read as a field – a process of production – charting Saddam Hussein's ascent to power and how oil enabled it. This stamp has borders around an expanse of a dense, interwoven visual pattern that looks like a field (or a sea) of infinite eyes. Within the inner frame of the stamp the figure on the right is younger, taller and over-powering in appearance. His face and features are visible, and his arm is extended around the President, the back of whose head is more visible than his face. While the stamp is celebrating another year of oil nationalisation after decades of exclusive possession and control by foreign corporations, Saddam's visible face and posture depict an impatient deputy head of state who has accumulated great(er) power (than his president) in the high echelons of the Iraqi government. The two bodies performing this transfer of power are not photographed, but illustrated. The field of interwoven eye-patterns that surrounds them further drives our attention towards a special feature of this stamp, Saddam's eyes: his experts, scientists, workers and informants who propped up his power.

This special edition stamp, *The Searchers* film and the clip of Al-Bakr's announcement of oil nationalisation were elements of national propaganda that aimed to foreground the main players in Iraq's new revolutionary happiness: oil, Saddam Hussein as deputy to the president, and a growing group of local (re)searchers. The latter were key to the making and receiving of the propaganda message. One can read the actions of the protagonists of *The Searchers* using the logic of the stamp. The two men who find the Rumaila fields in the film are the President and his deputy, or perhaps one represents the presidential couple combined and the other embodies the researchers as the carriers of oil knowledge and technical logic.

Nuha al-Radi's Ceramic Mural

One month before the nationalisation of Iraq's oil companies in 1972, the Iraq Petroleum Company (IPC) published the last issue of *Al-ʿAmilun fil Naft* (The Oil Workers) (Figure 5.3), a magazine that had circulated since 1961.[5] Distributed free to IPC employees, the magazine also had a literary and artistic mission. In addition to offering regular news on the oil industry, archaeological findings, fashion and sports highlights, the magazine sourced literary works from young and upcoming writers, and published reviews and images of Iraqi art as well as interviews with writers and artists. Each issue had an Arabic front cover, often showing images of art works by Iraqi artists, oil fields or refineries, and a rear cover with photos taken at one of the workstations of the various IPC's subsidiary companies operating in Iraq. The magazine was published under the supervision of the Iraqi Ministry of Oil and in many ways encouraged readers across Iraq to identify with the locations, technology and faces that brought wealth to their country. By the time the magazine ceased to exist in 1972, the careers of many of the artists it showcased had become established.

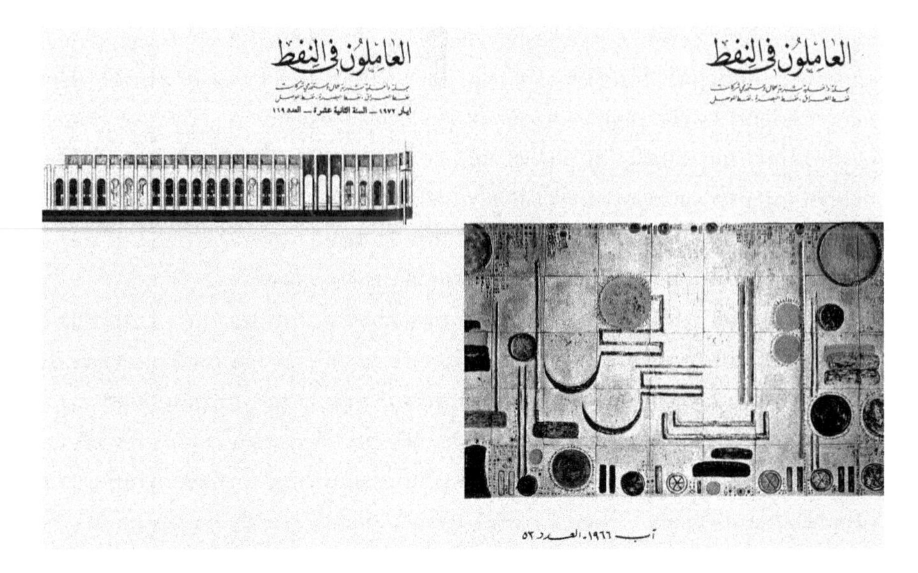

Figure 5.3 Two reproductions of *Al-ʿAmilun fil Naft* magazine covers. Right: Issue 53, August 1966, featured a ceramic mural by Nuha al-Radi, commissioned by IPC for its headquarters in Baghdad. Left: An imagined cover for the magazine's last issue, 119, which came out in May 1972. This issue featured the work of two architects before they led the Haifa Street development projects less than a decade later [*Al-Bahithun* (2018) by Ala Younis].

Among these talents was Nuha al-Radi (1941–2004), an artist whose ceramic mural was featured on the cover of the August 1966 issue of *Al-ʿAmilun fil Naft* after having been installed at the entrance to the IPC headquarters in Baghdad. Inside the issue, an article expanded on this mural and how its 25-year-old author had been the first in Iraq to revive the local ceramic heritage as art. It also mentioned how al-Radi's moderately priced works had earned her the IPC commission:

> Although the work was generally abstract in design, the artist based it on the concept of oil. Looking for a phrase connected with Iraqi oil, the Arabic lettering of which would provide the centre of her design, she hit upon the words 'Baba Gurgur', the name of the place near Kirkuk where oil was first struck in 1927, which formed the nucleus of one of the richest oil fields in the world. The two words were so designed as to combine Arabic letters, crescent shapes, and motifs resembling storage tanks and pipes. Decorative details were also based on oil motifs – valves, pipes, resembling towers, etc.

> with a border displaying traditional Iraqi ornamentation. When firing the tiles, she [al-Radi] minimised the glazed parts, in keeping with an ancient Sumerian practice, and did the background in an earthy yellow interspaced with brilliant suns suggestive of the visual effect of the locations where oil was usually found.[6]

A decade before the film *The Searchers* was made, therefore, Nuha al-Radi attempted to translate the sounds of the names of the oil fields in Arabic letters into ceramic patterns. In the IPC mural, she disconnects the letters to modernise *Ḥurufiyya*, an art style based on Arabic calligraphy and letters that was emerging in Baghdad at the time.[7] Al-Radi revives the Iraqi pottery heritage as new tableaux in the shape of a tiled mural which features not one but many suns. With the inclusion of the name of the oil field she acknowledges the geographic and geological features of a national treasure, despite the fact that oil revenue at the time was only partially accessible to the Iraqi government. The sun in the mural awakens the viewers' senses to the colour and heat felt by the workers in the oil fields. It is interesting to think of al-Radi's use of clay cooked with fire to produce an art work while under Iraq's very hot sun (Figure 5.3).

In *Al-Bahithun*, I leave out the Arabic letters but take Nuha al-Radi's choice of ceramic as a medium. I think of oil as it springs from under the feet of the workers in the oil fields, not of oil as in the mural or the film made in the 1960s and 1970s respectively, but of oil as experienced in Iraq in more recent years. I acknowledge the tiles as pieces of research culled from several resources and thus influenced – and perhaps sometimes sent off-track by – my narration of the multi-layered oil histories in each of the elements that I put together in this project. Oil, when imagined, prompts a variety of visuals: black liquid, bursting out of earth, running in pipes, indicating on a world price ticker, smell-spreading at gas stations, and seducing/begging/spying eyes as it morphs into lives, vehicles, buildings, wars, lootings, and so forth. While oil's imagination rattles between one that nourishes and one that diminishes, it is this morphing image of petroleum that leaks from between these fragments of stories: crude, thick, heavy, opaque, inflammable and ambiguous. How do I reproduce oil as a morpheme in an art installation, one that is part of a larger meaning and historicity, that can also change its meaning in accordance with the size or context of the story it has created, while having a stable connotation across all these stories? In my installation, one ceramic piece is made in the shape of a bubble that resembles oil gushing out of the earth dug out illegally by smugglers and militias. Another piece is presented as a leak that comes out of an oil dictionary pointing to the seepage of expertise funded by Iraqi oil (Figure 5.5), also with reference to the wave of assassinations that targeted scientists and researchers in the aftermath of the fall of the Ba'ath regime in 2003. The latter piece makes us tread these 'waters' more cautiously, as I consider how to speak of scientists and artists as products of an oil era, but also of their own scientific or creative capacities. Here, I hear oil, in a multiplicity of Iraqi Arabic and Iraqi English accents uttering technical terms, negotiating with management, conversing anxiously

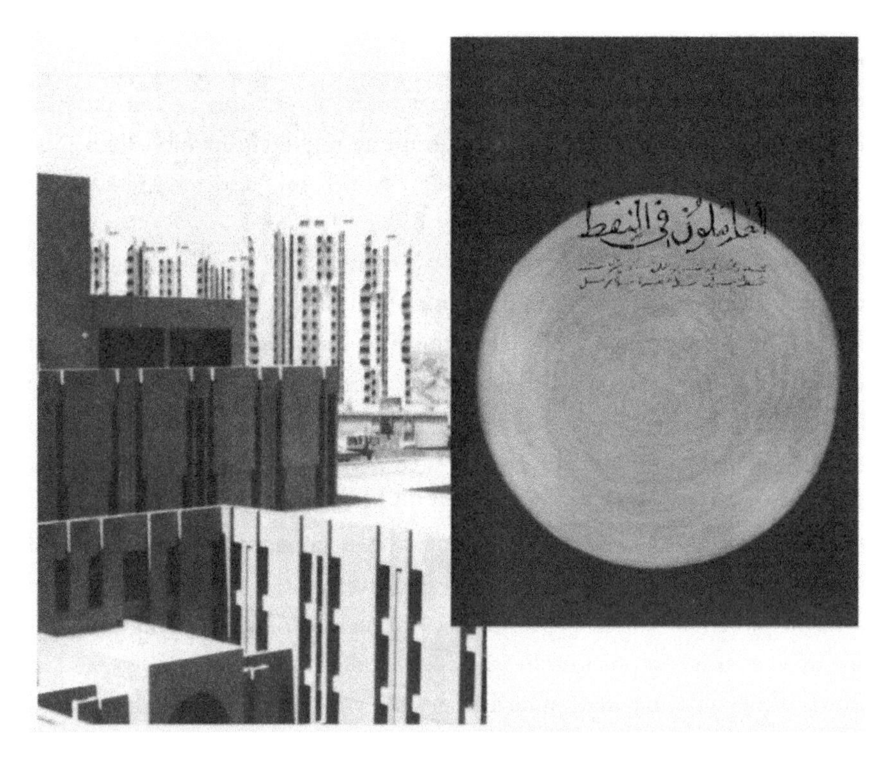

Figure 5.4 Left: Saddam Centre for the Arts in the foreground and the Dutch Buildings in the background of Haifa Street. Right: An invented cover for *Al-'Amilun fil Naft*, using the same name calligraphy juxtaposed with a drawing of the sun [*Plan for Greater Baghdad* (2015) and *Al-Bahithun* (2018) by Ala Younis].

or excitedly with colleagues on pertinent issues, or casually greeting a neighbour on their way back from buying their groceries. How far from and close to the oil fields were they, how did they endure a form of life generated by a tight grip on power, and how dangerous did their careers become (for them) on the demise of this power? It is this shifting headlight that renders the multiplicative presence of Nuha al-Radi's suns. There is no single but many (oil truths under the) suns.

The Researchers

Approximately fifteen years after her work featured in *Al-'Amilun fil Naft*, Nuha al-Radi produced another art piece with two components: a model Mercedes with human brains oozing from its windows, and the same brains flying Mercedes flags. This was the artist's critique of the new social trend that had emerged as a result of the state granting privileges to college graduates and thus encouraging more Iraqis to pursue a research or academic career in order to foster national knowledge and sovereignty. Graduates were offered, in addition to subsidised land or housing, good tax exemptions on cars.[8] These exemptions made many of them opt for Mercedes-Benz Sedans, which became a familiar sight in the neighbourhoods where academics lived.[9] I was searching for any images or further descriptions of this art piece by al-Radi, but to no avail, until I spoke with two of her close friends who confirmed that they had never seen or heard of this piece. Yet its description is included in an obituary written by another close friend of the artist.

In my project *Plan for Greater Baghdad* (2015), I research the story of commissioning, delaying and building a gymnasium in Baghdad over twenty-five years that witnessed shifts in political powers. My research was obsessed with finding clues in printed and oral narratives, many of which lacked not only pictures but also witnesses that could confirm them. The project's methodology was to extract elements from several materials and combine them together to produce non-existent documents on architectural (de)feats and the characters that shaped them. This combination produced an aesthetic of the time: elusive, faint, hesitant and, again, opaque. Oil ran through the stories as a side character, an enabler but not at the centre, because the events represented a timeline of happenings taking place in Baghdad, where oil revenue was spent. In 2018, I presented the timeline once again, from the perspective of the 'personas' who were entangled in these shifts, but who had no control over them. Nuha al-Radi was one of these characters, through her Baghdad Diaries which she penned during the First Gulf War and in the following years. She writes of how lifeless the infrastructures of their lives became once the oil that Iraq had always sold became of no value. Buildings became powerless, the state helpless, and knowledge was suppressed. Her nightmares were of US soldiers taking over Baghdad's most modernised neighbourhood, Haifa Street.

I consider this nexus of the knowledge economy and oil in *Al-Bahithun* as well. In a plastic-sculpture part of my installation *Al-Bahithun*, a Mercedes comes out of a housing block that resembles any of the nine identical buildings, informally known as the 'Dutch buildings',[10] that were erected on Baghdad's Haifa Street between 1980 and 1984 (Figure 5.4). The fifteen-floor buildings, which cost 17.5 million Iraqi dinars (c. 50 million USD) and consisted of two- and three-bedroom flats, were designed by the design department of Baghdad's Municipality which was established to produce in-house architectural plans for state projects. This is another example of the nationalisation of services that had previously been outsourced to international contractors. The flats were sold to academics and graduates at high but subsidised prices (37,000 Iraqi dinars, a substantial amount equivalent to approximately 100,000 USD). The cost of designing, supervising, building and subsidising these blocks and paying the wages to their tenants came from oil revenues. The same revenues paid for the Iran–Iraq war (1980–8), which was waged at the same time as their construction. In *Al-Bahithun*, a plastic sculpture of the building and the car sit on a strange water tractor, a miniature of the model used in *The Searchers* film. In the film, the tractor shuttled the searchers between their work and their accommodation unit on the river. In *Al-Bahithun* the tractor is carrying the building and the car, all dipped in black. Stuck to its wheel is a stack of books which are suspended there, in a state of tensed fragility.

While, up until 1972, the oil-funded *Al-'Amilun Fil Naft* had offered artists and intellectuals a venue for showcasing their talent and expressing their voices, under the Ba'ath regime access to books was a privilege for academics, who also lived under strict mobility regulations and censorship. This strict travel policy was part of a national strategy that aimed at preventing brain drain during the Iraq–Iran War. In this period, a surveillance regime scrutinised the activities of local experts (scientists, architects and others). The invasion of Kuwait in 1990 led to another war and economic sanctions.[11] Iraqi oil barrels could only be exchanged for a limited list of goods and the national currency was consequently devalued. Many academics sold their collections of books in the 1990s, one after the other as a consequence of the devaluation of their state-paid salaries. They also sold their Mercedes when these salaries could no longer pay for the fuel, or for spare parts. Mercedes thus exited the life of researchers while the housing blocks in Haifa Street sat still.

In my *Al-Bahithun* installation, a ceramic oil-bubble (Figure 5.5) sits close to the tractor-ensemble. I took this image from footage from a mobile camera (dated 2011) that I found on YouTube that shows oil flushing out of the earth as the pipes that carry it are pierced by a group of men. Close to the men is a tractor – not a water one – that de-earths the site while a line of oil

Figure 5.5 Left: Ceramic piece representing an oil bubble from YouTube footage documenting the illegal appropriation of crude oil in Iraq. Right: Another ceramic piece in the shape of an oil seep leaking out of Dictionary of Petroleum [*Al-Bahithun* (2018) by Ala Younis].

tanks awaits close by to collect the stolen crude, which is presumably going to be sold to secure funds and weapons. I hear the oil bubbling and therefore include in the installation a 3D effect of it in the shape of smaller bubbles fountaining in the sculpture.

Alongside the ceramic oil bubble stands another black ceramic piece. This resembles a leak coming out from the pages of an oil dictionary, like an oil (or blood) seepage (Figure 5.5). The *Dictionary of Petroleum* (*Muʿjam al-Naft*) had begun as a list of words compiled by Saudi Aramco – the American oil company that controlled oil extraction in the Kingdom of Saudi Arabia – that was then given to the Arabic Language Society in Baghdad, which organised an oil symposium in the second half of 1970s. In this thick volume, every English technical term used in the oil industry up until 1993 is rendered in Arabic. Since Iraq was not on good terms with most other Arab states in the aftermath of the First Gulf War, no Iraqi experts took part in the seminar organised by the Arabic Language Union of Societies in Damascus in 1994 to discuss the translation of the oil terms included in this dictionary. Over three days, participants from Syria, Egypt, Jordan and Tunisia scrutinised every page of the dictionary. I think of the sounds of war in parallel with the sounds of compiling, translating and uttering these Arabic terms. I am reminded of the indecipherable sounds that called men, artists, writers, graduates, scholars, teachers, engineers and foreign forces to speak in the oil fields using these terms. I make the ceramic spill drip out of the book for two reasons: thinking of the oil seepages caused by the two conflicts (the Iran–Iraq War and the First Gulf War), and the blood of the researchers and academics who were assassinated in the vicious decade following April 2003.

Architectures of Oil: The Plan for Greater Baghdad

In 1986, Nuha al-Radi completed a piece of public art on Haifa Street representing the fishermen that had their businesses close to the river-front near the Dutch development project (Figure 5.4) built by the government in the 1980s. This mural was one of the many projects commissioned by the government to mark the street as the new centre of Baghdad's intellectual life. In close proximity stood the Saddam Centre for the Arts. When this grand museum complex was completed in 1986, like al-Radi's mural, several thousand modern and contemporary art works from artists' studios and museums of modern art across Baghdad were relocated there. Undoubtedly, it was oil revenue that fuelled (literally and metaphorically) machines of construction, production and acquisition.

Art and monuments are thus also contaminated by the history of oil. As Iraq continued to flare Iranian oil fields in the south, Iranian missiles were falling on Baghdad. War did not interrupt the city's daily life, the construction schedule of Haifa Street or the erection of the Martyr's Monument, designed as a circular platform 190 metres in diameter at the centre of an artificial lake. The monument had a forty-metre-tall glazed turquoise ceramic dome that was split into two halves. At the centre of the two half-domes stood a flagpole and a spring of water symbolising the blood of the fallen. It cost the Iraqi government a quarter of a billion US dollars, at a time when one Iraqi dinar was worth over three US dollars.[12] The names of dead soldiers were to be inscribed on the inner walls of the museum that lies below the monument. At the time, however, the architects were unsure how to calculate space for the casualties of an ongoing war. Inscribed seamlessly in the texture of the stone cladding, the monument should have memorialised the several hundred thousand names of those killed in battles for the annexation of more oil fields. The sounds of oil here are not of bubbly crude or wiggly science-fiction signs, but sounds of authority, exchange of bullets, and (again) heavy vehicles.

These and the other stories that constitute *Al-Bahithun* represent a stream of thought drawn from a larger research project titled *Plan for Greater Baghdad*.[13] The project, as noted above, is a timeline of monumental, architectural, artistic and political interventions in Baghdad from 1956 up until today (Figure 5.6). When, in 1952, Iraq started to negotiate with IPC for a larger share of oil wealth, the increase in revenue led to the establishment of the Iraq Development Board which became legally entitled to 70 per cent of all oil revenue. This wealth allowed the country to invest in infrastructure, invite star architects like Le Corbusier, Walter Gropius and Frank Lloyd Wright to visit Iraq and design projects, and enabled young Iraqi graduates to participate in the infrastructural development of the country. These international guests were hosted by local artists and architects whose own experimental work started to mushroom in Iraqi cities. These infrastructural works were redrawn and modified by continuous political fluctuations as Iraqi artists and architects had to negotiate their projects with the regime.

Figure 5.6 Ala Younis, *Plan for Greater Baghdad* (2015), two- and three-dimensional prints, drawings, archival and found materials, and model. Image: Alessandra Chemollo. Courtesy of the artist and La Biennale di Venezia.

Le Corbusier's first set of designs for a Sports Complex in Baghdad (which included a stadium, gymnasium and swimming pools) were approved by the Iraqi Government two days before the 14 July 1958 revolution overthrew Iraq's monarchy. The Portugal-based Calouste Gulbenkian Foundation had offered to sponsor the construction of the Baghdad Stadium (also part of Corbusier's commission) on condition that a Portuguese architect designed it instead. This offer of funding came directly from Calouste Gulbenkian, who had brokered Iraq's original oil agreement in the 1920s and was committed to paying back his 5 per cent commission on Iraqi oil sales. Oil was a protagonist in a complex set of histories linking architects, presidents, monuments and budgets. In addition to the stadium that was inaugurated in 1966, the Gulbenkian Foundation sponsored the museum of modern art in Baghdad, which was completed in 1962 and then moved to the Saddam Centre for the Arts in 1986.

Le Corbusier's commission for the rest of the Sports Complex, on the other hand, remained dormant until an influx of capital from the oil crisis of 1973 infused new life into the project. The construction of the new gymnasium did not start until 1978, and when it opened in 1980 it was named after Saddam Hussein. Its completion came at a time when the government was developing a more ambitious plan of designing and building the intellectual district of Baghdad on Haifa Street. The President himself was sitting in architectural conferences discussing how Iraq's heritage could be incorporated into this new architecture. It is in this period that the President announced the outbreak of the Iran–Iraq war over the oil fields of southern Iraq. Architects who sat in these conferences did not realise that they were imagining an architecture of (oil) war. Oil became a contamination of architecture, as much as it attempted the containment of knowledge. These conferences, just like the web of monuments that were built afterwards, remind me of the field that surrounds the presidential embrace in the 1976 stamp.

In a film titled *Love in Baghdad* (1987), a man from the city loses his memory and is taken care of by a family who live in the countryside. He regains his memory as he returns to Baghdad wandering through the new Haifa Street. He cannot believe that the Baghdad he left has now become the paradise he had been promised. The sounds we hear are of astonishment.[14] A decade after the *Al-Bahithun* film, the promised paradise is witnessed in Baghdad's new neighbourhoods housing Iraq's scientists; oil sounds are close and loud, as if the fields have moved to the city.

The Oil Spill Song

I was in early elementary school in Kuwait when, in 1983, a locally-produced song called *The Oil Spill Is Approaching Us*, about an encroaching oil spill in the Arabian Gulf, became popular. Local news reported that oil was seeping from Iranian oil platforms that had been attacked by Iraqi missiles. Too young to understand the threat posed by oil spills to the surrounding marine environment, beaches and the lives of birds now dipped in black, we recited the song with amusement. We heard it on the radio as we moved around in our Mercedes Sedan through the streets of Kuwait or danced to it when it was featured in the soundtrack of Kuwaiti television dramas. I was thinking of the song and the faint memory in my head of an encroaching oil spill as I was looking at the weird tractor carrying the searchers/explorers in the film becoming smaller and smaller, a receding spot in the river. How fragile, evanescent and entangled their lives could have become! Sounds of oil fields were dominating everything.

Notes

1. Directed by Mohammad Yousef Al-Janabi and produced by the Iraqi state's Cinema and Theatre Department. I found a reference to this film in a catalogue published by this Department in the 1980s, then located a copy on Youtube.com.
2. The quotation is taken from the film.
3. Michael E. Brown, 'The Nationalization of the Iraqi Petroleum Company', *International Journal of Middle East Studies* 10, no. 1 (1979), 120.
4. John Galvani, 'The Baathi Revolution in Iraq', *MERIP Reports*, no. 12 (1972), 15–16.
5. An earlier IPC magazine was published in the 1950s called *Ahl al-Naft* (Oil People). It ceased publication after the 1958 Revolution. *Al-'Amilun fil Naft* was its replacement which started to be published in 1961 with a circulation of 8,000 copies, selling for 0.25–0.30 Iraqi dinars, and offering subscriptions. Its editorial office was based in Baghdad, unlike that of *Ahl al-Naft*, which was based in Beirut.
6. *Al-'Amilun fil Naft* no. 53 (August 1966).
7. An aesthetic movement that emerged in the second half of the twentieth century among Arab artists who experimented with traditional Arabic calligraphy as elements of their modern art.
8. Hayat Sharara, *When Days Dusked* (Beirut: Arab Institute for Research & Publishing, 2002), 183. Translated by the author.
9. Since the tax was almost three times the price of a car, investing in Mercedes cars that could be sold in the future at their market price was worthwhile.
10. Named after the nationality of the contracting company that built them.
11. Following its defeat in the First Gulf War, Iraq saw a near-total financial and trade embargo imposed by the United Nations Security Council to pay reparations and disclose and eliminate any weapons of mass destruction. The sanctions led to a severe drop in value of the Iraqi dinar. The salaries of the academics remained the same in the first years of the sanctions but afforded only one tenth of their original purchase power. The sanctions also led to spare car parts becoming scarce and expensive.
12. Samer al-Khalil, *The Monument, Art, Vulgarity and Responsibility in Iraq* (London: André Deutsch, 1991), 22–3.
13. Commissioned and premiered in the context of *All the World's Futures*, the 56th International Art Exhibition – La Biennale di Venezia, curated by Okwui Enwezor, 2015.
14. Directed by Abdul Hadi Al Rawi in 1987. Available on Youtube.com

PROJECTING FUTURES

6

EMPATHY FOR THE GRAPH

Arthur Mason

> . . . the nature of our civilized minds is so detached from the senses, even in the vulgar, by abstractions . . . that it is naturally beyond our power to form the vast image of this mistress called 'Sympathetic Nature.'
>
> Giambattista Vico[1]

As an inscription device, the energy graph reflects a style of hydrocarbon aesthetics that celebrates abstractness. The graph is an abstract image that does not make a great deal of sense on the surface of things. In fact, as I argue below, abstractness in graphical representation is analogous to the contemporary art image – the latter, a visual representation whose recognition reflects the consecration of an effort. Pierre Bourdieu notes that contemporary art invites appreciation through an imposition of refinement of taste – what he calls the Kantian aesthetic.[2] As a form of taste, the Kantian aesthetic functions to endow populations with a feeling for the world that favours distance. Peter Sloterdijk employs a similar aesthetic regime, which he labels cynical reason.[3] Both the Kantian aesthetic and cynical reason tend towards a rejection of the obvious in favour of abstractness. By applying these discourses in this instance, in this chapter I draw attention to the way in which graphical representation in energy development (and in climate change) favours distanced reflection as a form of aesthetic appreciation.

In *The Pulse of Modernism*, the historian Robert Brain notes that the origin of the modern graph lies in graphical inscription instruments invented in the nineteenth century which opened new ways of moving from materiality to semiotics and established a desire for linear temporality.[4] For example, in their attempts to modify the work of steam engines, engineers such as James Watt introduced graphical copying processes for mechanically tracing the movement of the piston inside the cylinder of an engine. Brain notes that such instruments, which began crudely by affixing a pencil to a piston rod and then to a registering apparatus made up in part of writing paper, transformed the piston's mechanical movement into graphical expression. Through this process, according to Brain, conceptions of productive labour shifted from an alignment with an idealised understanding of the regularity of natural objects (movements of the sun, the stars, the earth)[5] and towards an obsession with the passage of linear time:

> The new graphical recording instruments implemented in myriad concrete measures the primacy of linear time in a new cosmos of irreversible forces, conversions, history, activity, progress, energy, and eventually, entropy associated with the [capitalist] steam-powered world. In nearly every science – thermodynamics, astronomy, political economy, philology, archaeology, evolutionary biology, and soon, physiology – linear temporality became the critical variable.[6]

Notably, Brain also calls attention to the graph's poetic function by its extension from the cognitive realm to the affective. The graph amplifies both cognition and emotion through what he calls a *double-reading*: first, the graph involves *intellectual calculation* associated with interpreting the scale of coordinates, time frames and ruptures that re-enact the event inscribed; second, the graph requires *sensual-intuition* associated with the rising or falling line, rhythm of direction and repetition.

The humanities scholar Heather Houser similarly points to a double-reading of the graph in her exploration of InfoVis or information visualisations on the internet.[7] InfoVis are skilful presentations of complex data sets that can bring viewers to the point of 'infogasm' through an allure of imagery that promises knowledge. The material performativity of the internet's speed of transferring signals and light through 'dispersed material infrastructures'[8] gives an illusion of immateriality that enables visuals to appear free from the chemicals, metal and plastics of its resource requirements. The popular *flattening the curve*, in reference to COVID-19 representations of infected people needing healthcare over time, situates the graph in a globalised mediascape where data transparency aligns with forms of erasure.

Recent anthropological discussion of the energy graph emphasises the popularity of oil projections in transforming feelings of present security into an uncertain futurity.[9] Elsewhere, I detail a historical shift towards favouring an image of the graph in visual communication.[10] In all manner of conversations – energy supply, CO2 emissions – the graph is a key tool for visualising the future. I explain the popularity of the graph by reference to key moments: first, its promotion by economists in the restructuring of energy and financial markets during the 1980s;[11] second, its adoption as science imagery by environmental activists during the 1990s.[12] Notably, these events signal a departure from earlier text-based and self-evident visual encounters. As such, viewer engagement shifts from favouring images composed of textual and realistic renderings to images where pleasure occurs from uncertain interpretation.

Refinement through the Energy Graph

Among the energy sciences, visualising through the graph is preoccupied with timelines that are inaccessible through direct human perception. Graphs that depict forecasts of liquid oil production, for example, are created from three distinct communities of practice: resource economics, petroleum geology, and energy and climate modelling.[13] Economists employ optimal depletion models aimed at determining levels of production and rate of depletion, thus identifying the maximum net present value of the resource. Petroleum geologists embrace the Hubbert methodology (Gaussian curve) assuming that oil extraction increases until the reservoir is half-consumed, at which point production decreases. Finally, climate analysts are interested in oil production because of carbon dioxide emissions from oil use. Their efforts to simulate the extraction of oil are based on simplified non-specific petroleum resource bases. These approaches to the future are captured, plotted and rendered through the graph. In other words, the graph is a readily acceptable visual in representing diverse activities that can transform information into knowledge that purports to have prognostic value.

Types of graphs used for visualising energy information include linear, bar and histograph. Image characteristics comprise colour, texture and edge style. These strong qualitative aspects draw attention to style classifications similar to those within the visual arts. The linear graph used in representing CO2 emissions employs a spontaneity of unrestrained lines which is a visual technique resembling Abstract Expressionism with its reductionist colour palette, vivid strokes and lack of emphasis on form. By contrast, the bar graph focuses on geometric shapes which, arguably, could be associated with Minimalism with its anti-humanist rupture of the picture plane. Visuals of this sort suggest a lack of resemblance to pictorial art that typifies well-formed

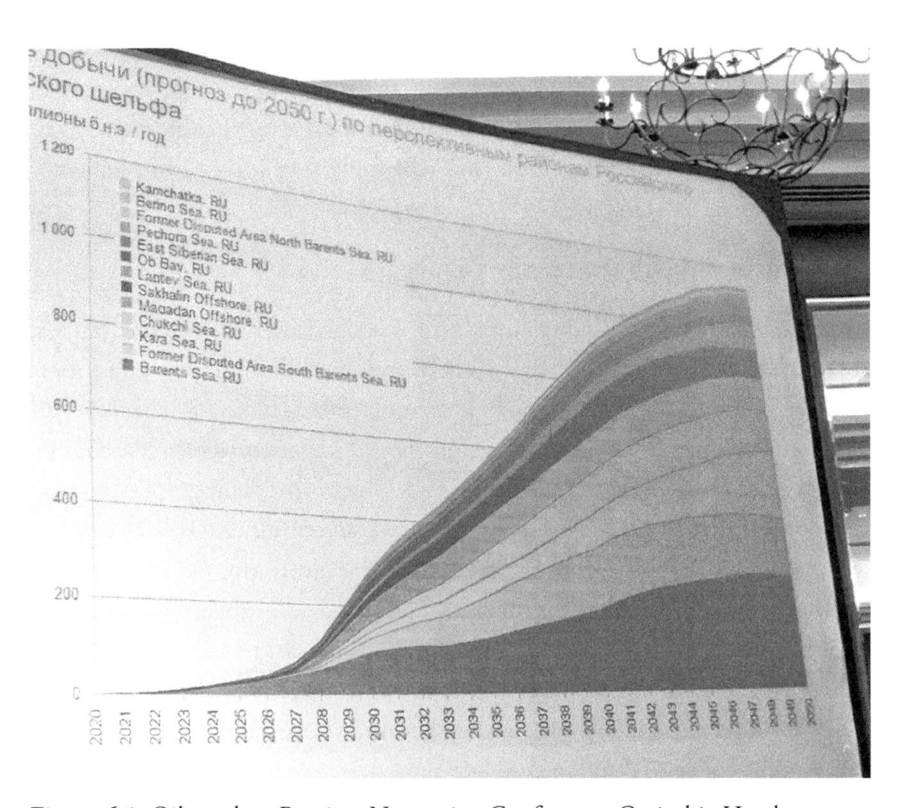

Figure 6.1 Oil graph at Russian–Norwegian Conference, Corinthia Hotel, St Petersburg, Russia. Photo by Arthur Mason.

subjects such as those found in figurative historical art styles or scenes of delicately rendered landscapes. The use of qualitative attributes such as line, shape and colour suggests that the graph can provoke an aesthetic experience that disrupts its content of meaning.[14]

The idea that the energy graph utilises artistic form came to me when I happened to pass through the mezzanine floor on route to an energy round-table inside the Corinthia Hotel in St Petersburg, Russia. Executives attending the oil conference repeatedly walked past a set of hotel paintings whose colour schemes and shapes struck a similarity with Arctic oil development graphs shown in an adjacent room. In one instance, the straight-line features of an oil painting mirror the block-shaped figures of a bar graph used to depict petro-geologic development futures in the Arctic. Placed side by side, the graph and the art painting seemingly collude in a form of representation whose visual resemblance gives no indication of which is a primary reference. In a second painting from the hotel, a colour palette of yellow and red creates

Figure 6.2 Mezzanine-level contemporary paintings at Corinthia Hotel, St Petersburg, Russia. Photo by Arthur Mason.

a contrast between a particular feature (red) and background state of affairs (yellow), providing a visual sensation of seeing-in. This painting mirrored an oil development image whose yellow/red contrast also gave a feeling of 'seeing-in' on directional ice-flows in the oil-rich regions of the Barents Sea. In this instance, the energy graph is a visual representation whose model of resemblance is the artistic oil painting.

Thus, what appeared to me initially as images created to ensure reliability in planning came to suggest a social habitus through semantic provisioning. Faced repeatedly with oil graphs whose resemblance to abstract paintings seemed made to order (and vice versa), I began to wonder if a veracity regarding planning is in fact an aesthetic preference for abstract expression. In this instance, the graph is a type of visual whose primary resemblance is an art image – the latter, a representation whose recognition reflects a stress on achievement. In this manner, abstractness in the graph, as with art, reflects a desire for knowledge expressed as uncertainty. By replacing recognition with uncertainty, the graph increases the difficulty and length of perception as a way of experiencing the artfulness of reality. Through this demand for uncertain interpretation, the graph invites a hedonistic reaction to objects of scientific judgement.

Not to be ignored, the Corinthia is one of the more luxurious hotels in St Petersburg. Its interior design, and the selection of paintings, fixtures and furniture, are original creations by the G.A. Group, a global hospitality specifier in London. In this setting, knowledge provisioning and incidental luxury become bound together, creating a link between aesthetics and credibility. Here, hotel extravagance enhances the possibilities of achieving knowledge acquisition whereby abstractness in art work represents a bid. Indeed, a substantial component of the allure of executive locations like the Corinthia is the way they utilise space to 'amplify and refine' an explicitly cosmopolitan identity.[15] In this way, the conception of abstractness in the graph is an aesthetic ideal that stands in for the appearance of a hydrocarbon production ideal.

The historian Robert Brain, mentioned above, notes a similar pattern of representation wherein art work is modelled from graphical expression.[16] Brain points to the well-known painting *The Scream* by Edvard Munch. In this image, a linear waveform flows from the perspectival back of the painting onto the humanoid figure, drawing the viewer into the painting's landscape and soundscape. According to Brain, the painting deliberately sets the conditions of its visuality to resemble the graphical expressions of psycho-physiological experiments, with which Munch was familiar at the time. *The Scream* is a visual representation whose model of resemblance includes the graph.

The resemblance between art work and graphical expression offers an additional route for considering the graph by reference to Bourdieu's regime of taste.[17] Bourdieu notes that perception is a process of interiorisation of signs through pedagogic work by the individual. Such work produces a durable training of recognition, or what he calls habitus. Largely through mapping differences between consumer goods onto differences between social groups, Bourdieu demonstrates that differences in habitus represent distinct cultural classification systems that appear as differences in taste. Arguably, the surface/depth dichotomy of his Kantian aesthetic regime is a simplification; the idea that groups identify with a sense of taste exclusively has a diminishing utility

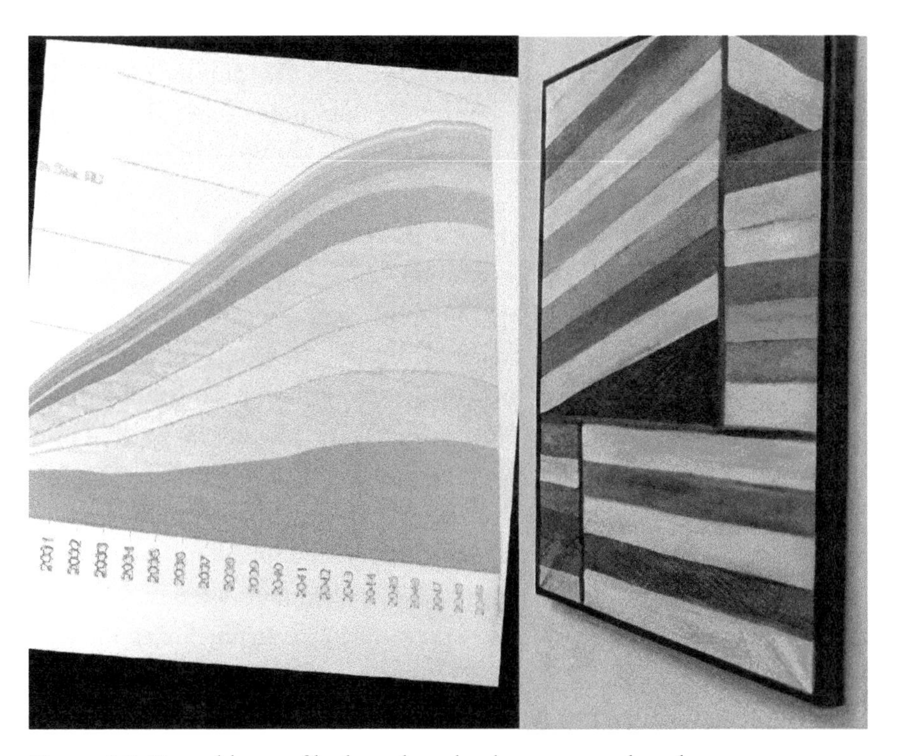

Figure 6.3 Resemblance of hydrocarbon development graph and contemporary painting. Photo by Arthur Mason.

with the intermingling of popular and high culture in the West.[18] Still, as I describe below, visualising through the graph emphasises an appreciation for abstractness where knowledge expresses uncertainty in temporal order and aesthetic judgement.

Clear Thinking through the Graph

In the Hollywood movie *Syriana*, the actor Matt Damon plays an energy analyst hired by a Saudi prince to serve as 'economic advisor'. According to the *Internet Movie Database*, the film is 'a politically-charged epic about the state of the oil industry in the hands of those personally involved and affected by it'. After weeks of globetrotting with the Saudi royal, Damon catches up with his wife, played by Amanda Peet, who expresses doubt over the new job, prompting Damon to explain, 'I know it seems like people sitting in hotel rooms, but that's what they do, that's how they do business.' Damon then rattles off a wish-list of events derived from sitting around talking business: 'We're talking about the world historical stage, with the delivery deals we can make with Europe, oil transport through Iran. The [Saudi prince] might be able to revolutionize not just his country but the whole region. He could be like Mosaddegh [Prime Minister of Iran 1951–3] – a real democracy rising up organically. If we can be a part of any of these countries getting a parliament, helping them find efficiencies . . .' As Damon's speech trails off, movie audiences confront Peet's sullen expression.

The exchange above aligns with my own impression of a global petro-power elite, many of whose members I met at first hand while attending executive energy roundtable events. At these meetings, I too became aware that experts, politicians and industry leaders sit around in hotel rooms immersed in 'Oil Talk'.[19] Following Peter Adey, such events are elite premium networked environments that take place in expensive hotels located in global cities.[20] I have, over time, attended energy roundtables in cities across the Global North, including Moscow, Oslo, London, Houston and Mexico City. At each roundtable, experts give individual talks lasting fifteen minutes each. The expert stands near a wall-screen onto which PowerPoint slides are projected. Attendees observe and listen, but also follow along in an agenda booklet that they are issued with upon arrival. This booklet contains reproductions of the slides that are being shown by the expert. Often participants scribble notes in the booklet, an activity that I have come to understand as an effort to elucidate the relationship between the printed material and its meaning as explained by the expert.

The global oil executive roundtable has various historical antecedents, including the club-style setting associated with collusion on price fixing among companies or by governments, whose origins can be traced to the first quarter of the twentieth century.[21] One example is the *As Is* agreement of 1928 in which leaders of Exxon, Royal Dutch/Shell and British Petroleum met at Achnacarry Castle, Scotland, to devise a collective strategy to defend their companies' profitability from problems of over-production and low oil prices.[22] Collusion in the oil industry in terms of 'a very exclusive club' may be initially attributed to Pérez Alfonzo in reference to his conception of the sovereign cartel, the Organization of Petroleum Exporting Countries or OPEC.[23] Often described as the 'founder' and 'architect' of OPEC, Alfonzo served as Venezuela's Minister of Mines and Hydrocarbons in the 1950s.[24]

In recent years, select global advisory firms including IHS, WoodMac and Cambridge Energy Research Associates (CERA) have sought to establish executive gatherings by reference to these early twentieth-century forms. During the 2000s, for example, CERA established the Global Power Summit, inviting industry leaders to take part in 'a private, neutral, club-style setting' to explore issues facing the international energy community.[25] The venue offered formal and informal exchange in a focused but relaxed atmosphere. All details – ground transportation, meals, internet access, language translations, entertainment – were arranged by CERA. One venue of the bi-annual Power Summit was the Westin Turnberry Resort in Ayrshire, Scotland, a luxury accommodation now owned by former US president Donald Trump and located not far from Achnacarry Castle.

CERA emerged in the 1980s during energy deregulation spurred by growth and innovation in activities related to energy marketing. The need for information created robust opportunities for firms which collect, analyse and distribute information of relevance to energy buyers and sellers, including information about weather, future prices, demand patterns and storage flows. This need was reflected in a conversation I had with a planning analyst for the natural gas marketing company KeySpan, who I met with at a CERA energy roundtable event:

> Our company needs to look at prices and know when prices will go up. We need to know that to let the customers know. We've been working with CERA for a long time. We always renew the contract to get the internet subscription and a spot to attend the Roundtables. They've been doing it for a long time, so people trust them. Because internally, you're interested in [gas pricing and storage] but you don't have the time to develop that, as a company you only have so much resources and time.

One of the more compelling uses of graphs that I have witnessed took place at a roundtable event in 2003, where CERA analysts invited clients to understand the process by which they predict market functioning. In this case, CERA analysts demonstrated how to identify when natural gas producers ramp up production in response to a rise in energy prices, what they call the *forward price curve*. CERA analysts described various 'mistaken results' that occurred while developing their findings. But they explained that by charting responses at various time differentials from the price rise, they could discover

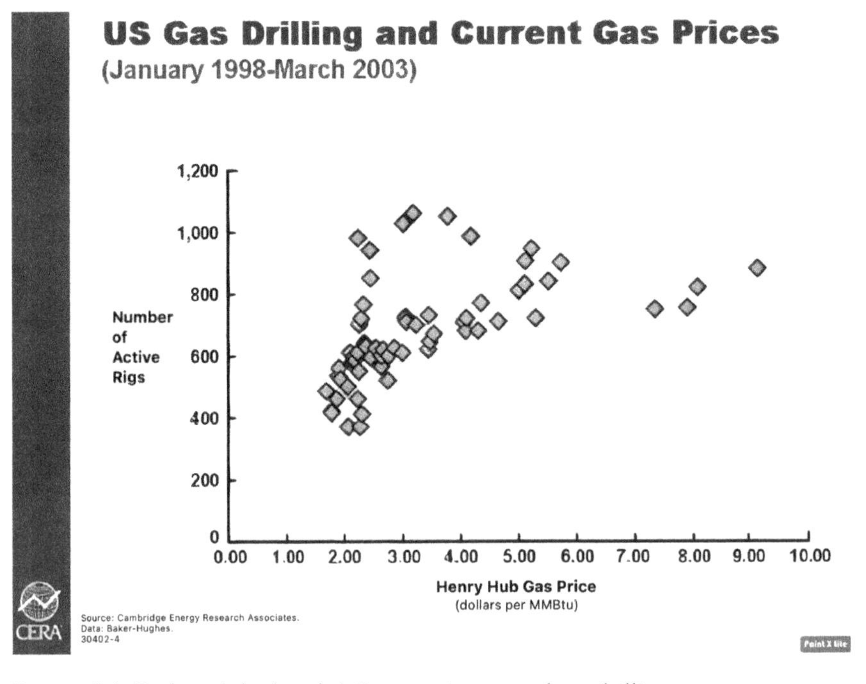

Figure 6.4 'It doesn't look right'. Powerpoint natural gas drilling response. Courtesy of PanArcticon Energy Archives.

the correct relationship of a drilling response to the rise in price. What the analysts discovered from these trials was a six-month lag between a rise in natural gas prices and an increase in gas-related rig drilling for the period January 1998 to March 2000.

In their demonstration, the early test results shown to the audience took the form of a scatter plot graph in which square figures were individually dispersed across the image and about which analysts stated 'You see, it doesn't look right'. Analysts then showed what they referred to as their correct results in the form of a different scatter plot. Here the individual squares appeared concentrated together, and about these the experts stated 'You see, it looks right'. CERA analysts presented this six-month lag to clients as a pattern discovered through trial and error. The pattern that did 'look right' offered a true representation of market functioning. The correct display suggests, following Edward Tufte, that 'excellence in the display of information is a lot like clear thinking'.[26] In this case, the sensorial image of graphical

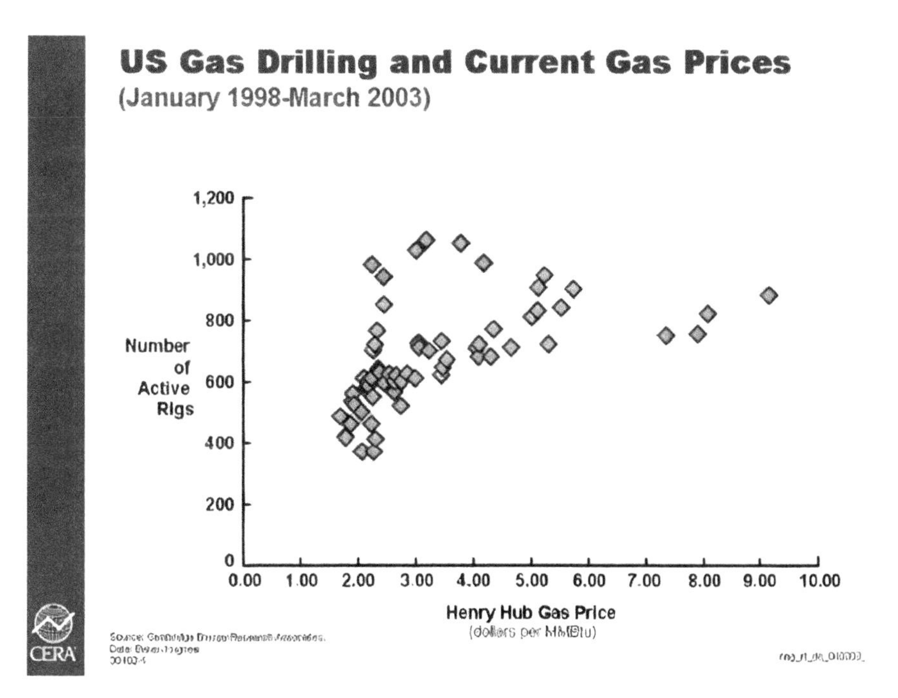

Figure 6.5 'You see? It looks right'. Powerpoint six-month time lag natural gas drilling response image. Courtesy of PanArcticon Energy Archives.

expression represents the abstract image of *market functioning*, which itself is an abstraction.

This pattern of market functioning presented by CERA analysts took the form of a proprietary discovery. I was surprised when, some weeks later, during a review of the US government's Energy Information Administration (EIA) website, I discovered a report produced two years earlier that offered the same six-month price lag. A similar type of graph used by CERA, a scatter plot with square figures, was employed by EIA. Moreover, the EIA graphs appear to have nearly the same configuration as the CERA image, but with different dates. The EIA report is titled *U.S. Natural Gas Markets: Mid-Term Prospects for Natural Gas Supply, December 2001.*[27] The graph in question states: 'Scatter Plot of Monthly Natural Gas Drilling Rigs Versus Wellhead Natural Gas Prices 6 Months Earlier, July 1992–September 2001' – a similar title to the one CERA uses about its discovery.

CERA relies on EIA for a substantial amount of information and often cites the organisation directly. The analysts who presented the price curve discovery had worked previously at EIA. CERA analysts with whom I spoke about the scatter plot provided responses similar to that of the analyst for

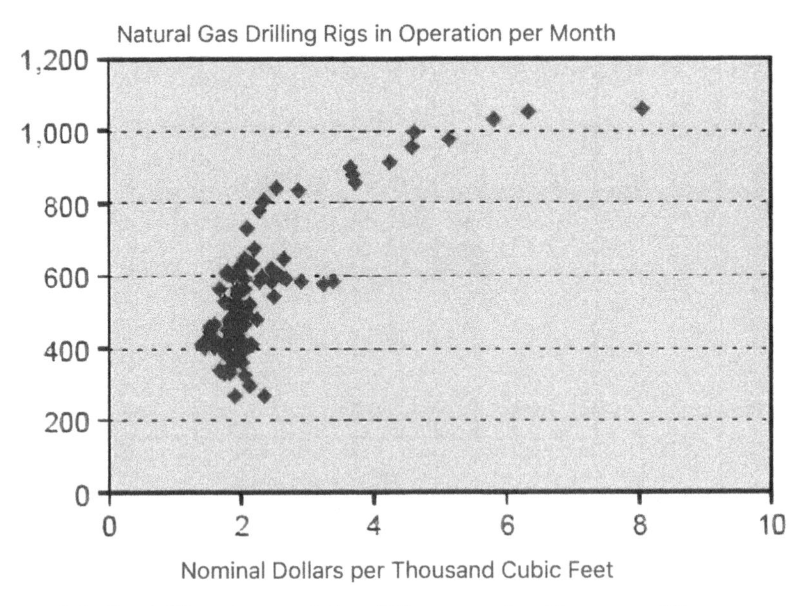

Figure 6.6 Gas drilling six-month lag, IEA scatter plot. Courtesy of Energy Information Administration.

Key Span quoted earlier, that companies do not have the internal resources to develop market understandings and rely on CERA to re-package hard-to-find analysis. Here, re-packaging may be considered a type of classical resemblance where representation transfers an original idea through a technique of family resemblance, such as substitution and imitation.

Vitality through the Graph

One unfolding structure of the energy future taking place at the turn of the millennium involved attempts to develop Arctic natural gas frontiers. In 2000, CERA consultants circulated the graph shown in Figure 6.7 at their executive roundtable events. The image depicts an eight-year span of natural gas prices, both historical and projected, as understood during the fourth quarter of 2000. The graphic appeared as a wall-sized PowerPoint image but also as an 8-by-11-inch illustration printed in an agenda booklet. The title of the graphic is an open-ended question: 'Natural Gas Markets: How Long Will High Prices and Volatility Last?'. Together, the graph and title provide an outlook of expectation on the future of higher natural gas prices in the North American energy market.

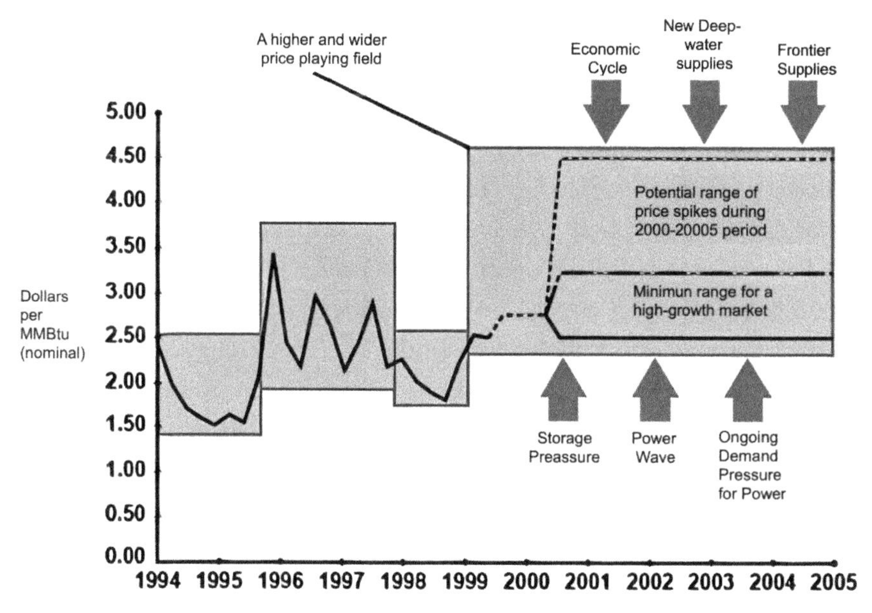

Figure 6.7 Natural gas markets: how long will high prices and volatility last? Courtesy of PanArcticon Energy Archives.

A likely entry point into the symbolism of the graph is the year 2000, which appears as at the mid-point of the timeline located at the bottom. At that time, the year 2000 was the temporal present and carried the weight of a self-evident fact. Thus, the figure '2000' gave its location an attention-grabbing significance as a departure point for exploration. To the left, the eye glides across a projection of six years then moves diagonally upward to a historical record of natural gas price. Created in late summer, the graphic was distributed to clients during autumn and discarded by late winter 2001.

In addition, the graphic served as an originary visual for an industry discourse about growth imperatives. It is a fragment of evidence supporting a highly regarded claim circulating at the time called *The Long Ascent*, an industry expectation of increasing natural gas consumption, then estimated at 22 trillion cubic feet (tcf) and moving to a 30 tcf market by 2010. The graph represented also a new contribution to industry restructuring. As noted above, the natural gas industry shifted to a liberalised market form beginning in the 1980s. Previously, government had provided companies with a structured risk environment that secured rates of profit over fixed periods. The shift to a liberalised market collapsed these agreements. Thus, the appearance of price projection in the graph represents a visual declaration that energy companies had entered into a more competitive arrangement. It suggests that critical knowledge was shifting from the regulatory (textual) to the economic (graphic) as part of a new strategic approach for demonstrating balances between increases to supply and incremental rises in demand.

In this way, the graph gives evidence of the disadvantages of destabilising balance whereby, for example, an abundance of supply destroys price. The destruction of price could be disastrous for new development projects whose investment recovery depends on the stability of long-term prices. By both forecasting the incremental amount of new energy additions that can satisfy demand and depicting the actual price these energy additions will fetch in the market-place, the graph serves as an indication of the changing politics surrounding the development of new supply sources. While, prior to restructuring, the political community established incentives for corporate decision-making, by the early 2000s fluctuation in price, as evidenced in the graph, established the risks by which industry sought political concession.

The graph also stands as a new determinant towards investment decision-making. It is a bit of knowledge in a stable stream of information that experts provide to clients about a system of pricing. In this case, it suggests that new instabilities have given rise to a field of consultant forecasters whose expertise is the announcement of a collective need within industry for knowledge on pricing, since knowledge of future price facilitates trade, provides longer-term signals that govern investments, and allows producers and consumers to manage risk.

Skipping ahead within the same agenda booklet, we see that another graph similarly depicts a decadal temporality. Instead of an eleven-year range, this graph depicts a fifteen-year trajectory. Instead of splitting time into a past, present and future, this graph confronts a singular future extending directly out of an isolated present. The graph provides a visual into the expected planning stages for a large pipeline construction project to bring Alaskan natural gas on-stream. The image portrays a series of dates and events and is a history

Artic Gas Supply Build - Supply Realignment

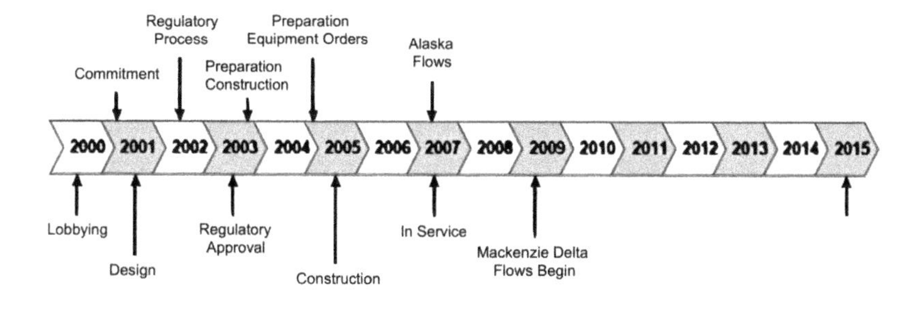

- Current supply disappointments accelerate Northern development
- Strong prices lead to pipe capacity commitments by the end of 2001
- Plans develop for a large project(s) to bring both Alaskan and Mackenzie gas on-stream
- Project size of 4 Bcf per day
- Regulatory and government requirements necessitate building pipe capacity to "market"
- New pipe to California and Pacific Northwest: 1.75 Bcf per day growing to 2.4 Bcf per day by 2015
- Total of 4.25 Bcf per day of supply by 2015

Figure 6.8 Arctic Gas Supply Build. Courtesy of PanArcticon Energy Archives.

lesson in the projected gas flow to North American markets by year 2007. The graph appears in the shape of a pipeline, and printed directly below in the booklet are specifications that provide information relating to expectations: capacity size, date of initiation and timing completion.

Importantly, the two graphs above were not expected to be viewed independent of each other. They are visuals as part of an installation within a series of graphs about energy future. Both graphs – Natural Gas Markets and Arctic Gas Supply Build – share a similar placement within the agenda booklet. They appear as the last graph in a sequence of graphs. They represent the final images of two separate presentations. They are concluding images. As such, they carry the drama of a finale. Their emotive power lies in their being presented as last (a lasting image), and in their capacity to represent a climax as the sum total of previous graphs – that is, to have greater meaning than the previous parts to which they owe their final existence. The Natural Gas Market graph appeared as part of a set of initial presentations titled 'Market Focus', which took place in Calgary on 2 October 2000. This graph sums up a story about market dynamics that is captured in a series of thirty graphs that precede it. The information of these earlier images recounts the content assumptions of *The Long Ascent*. In short, these previous graphs help build up information for the final image, but also become inscribed into the final image. The graph titled 'Arctic Gas Supply Build' appeared in a subsequent presentation titled 'Arctic Gas', which, according to the same agenda schedule, was the last presentation in the two-day executive roundtable session. Thus, the Arctic Gas Supply Build was the final image of all the previous graphs combined.

The terms and shapes generated by these two graphs are worthy of comparison for the vitality affects they aim to generate. Figure 6.7 portrays a stratified harmony of points. By contrast, Figure 6.8 presents a singular line rhythmically punctuated with arrows, like staccato emphases. In the former, a constellation of rhythmic movements defines a market characterising creative risk: drivers, ongoing demands, pressures, cycles, playing fields, frontiers, minimum and potential ranges, high growths, spikes. The jagged lines guide the eye from left to right towards a finality of parallel lines that present a two-part illusion of progression in which time accelerates towards a finale where tempo decreases, becomes more powerful, rises higher and is then set right. By contrast, the second graph emphasises a staid and conservative set of terms: commitments, design, approval, service, flows, full capacity, preparation, process, orders. Such terms frame movement with a sense of stability, finality, and the feeling of inevitability and commitment to procedure. It is a step-by-step aesthetic for achieving full capacity placed into linear visual style. In these graphs, the exteriority of the written and figurative element symbolises the penetration of discourse into the form of things whereby representation and verbal sign are inextricably bound, revealing discourse's ambiguous power to redouble.[28]

Thus, both graphs suggest that inscription by itself does not carry the burden of explaining the power of futurology. Following Bruno Latour, inscription serves as 'the fine edge and the final stage of a whole process of mobilization'.[29] Certainly, the graphs excite yet provide little of the relations between companies, governments or other stakeholders. This kind of minimalism creates a tension within the graphic image: the impulse for narration. Thus, the challenge of energy graphs is to uphold their 'simplicity, clarity, and instantaneity, while elaborating data's intricate stories and sparking a craving for more'.[30]

In these instances, then, the energy graph is a 'virtual entity' capable of condensing relations between materials, technical processes and economic conditions, thereby enabling stakeholders to extend their decision making into the future by envisaging actions without 'necessarily ever knowing precisely what exists in the present'.[31] By simplifying complex systems into an abstract image, the graph channels disruptive activity, governs transparency and generates accountability.[32] In sum, these graphs are teleological; they not only telegraph the future but also attempt to manage it.[33] Or, following Karin Knorr Cetina's metaphor of knowledge construction as a form of empiricism, the energy graph is an object made to facilitate empirical success, where *representation is a model of adequate phenomena* whose conceptual entities are seen in direct relationship to objective realities of investment decisions.[34]

Empathy for the Graph

Today, images that connect the immediacy of identification to the weight of a temporal present are referred to as *special effects*. The Hollywood movie industry and news format of re-enactments have a monopoly over the legitimate production of the realistic (simulacra) aesthetic. These are the *dream factories* that merge the future (*Mad Max*) and the past (*Jurassic Park*) with a self-evident temporal present. The entire premise of reality programming with its emphasis on the staging of verifiable truth through unscripted formats was, initially, an attempt to reinstate authenticity in an anti-Kantian realistic aesthetic.[35] But such imagery, while immediately identifiable, can no longer verify what is authentic to reality.

By contrast, the Kantian aesthetic is produced by artists and (energy) scientists. Art and science have a monopoly over the legitimate production and interpretation of Kantian images. Images in science are not immediately recognizable, but still represent an apparently trustworthy verification of reality. Such images entertain, but the drama of their style classification emphasises a visual encounter that favours an achieved understanding. Yet, while the Kantian aesthetic is difficult to interpret (and inclusive of practices of erasure), Kantian images serve to verify, in contrast to imagery whose immediate identification has come to signify artificiality.

It is because of the graph's independence in verifying *as an image*, that is, in circulating as a free-floating signifier of energy development and planetary crisis, that I argue that mass populations develop an empathy for it. Empathy is a quality of identification with the emotional life of another, and in this case a process of projecting feeling into aesthetic production. A sculpture, a figure in a painting, a literary or musical composition, all may evoke feelings of empathetic engagement.[36] Graphs too, provoke embodied experiences of empathy. The graph is an abstract visual whose colour, shapes and patterns provoke elusive qualities that may be described kinetically, such as surging, fading away and crescendo, and also thermally, such as cooling and warming. These vitality affects[37] may be experienced as empathy for example, by the way Cartesian graphs divide the plane into quadrants, distinguishing up and right with positive affects and down and left with negative affects.[38]

Perhaps there is no better example of an empathy for the graph than the popular reception of the movie *An Inconvenient Truth*.[39] In this environmental-activist film, imagery no longer relies solely on a realistic aesthetic to create a sense of urgency for saving the planet. In the movie's most dramatic scene, former Vice President Al Gore traces 650,000 years of CO2 levels and temperature variation in a visualisation derived from paleoclimatologist Lonnie Thompson's research. Dramatising the expression 'off the charts', Gore climbs into a cherry picker that carries him up a giant screen to trace the staggering projections for CO2 and temperature increase over the next fifty years. Such abstract visuals, including the controversial 'hockey stick' graph of climatologists Michael Mann, Raymond Bradley and Malcolm Hughes, have visually shaped current environmental debate.[40] That is, empathy for the planet is now based on a graph, an image that substitutes classical notions of resemblance through simulated renderings with an abstraction.

This shift – *from deriving empathy for environmental crisis through realistic representation ('seeing a whale harpooned'), to having empathy for environmental crisis from seeing a* graph – is what I call a transfer of the realistic or anti-Kantian aesthetic to the Kantian aesthetic. The whale and the graph are two types of images. Appreciation of environmental crises from these images requires two separate ways of seeing things. In the former, empathy involves classical notions of resemblance whereas in the latter, the underlying trait of resemblance lies in *transposition* or Epiphora (in Aristotle's sense of the term) where assimilation occurs between alien ideas, distant from one another. In this sense, the whale and the graph are two separate types of synecdoche for classifying knowledge, that is, two separate ways of discerning both what is good and what is in good taste.

Actually, the staging of Gore's empathy for the graph involves a transfer to Kantian aesthetic in real time. In the movie, Gore presents climate change initially through photographic comparisons of glacial melt over a hundred-year time span. This diptych imagery presents climate change in the form of a family resemblance. That is, glacial melt is presented through a comparison of two photographs whose perspectives, though taken nearly a hundred years apart, nevertheless replicate the same position, angle and elevation. After displaying these simulated resemblances, Gore then introduces the abstract graph to show timelines measured in thousands of years.[41]

Conclusion

For Pierre Bourdieu, the Kantian classifying scheme derives from a life distanced from economic necessity and favours distance in all manners: economic, and also temporal, spatial, aesthetic.[42] What makes the Kantian perspective especially political in the context of this chapter is the imposition on mass populations of a sense of livelihood peculiar to the artistic and intellectual professions, both of which seem to share an idealised image of uncertainty.

A preference for the graph marks a refinement of perception. It draws attention to a growing unease among those who rely on images that this refinement seeks to replace. Such a shift is not unlike the history of manners, which Norbert Elias[43] describes, in which European society modelled itself on the etiquette of the noble court. Perhaps the denial of current events whose authenticity relies on the Kantian image is not merely a rejection of the basic facts but instead a rejection of how such facts are made social. After all, climate change seldom receives protest when viewed through the anti-Kantian aesthetic, that is, as special effects in the form of a Hollywood movie. The Kantian graph, by contrast, is a move towards refinement and an enforcement of visual communication whose mode of perception seems to suggest a new style of manners. Empathy for the graph is an acquired taste. But it has also become immediately identifiable, not for its content of meaning but for what it says to the viewer as an arbiter of his or her own sense of taste.

Notes

1. Giambattista Vico (trans. Thomas Bergin and Max Fisch), *The New Science of Giambattista Vico* (Ithaca: Cornell University Press, 1988), 118.
2. Pierre Bourdieu, *Distinction* (London: Routledge, 1984).
3. Peter Sloterdijk (trans. Michael Eldred), *Critique of Cynical Reason* (Minneapolis: University of Minnesota Press, 1987).
4. Robert Brain, *The Pulse of the Modern* (Seattle: University of Washington Press, 2015).
5. Mikhail Bakhtin (ed. Michael Holquist), *The Dialogic Imagination* (Austin: University of Texas Press, 1994), 209; Norbert Elias, *Involvement and Detachment* (Oxford: Blackwell, 1987), 42.
6. Brain, *Pulse*, 14–15.
7. Heather Houser, 'The Aesthetics of Environmental Visualizations: More than Information Ecstasy?', *Public Culture* 26, no. 2 (2015), 319–37.
8. Jennifer Gabrys, *Digital Rubbish* (Ann Arbor: University of Michigan Press, 2011), 58.

9. Mandana Limbert, 'Reserves, Secrecy, and the Science of Prognostication in Southern Arabia', in Hannah Appel, Arthur Mason and Michael Watts (eds), *Subterranean Estates* (Ithaca: Cornell University Press, 2015).
10. Arthur Mason, 'Graphical Representation' (paper presented to 'Team Society', NTNU, 27 January 2021).
11. Arthur Mason, 'The Rise of Consultant Forecasting in Liberalized Natural Gas Markets', *Public Culture* 19, no. 2 (2007), 367–79; Caitlin Zaloom, 'How to Read the Future: The Yield Curve, Affect, and Financial Prediction', *Public Culture* 21, no. 2 (2009), 246–68.
12. Paul Wapner, 'Politics Beyond the State', *World Politics* 47, no. 3 (1995), 311–40.
13. Alex Farrell and Adam Brandt, 'Risks of the Oil Transition', *Environmental Research Letters* 1, no. 1 (2006), 1–6.
14. Donna Rosato, 'Worried about Corporate Numbers? How about the Charts?', *The New York Times*, 15 September 2002, Business Section.
15. Caroline Levander and Matthew Guterl, *Hotel Life* (Chapel Hill: University of North Carolina Press, 2015), 6.
16. Brain, *Pulse*, 182–5.
17. Bourdieu, *Distinction*.
18. Colin MacCabe, 'Preface', in Fredric Jameson (ed.), *The Geopolitical Aesthetic: Cinema and Space in the World System* (Bloomington: Indiana University Press, 1992), xiii.
19. Hannah Appel, Arthur Mason and Michael Watts, 'Oil Talk', in Hannah Appel, Arthur Mason and Michael Watts (eds), *Subterranean Estates: Lifeworlds of Oil and Gas* (Ithaca: Cornell University Press, 2015).
20. Peter Adey, 'Security Atmospheres or the Crystallisation of Worlds', *Environment and Planning D: Society and Space* 32, no. 5 (2014), 834–51.
21. Morris Aldeman, *Genie out of the Bottle: The World Oil Since 1970* (Cambridge, MA: MIT Press, 1995).
22. Daniel Yergin, *The Prize: The Epic Quest for Oil, Money, and Power* (New York: Simon & Schuster, 1992), 263–9.
23. Anthony Sampson, *The Seven Sisters: The Great Oil Companies and the World They Shaped* (Toronto: Bantam Press 1975), 4.
24. Fernando Coronil, *The Magical State: Nature, Money and Modernity in Venezuela* (Chicago: University of Chicago Press, 1997), 353.
25. GPF, Cambridge Energy Global Power Forum Summits, 19 October 2008.

26. Edward R. Tufte, *Visual Explanations: Images and Quantities, Evidence and Narrative* (Cheshire, CT: Graphics Press, 1997), 141.
27. Energy Information Administration (EIA 2001), *U.S. Natural Gas Markets: Midterm Prospects for Natural Gas Supply, December 2001*, https://www.eia.gov/naturalgas/articles/storageindex.php (last accessed 22 February 2022).
28. Michel Foucault, *This Is Not a Pipe* (Berkeley: University of California Press, 2008).
29. Bruno Latour, 'Drawing Things Together', in Michael E. Lynch and Steve Wooglar (eds), *Representation in Scientific Practice* (Cambridge, MA: MIT Press, 1990), 42.
30. Houser, 'The Aesthetics of Environmental Visualisations', 329.
31. Andrew Barry, *Material Politics: Disputes along the Pipeline* (Chichester: Wiley Blackwell, 2013), 14.
32. Timothy Mitchell, *Carbon Democracy: Political Power in the Age of Oil* (London: Verso, 2011).
33. Arthur Mason, 'Images of the Energy Future', *Environmental Research Letters* 1, no. 1 (2006), 1–8.
34. Karin Knorr Cetina, 'Strong Constructivism – from a Sociologist's Point of View: A Personal Addendum to Sismondo's Paper', *Social Studies of Science* 23, no. 3 (1993), 555–63.
35. Steven Brown, 'Experiment: Abstract Experimentalism', in Celia Lury and Nina Wakeford (eds), *Inventive Methods: The Happening of the Social* (London: Routledge, 2014), 61–75.
36. David Freedberg and Vittorio Gallese, 'Motion, Emotion and Empathy in Esthetic Experience', *TRENDS in Cognitive Sciences* 11, no. 5 (2007), 197–203.
37. Susan Lanzoni, 'Empathy in Translation', *Science in Context* 25, no. 3 (2012), 301–27; Elizabeth Lunbeck, 'Empathy as a Psychoanalytic Mode of Observation: Between Empathy and Science', in Lorraine Daston and Elizabeth Lunbeck (eds), *Histories of Scientific Observation* (Chicago: University of Chicago Press, 2011), 255–75.
38. Susan Gerofsky, 'Ancestral Genres of Mathematical Graphs', *For the Learning of Mathematics* 31, no. 1 (2011), 14–19.
39. David Guggenheim, *An Inconvenient Truth* (Documentary Film, 2006).
40. Houser, 'The Aesthetics of Environmental Visualisations', particularly 322.
41. In this instance, while the graph aims to represent, its work of resemblance lacks a primary visual reference. On closer inspection, however, resemblance is evident in the CO2/temperature graph, where it takes the form of two identical

timelines internal to the graph itself – the CO2 timeline mirrors the temperature timeline. As shown in the movie, the CO2 timeline is unveiled *after* the original appearance of the temperature timeline below it, lending the former a classic representational character, a model that presupposes an original whose resemblance is a less faithful copy struck from it. In fact, in the movie, the CO2 timeline is unveiled in a kind of anti-Kantian dramatic flourish whereby the audience gasps at its immediately recognisable appearance. In this way, the upper line (CO2) is a recognisable image by its resemblance to a primary reference – that is, the Kantian-line directly below it (temperature). With the graph, as with much of modern art, representation is highly self-referential.

42. Bourdieu, *Distinction.*
43. Norbert Elias, *The Civilizing Process* (London: Routledge, 1972).

7

AFTER OIL

Rania Ghosn

The 1970s marked the end of cheap, abundant and guilt-free petroleum.[1] In the USA, long queues at the pump materialised the threats of foreign oil dependency following the 1973 Arab oil embargo. Since then, the surging global demand for energy, projections of oil depletion, and the need to manage climate change have brought forth the necessity for long-term structural change in energy systems. Most energy transitions, however, inherently become a problem of carbon, the response to which is low-carbon energy sources and a series of techno-fixes: carbon sequestration, carbon credits, carbon markets. The euphoric tone of low-carbon or carbon-free narratives is uncanny; it purges, or at least masks, carboniferous matters while perpetuating a series of myths, notably that 'any newly discovered source of energy is assumed to be without faults, infinitely abundant, and to have the potential to affect utopian changes in society. These myths persist until a new source of energy is used to the point that its drawbacks become apparent and the failure to establish a utopian society must be reluctantly admitted.'[2]

The project *After Oil* by the architectural collaborative DESIGN EARTH presents a speculative future of the Arabian Gulf when the world has transitioned away from fossil fuels. In a series of three triptychs, *After Oil* renders visible the embeddedness of petroleum in a region of many oil-producing economies by charting matters of concern at three nodes in the system: an offshore oil extraction and processing facility (Das Island); a transit chokepoint (Strait of Hormuz); and the site one of the largest oil spills in history (Bubiyan Island). To imagine a society after oil foregrounds the past and present geographies of oil. Such an extrapolation of issues critical for today's oil landscape favours an approach to energy transitions that upholds the geographies of technological systems – their processes, sites, objects and externalities.

Figure 7.1 Das Island, Das Crude. The section indexes the increasing heights of architectural landmarks in relation to the depths and geological periods of extraction of subterranean fields. Courtesy of DESIGN EARTH. Project Team: Rania Ghosn and El Hadi Jazairy; Rawan Al-Saffar, Namjoo Kim, Hsin-Han Lee, Kartiki Sharma, Jia Weng, Sihao Xiong.

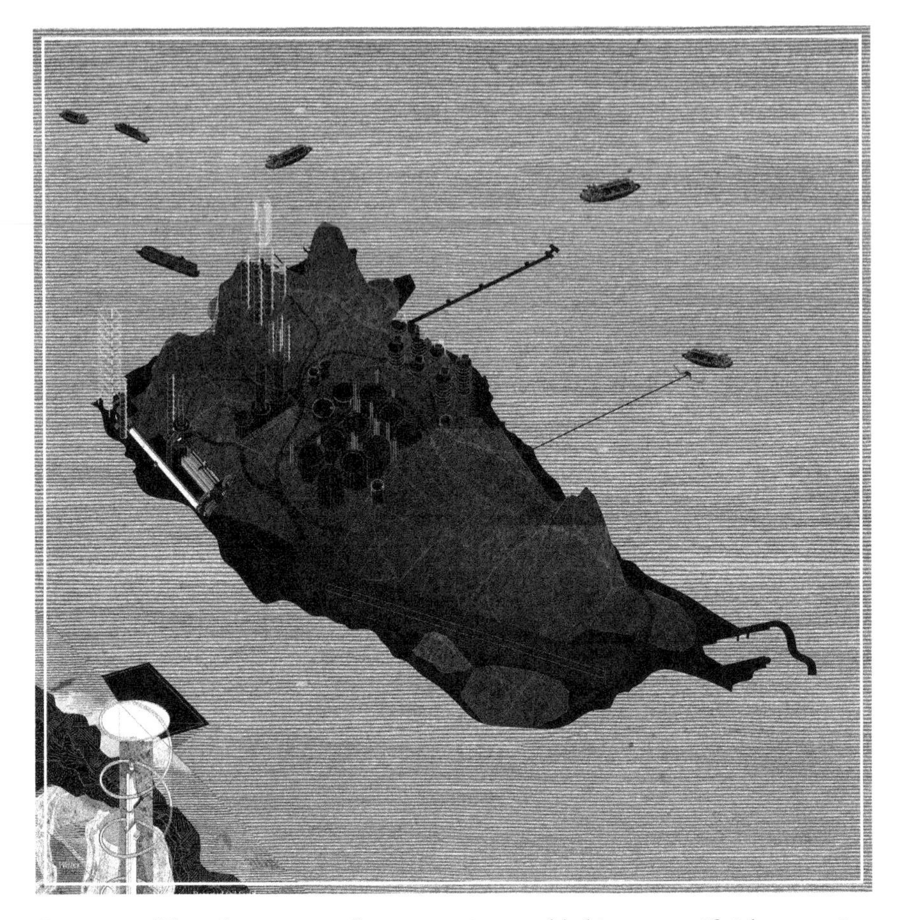

Figure 7.2 The refuse matter of extraction is assembled into an artificial mountain, a landform monument to the age of oil. Courtesy of DESIGN EARTH.

Das Island, Das Crude

Das Island is a major offshore Emirati oil and gas industrial facility. Oil production began following prospecting during the 1950s and has since financed the urbanisation of Dubai and Abu Dhabi, with many of the country's iconic buildings built from oil wealth. Although such territories of

Figure 7.3 The image of the city as seen from the offshore territory. Courtesy of DESIGN EARTH.

extraction are critical to urban development, they are out of sight and often external to urban representation. This architectural section makes visible the displacement of value in oil urbanism by situating architectural landmarks within exploited oil reservoirs – it collapses the surface and the underground. The drawing also indexes the increasing height of such landmarks in relation to the depths and periods of extraction of the subterranean fields, weaving

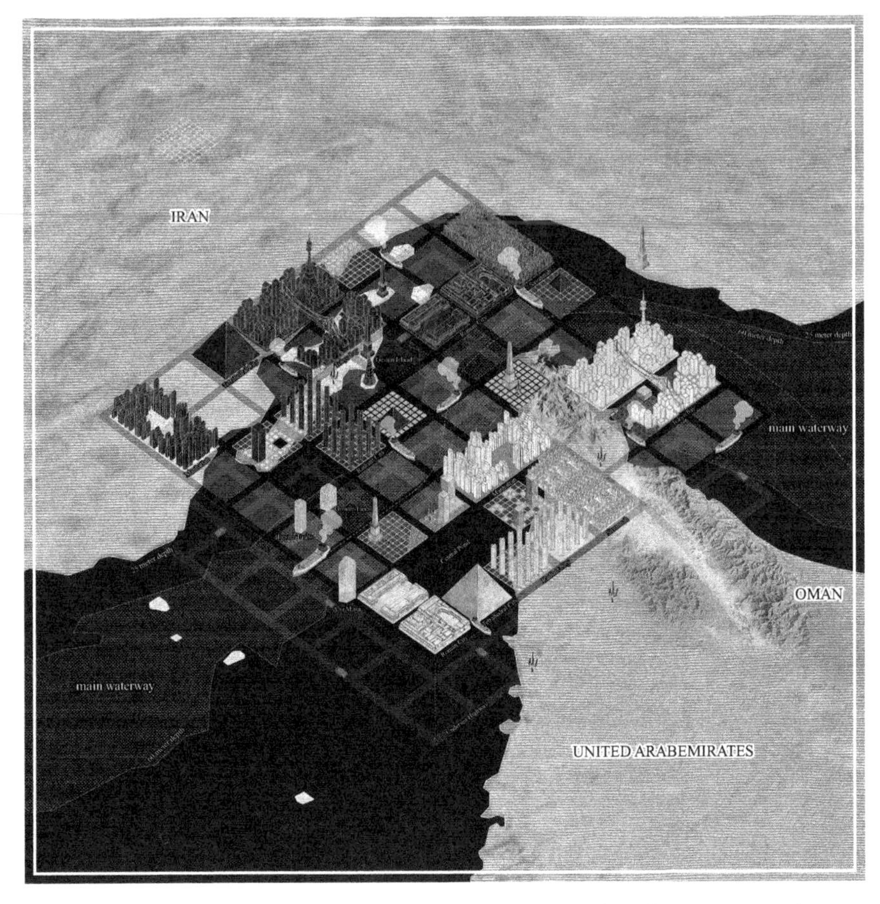

Figure 7.4 Oil Futures gameboard where players buy and trade iconic speculative projects. Courtesy of DESIGN EARTH.

together the architectural and geological attributes of the territory. The material by-products that accompany oil drilling – volumes of excavated soil and stone – are assembled into an artificial mountain, a landform monument to the age of oil.

Figure 7.5 Section through chokepoint, which is strategic to world oil supply. Courtesy of DESIGN EARTH.

Strait of Hormuz Grand Chessboard

The Strait of Hormuz is a critical oil transit chokepoint where 20 per cent of world oil trade moves across the 34-mile-wide passage. The strait was never actually shut down in spite of persistent geopolitical anxiety over territorial disputes, notably the disagreement between the UAE and Iran over the three islands of Abu Musa, Greater Tunb Island and Lesser Tunb Island. The

Figure 7.6 Underwater view of things – drones, submarines and dolphins – that conduct mine countermeasures operations in the narrow strait. Courtesy of DESIGN EARTH.

Grand Chessboard repurposes the strait into a territorial real estate game between traditional adversaries across the Gulf. Financed by Oil Futures, which is the legal contract name for the trade of oil, the chessboard echoes a long history of speculation – both in the sense of the practice of buying and selling land and the speculative nature of urban projects in the region, some of which render concrete visionary projects, such as Le Corbusier's Radiant City, Buckminster Fuller's Pentahedron City and Kenzo Tange's Tokyo Bay.

Figure 7.7 The slow violence of fossil fuel produces a geography of submerged islands. Courtesy of DESIGN EARTH.

The chessboard also absorbs the three disputed islands in the field of black and white utopian projects.

There Once Was an Island

The end of the Persian Gulf War in 1991 was accompanied by the world's largest oil spill, which drastically affected Kuwait's costal environment.[2] Beyond the apocalyptic intensity of such a geo-traumatic event, the everyday

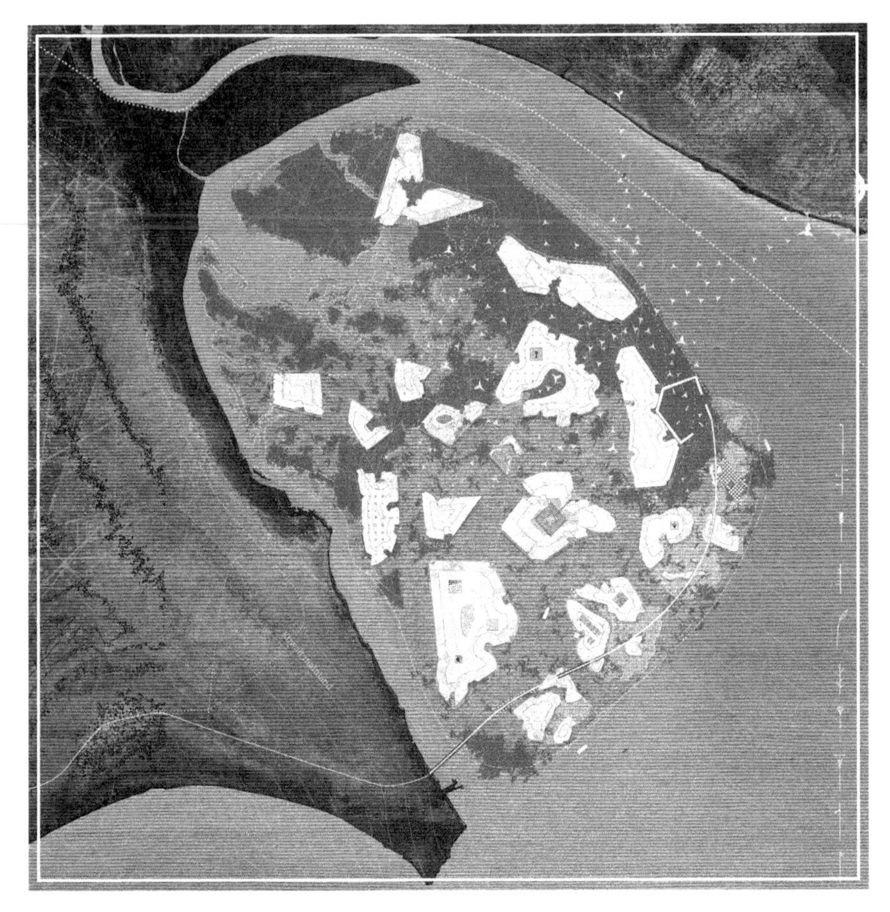

Figure 7.8 Archipelago of sixteen highest mounds. Courtesy of DESIGN EARTH.

business-as-usual oil industry, with its increasing rates of carbon emissions, is a slower form of violence. Kuwait's flat and low-lying Bubiyan Island is vulnerable to the anthropogenic sea level rise that the oil industry contributes to. *There Once Was an Island* gives forms to the slow violence of climate change and figures the gradually transforming landscape. A series of *nabkha*, vertical pole elements, are inserted into the highest sixteen elevation mounds of the

Figure 7.9 Section of *nabkha* landscape to stabilise the ground. Courtesy of DESIGN EARTH.

island to stabilise its shrinking shoreline into an archipelago of edenic islands, home to sea turtles, Arabian oryx and other forms of wildlife that acclimate to the increase in temperature and acidity of the water. Such a geographical holding pattern also freezes development plans to transform the island, uninhabited by humans, into a major container port and tourist resort.

Notes

1. This chapter is reprinted, with minor changes, from Rania Ghosn and El Hadi Jazairy (eds), *Geostories: Another Architecture for the Environment* (Barcelona: Design Earth, 2018).
2. George Basalla, 'Some Persistent Energy Myths', in George H. Daniels and Mark H. Rose (eds), *Energy and Transport: Historical Perspectives on Policy Issues* (Beverly Hills: Sage, 1982), 27–38.

PART III

OIL SUBJECTS

ECOLOGIES AND GENEALOGIES OF PETROLEUM KNOWLEDGE

8

SPECULATIVE MATTERS: THE PASTS AND PRESENTS OF OIL IN TURKEY

Zeynep Oguz

Absent and Present, Visible and Invisible

Intangible, ubiquitous and elusive at the same time, oil in Turkey is a substance infused with much confusion, speculation and imagination. Due to Turkey's particular geological history and setting, oil is concentrated in very small oil traps. Today, Turkey's limited domestic oil production covers only 7 per cent of its demand. All of this domestic oil, however, is extracted in Turkey's Kurdish-populated south-east, a region characterised in the past century by armed conflict, emergency rule and military occupation. Many do not understand why Turkey has so little oil when its close Middle Eastern neighbours have so much. Some have no idea that oil is being extracted in the south-eastern parts of the country. And yet others think that there is so much more oil than is being revealed, and that sinister powers are obstructing oil's extraction. In the midst of so much speculation, petroleum geologists – and anthropologists of oil – in Turkey are met with puzzled reactions and curious questions that often lead to further speculation and uncertainty.

Faruk, an exploration geologist employed at the state-owned oil company Turkish Petroleum, for instance, often complained about the questions he always got. 'When I tell people what my job is, I often get this reaction: "What, oil? Is there oil in Turkey? I had no idea!"' That many people who were not part of the energy industry were not aware of the existence of any domestic production in Turkey was for Faruk an unavoidable consequence of a variety of geological factors that have determined the specific characteristics of oil in Turkey. Yet for a considerable number of people, there are political, not geological forces behind Turkey's low domestic production. For them, the question is not one of geological setting and, thus, of the real absence of large oil reserves. The issue, for them, is that Turkey has, in fact, abundant oil deposits under the ground, but that this fact is hidden by the government, foreign powers or secret international treaties.

Oil has been a key material in the reproduction and mediation of social and political life in the twentieth century. Oscillating between instant wealth, growth and freedom, on the one hand, and political conflict, corruption and inequality on the other, oil has been fetishised in petro-states around the world.[1] In oil-rich petro-states, for example, oil has been considered as a 'magical' commodity and cast as an integral part of the body of the nation, while sustaining the power of the state.[2] In contrast to these tales

Figure 8.1 An exploratory drilling rig and container camps near the province of Siirt, Turkey. Photo by Zeynep Oguz.

of abundant oil that brings about fortunes and curses, Turkey, having very limited oil reserves of its own, could never become a petro-state. Turkey emerged from World War I having lost significant territories in what used to be the Ottoman Empire. These included the oil-rich regions in present-day Iraq and Iraqi Kurdistan. For Turkey, not having territorial sovereignty over such abundant reserves, oil could not establish a smooth link between the nation's natural and political bodies. This, however, was not the end of Turkey's oil story.

In this chapter I trace conversations with geologists, petroleum engineers, police officers, village guards and other actors around the confusion and speculation surrounding oil that were evident in Turkey during 2016–20. This period was characterised by a failed coup attempt, the national emergency that lasted for three years, a constitutional referendum on whether to grant immense powers to the president, and their aftermath. Oil's intangible yet elusive presence, and the modes of expectation and anticipation that surround it, were a characteristic feature of this tumultuous period. Prospects for oil discovery not only helped grant legitimacy to the increasingly authoritarian and nationalist government trying to re-consolidate its power in the aftermath of the coup attempt and decreasing popular support, but also exposed many of the fault lines upon which state power and national politics were built in Turkey.

Here I build on the recent scholarship in anthropology, geography and adjacent fields that understand matter as an elemental or earthly force in the constitution of the world, rather than a mere backdrop.[3] Revisiting the traditional questions around the relationship between nature and culture, these works have followed a trajectory of writing about 'geosocialities',[4] 'geopower'[5] and 'geontologies'.[6] These works stress the multi-directional relationship between earthly and socio-political formations. Yet the specific histories and localities through which this relationship occurs, and empirically grounded anthropological-historical accounts, are necessary if we are to fully understand these workings. How, when and where do earthly materialities and socio-political worlds relate to each other?

From this perspective, the absent presence of oil becomes a precisely productive vantage point from which to explore the social and political worlds it has nevertheless composed and continues to bring about. Oil has acquired an elusive materiality in Turkey, situated between presence and absence, visibility and invisibility, abundance and lack, loss and desire. It is in the space between these oscillations that national pasts and presents are reconfigured, repressed histories of violence and resistance are conjured up, and state power is both reproduced and exposed as highly fragile and fragmented.

Populism and Oil: AKP Elections, Declining Production

One of the most significant political events in Turkey's last decade has been the failed coup attempt of 2016 and its political aftermath. On 15 July 2016, a faction of the Turkish Armed Forces unsuccessfully attempted to seize control of the government. Not being able to garner popular support or mobilise other factions within the state, the coup attempt failed. Immediately after the coup attempt, the Justice and Development Party (AKP) government declared a state of emergency, which enabled the president and cabinet to bypass parliament when drafting new laws and to restrict or suspend constitutional rights and freedoms.[7] Inadvertently later described by Erdoğan as 'God's blessing', the 15 July coup attempt became a new milestone in the authoritarian Erdoğan–AKP regime, who used it as a tool to consolidate its base, eliminate the opposition, speed up the transition to a new regime, and intensify repressive politics through populist strategies, continuous war and polarisation.

Following the failed coup attempt, the government quickly set a date for the presidential referendum, which was viewed by the opposition as the final move to dismantle the democratic institutions of the Republic. As the 16 April 2017 referendum on the transition to an authoritarian presidential system approached, the government strove to gain supporters through further cultural polarisation and a rhetoric of internal and external enemies, labelling those who opposed the impending regime change as 'inauthentic' (*gayri milli*) and, therefore, enemies of the pious 'people' (*millet*), of whom Erdoğan claimed to be the true representative.

In the weeks leading to the referendum, I spent most of my time on the oil production sites with petroleum engineers in the provinces of Diyarbakır and Batman. On one of these days, we all sat down for lunch once the engineers were done with their routine well checks at the Garzan Production Camp. Glancing at the midday news on TV, Hakan, a young petroleum engineer, remarked 'Another election eve, another oil discovery', in his usual cynical way. On the screen, there were six or seven men in suits and protective gear standing in front of a drilling rig and a giant Turkish flag. 'What do you mean?', I asked Faruk. 'Well, the geological limitations of Turkey are clear. The possibility of a massive oil field located under our ground is almost absent. This news about "the largest onshore oil discovery in Turkey" and so on are just ploys to win people and secure votes. Before each election or referendum nowadays, these kinds of news surface, and then they suddenly disappear after voting day. You'll see, that's what's going to happen in this case too', he said, and a few other engineers sitting around the table nodded.

As Faruk said, before each local or national election, stories about oil exploration or discovery resurfaced in Turkey. Yet it was very difficult to see if oil production started. 'People who watch these kinds of news assume that somehow we are sitting on massive oil fields. And when there is no follow-up to these stories, they believe that somehow, their extraction is being

Figure 8.2 Lunchtime with petroleum engineers Hakan and Faruk at the Garzan camp. CNN Turk's News ticker reads: 'Towards the April 16 Referendum: President Erdoğan Speaking'. Photo by Zeynep Oguz.

obstructed or something sketchy is going on', said another engineer, Aslan. 'But there is some truth to such news about oil – they are not just making things up entirely, right?', I asked. 'Of course', Aslan said, 'there are trace amounts of oil in small traps, but many of them are not commercially viable. This is probably already known by geologists in the field, our colleagues, but politicians nevertheless send them and stage these ploy exploratory drills until the elections pass.' Aslan referred to the 'heavy' quality of oil in Turkey's south-east, meaning it had high viscosity and thus its density and gravity were significantly higher than those of light crude oil. These properties of oil in Turkey's south-eastern regions made it hard to get to flow to production wells and thus made drilling too costly.

According to data provided by the Turkish Petroleum and Natural Gas Platform Association, Turkey's domestic oil consumption increased from 38 million tons in 2015 to 42 million in 2016.[8] Turkey imports most of its energy and its domestic resources only account for 26 per cent of its total demand. Turkey's annual energy consumption is expected to more than double over the next decade. With its increasing energy demand and dependence on imports from Azerbaijan and Iraq (oil) and Russia and Iran (gas), two main characteristics dominate the Turkish energy market: growing energy demand and dependence on imports. With no new large oil field discoveries in recent years, and with the existing fields maturing, Turkey's oil production tends to fall each year as demand increases. To reduce its heavy reliance on imported fossil fuels, the AKP government launched a bold national energy plan in 2015. The plan pushed for energy diversification (including three nuclear power plant projects), a heavy reliance on domestic resources such as coal and hydropower, and increased hydrocarbon exploration on- and offshore.

Yet, as almost every geologist I spoke to told me, onshore oil prospects in Turkey were bleak, while offshore exploration was extremely costly and geopolitically sensitive. In this context, rather than significantly contributing to meeting the growing domestic energy demand or ensuring 'energy security', oil has been doing other kinds of things in Turkey. It has been operating as a material that generates consent before elections, with its power to put forward abundant and sovereign energy futures. Oscillating between visibility and invisibility, oil has also worked as a temporal device that reconfigures histories. The kinds of futures prolonged by the target year '2023', and the reconfiguration of Turkey's foundations, exemplify the multiple life worlds of oil in Turkey over the past decade.

Mighty Pasts and Futures: From 1923 to 2023

The then Minister of Energy and President Erdoğan's son-in-law Berat Albayrak stated, during a press conference in 2017, that voting Yes in the upcoming presidential system referendum in April 2017 also meant endorsing the government's ambitious energy future. This future revolved around a particular date: 2023. In 2011, the AKP government announced what it called 'The 2023 Vision', a list of ambitious economic growth goals to coincide with the centenary of the founding of the Republic of Turkey in 2023. According to this, Turkey will be an economic powerhouse by that date. Other goals included in the vision are increased energy capacity and the commissioning of three nuclear power plants, as well as ambitious infrastructure and transportation projects. However, the designation of the year 2023 also entailed a series of speculations: that there were secret and hidden powers out there actively obstructing oil discovery and extraction in Turkey, that there is the possibility of a subsoil that is abundant with oil, or that 'more oil is hidden', as has been the case in Oman.[9]

Figure 8.3 'Lausanne Will Be Over: Goal 2023'. An unofficial graphic designed for the AKP's presidential referendum campaign. Source: https://www.youtube.com/watch?v=0-hD1lcT44o.

Whenever I mentioned my research to someone in Ankara or Istanbul – where oil extraction and production is not as visible as it is in the Kurdish cities of south-eastern Turkey such as Adıyaman, Batman or Diyarbakır where oil fields are located – I would almost always get the same response: 'Are the rumours about 2023 true?' I would ask them which rumours they were talking about. 'Don't you know? In 2023, the Treaty of Lausanne expires, and we'll finally be able to extract our oil below the ground.' One day, in April 2017, only a week before the referendum I encountered and spoke to a young cleaning woman at the Ankara Headquarters of Turkish Petroleum. Ayten was born in the Black Sea region of Turkey and migrated to Ankara in the late 1980s with her husband, hoping to achieve upward class mobility. She had just started working at Turkish Petroleum, as the company had fired hundreds of employees for their suspected connections to the alleged perpetrator of the coup attempt. She told me she knew that every engineer and manager at the company were doing their best to discover oil, but that 'their hands are tied'. 'How so?', I asked. 'Well, you know, they can't do anything until 2023, when Lausanne expires', she replied, as if she were stating an obvious fact. I countered by saying, 'As far as I know, the Treaty of Lausanne does not have an expiration date'. My words were met with laughter. 'Of course, none of the official sources will tell you that. There are secret clauses in the treaty. And you can't find them, because they're secret!'

Speculation about Turkey's obstructed oil and 2023 centred around the iconic Treaty of Lausanne. Signed in 1923, following the collapse of the Ottoman Empire after World War I, the Treaty of Lausanne constitutes the foundational text that has recognised the territorial unity and sovereignty of the Republic of Turkey up to this day. After defeating the Ottoman Empire, Britain, France, Italy and Greece planned to occupy Anatolia. The national resistance movement thwarted this attempted division through shrewd diplomacy and several years of war. Subsequently, the 1923 Treaty of Lausanne recognised Turkey's victory and established the borders of modern Turkey. The believers in the conspiracy theory claimed that the Treaty of Lausanne was going to expire on the hundredth anniversary of the founding of the Republic of Turkey. For them, when it does expire, Turkey will finally have full sovereignty over its petroleum reserves. Between 2015 and 2018, this conspiracy theory featured in various talk shows, magazines and everyday conversations. It also figured on a daily basis during petroleum geologists' and engineers' interactions with other people.

One day, in January 2017, we were stopped at a police security checkpoint on our way to Diyarbakır from Batman. I was with Kemal, a Kurdish exploration geologist and his team (young Turkish geologists). The police officer collected our IDs and asked us what our business in Diyarbakir was. 'We are geologists, working for the state, see the sign?' Kemal said, meaning the Turkish Petroleum logo on the back of our truck. 'Turkish Petroleum?', the officer said, raising his eyebrows. He must have recently been assigned to Diyarbakır from somewhere else, otherwise he would be familiar with seeing TP and other oil companies operating in the area on a daily basis. 'Yes, we are from the exploration department, it's on our IDs.' More than security, the police officer was interested in talking to us out of personal curiosity. 'I thought we were not allowed to extract our oil reserves', he said. 'Oh yes we are, why would a sovereign state be unable to seize the resources in its territories?', the geologist replied. 'I don't know, because *they* don't want us to be powerful.' Kemal laughed. 'No, no, this is exactly what we are trying to find out, if there are more fields in the area, officer.' 'So, we don't have to wait until 2023? OK', the officer said. From the look on his face, it did not seem he was convinced by our arguments. 'Well, good luck gentlemen', he said, as he handed our IDs back to us.

I was curious as to what the geologists thought about the conspiracy theory. While many of the young petroleum engineers I met at production fields – like Hakan – dismissed them, others, while accepting the geological limitations of Turkey and thereby its relative lack of oil, pointed to a 'truth' behind these conspiracies. Yusuf, a middle-aged exploratory geologist based at the Batman headquarters of Turkish Petroleum, was among them. Yusuf considered himself religious and was a supporter of the AKP government. Many young engineers who opposed the government's policies and ideologies believed that Yusuf's connections to the government were the main reason behind his managerial position. For Yusuf, it was not a matter for debate whether there were hidden or obstructed oil reserves in Turkey. There were not. What the conspiracy theories correctly pointed out for him, however, was that Western powers had been obstructing Turkey's – and the Ottoman Empire's – might and progress for centuries. For Yusuf, the modern borders of Turkey that were established after the collapse of the Ottoman Empire intentionally left out the oil-rich regions of Mosul or Kirkuk. European states and their geologist informants, for Yusuf, knew all too well where large oil deposits could be located, and the Treaty of Lausanne and the founders of the republic left them out of the Republic of Turkey's borders.

Speculation over the expiry of Lausanne in 2023 stuck in public life. They often worked to ignite nostalgia about the extended borders of the former Ottoman Empire. They fuelled desires for territorial expansionism that characterise neo-Ottomanist nostalgia politics in contemporary Turkey. President Erdoğan, for example, often alluded to the former or 'lost' territories of the Ottoman Empire and the potential malleability of the Treaty of Lausanne as legitimising factors in Turkey's military operations in Iraq and Syria, and its irredentist aspirations in the Middle East. Almost every day between 2016 and 2017, prominent historians rushed to TV programmes to persuade the public that the agreement in fact did not have an expiry date and that opening the agreement up for discussion would be 'extremely dangerous for Turkey'. Historians, columnists and other experts wrote 'Lausanne is the title deed of the Republic of Turkey', implying that discussing Lausanne also meant dangerously placing the Turkish state's territorial unity and sovereignty on the table.

Oil re-arranges pasts and futures. As the above paragraphs suggest, it particularly re-configures neo-Ottomanist nostalgia in Turkey. It mediates desires for territorial expansionism that aspire to reclaim the territories deemed lost by the Ottoman Empire in the aftermath of World War I. Proponents of the conspiracy theories that revolve around oil and the year 2023 hope that the speculated-about expiry of the Treaty of Lausanne will give Turkey a second chance to negotiate or reclaim its political borders in the future. Others, however, who are not AKP supporters, cautioned that one must be aware of 'what they wish for', as the cancellation of Lausanne may also mean a possible foreign occupation since the treaty is the deed of Turkey's international territorial unity and sovereignty.

Both positions, however, point to the postcolonial and unequal place that Turkey holds in the world system, exacerbating fears of foreign occupation on the one hand, but also desires for lost power and fears about territorial disintegration, to which debates about Lausanne are deeply attached.

These fears, addressed in the next section, are also linked to the century-old Kurdish Question and the protracted war between the Kurdistan Workers' Party (PKK) and the Turkish Armed Forces. As oil's oscillations between absence and presence, visibility and invisibility, continue to mark the tensions of national politics in Turkey, pro-state actors utilise them to erase or counter past and present conflicts. Yet, as the following sections demonstrate, these projects are far from complete. Below, I tell a story about a geological field trip in south-eastern Turkey and briefly summarise the history of the Kurdish Question to explore how oil's oscillations both mark and fuel ongoing fault lines in Turkey – especially the ongoing legacies of imperial collapse, genocide and coloniality in Turkey's Kurdish south-east.

Colonial Pasts and Liberated Futures: The Kurdish Question

After the more than three-decades-long war between the Turkish armed forces and the PKK, in 2013 the Justice and Development government formally announced that negotiations with the leaders of the organisation were being carried out and a truce was instigated. In the summer of 2015 the peace process collapsed, and full-scale war resumed. As the war shifted to urban spaces, the military reclaimed some of the rural and mountainous areas that had once been controlled by the PKK, aspiring to create new resource frontiers in the previously inaccessible and unexplored terrains of the south-east.

As a part of this new initiative, I joined Turkish Petroleum's geologists for an exploratory geological fieldwork trip in Eruh in 2017. Eruh is a highly significant place in the histories of Turkey and Kurdish insurgency. A mile away from the Eruh outpost, where geologists and I met with Captain Enis, the very first PKK attack took place. In 1984, guerilla forces attacked the gendarmerie station in Eruh, killing one gendarmerie soldier and injuring nine people. The Eruh attack marked the beginning of the Kurdish armed resistance movement against the Turkish state. The group of geologists' first stop was the local military outpost to make sure we had the right permissions and 'protection' for our field trip. Captain Enis, a Turkish officer in his mid-thirties, was the commander of the local outpost in Eruh. For him the Kurdish Question was part of the same ploy to divide Turkey and obstruct its progress.

As Captain Enis went through the geologists' plans about their prospective route for fieldwork, he asked them if they were hopeful about the results. Kemal expressed cautious hopes. 'See', Captain Enis replied, 'the entire purpose of this war was to obstruct us from exploring oil. To me, there is no Kurdish Question, but the provocation of Kurdish people by bigger powers, do you agree?' I knew he did not, as Kemal himself was a left-leaning Kurdish democrat. Yet he had mastered the art of code-switching, but also kept his integrity while doing it. He nodded his head, and replied, with his subtle sarcasm, 'Of course, of course!' We made eye contact for a split second.

The origins of the Kurdish Question date back to efforts by the Ottoman Palace to centralise the empire in the nineteenth century. With the Ottoman Empire's integration into the capitalist world economy and the growing influence of European nation-states, the Empire's 'a-national and de-central structure'[10] started to become centralised. This symbolised the end of a period of de facto consensus between the Kurdish tribes and the Ottoman Palace, during which the tribes enjoyed relative autonomy. The Kurdish tribes' political power decreased dramatically after a series of administrative reforms in the first half of the nineteenth century aimed at destroying the autonomy of peripheries and centralising the economy, politics and administration.

The dissolution of the Ottoman Empire in the aftermath of World War I and the subsequent formation of the Republic of Turkey in 1923 did not put an end to the oppression of the Kurdish people. With the formation of a nation-state and other reforms initiated by the founders of the Republic, Kurdish tribal formations not only lost the little autonomy they had left, but also faced increasing pressure from the centralised government to accept the political and cultural system of the new nation-state that heralded a new territorial order and redefined relationships between the state and the people. Violently suppressing three major Kurdish rebellions in the first decades of the Republic,[11] the Turkish state has officially denied the existence of Kurdish identity, governing the region under martial law and emergency regulations, focusing on the aggressive displacement and resettlement of Kurdish populations in the 1920s and 1930s.[12] When violent strategies seemed to stop working or were deemed not preferable for various political and economic reasons, the state turned to policies of assimilation through forced resettlement, language instruction and education in Turkish history, politics and culture.

The Kurdish people were not passive recipients of such policies. Throughout the 1960s and 1970s, Kurdish students, intellectuals and politicians formed various platforms and criticised the colonial relationship between the Turkish state and Turkish Kurdistan. Finally, in the late 1970s, the PKK emerged as a Marxist-Leninist organisation aiming to establish an independent socialist state based on self-determination. The PKK adopted armed struggle as a revolutionary strategy. Following the PKK's initial attacks in 1984, the Turkish state retaliated by declaring a state of emergency in eleven provinces in 1987, recruiting and arming Kurdish peasants to serve as 'village guards', a paramilitary force, and conducting massive military operations in mountain hideouts, villages and cities.[13]

I met four Kurdish village guards, Mahmut and his brothers and cousins, during the geological fieldtrip in Eruh in 2017. After walking in the terrain of Eruh for a day, we finally climbed a steep hill, and met with the village guards who had been watching us and the broader terrain against a possible 'enemy' the whole time. These young Kurdish men were assigned as a protection unit to our team of geologists by the local military outpost office. We joined the village guards in the little camp area they had set up. As I snacked on the bread, black olives, tomatoes and cheese, Mahmut's younger brother Kenan talked about himself, recounting how being a village guard had been a tradition in their entire family, how they were all related, and their precarious work conditions. He talked about how they were viewed 'as traitors by the guerillas' and as 'errand boys' by the Turkish state. In no story could they be depicted as heroes or have the moral high ground. As paramilitary agents, they were caught in between. Ultimately, the conversation came round to the subject of the referendum. To my surprise, Kenan said he had voted for the pro-Kurdish left party HDP and that he would definitely be voting No in the upcoming presidential referendum.

The village guards were very curious about the activities of the geologists and prospects for oil in the area. One of them asked the geologists if they had found anything. The geologist explained to him that 'things don't work like that' and that many other studies had to be made to establish for sure that the terrain could have some oil potential. 'Oh, I wish', he said, with glowing eyes. 'How great it would be for us, the Kurdish people.' His brother jumped in: 'Kemal Bey, please find the oil for us. Once we are a free nation, we will take over the oil fields from you', he said, grinning playfully. Only Kemal and I heard him. Kemal nodded, half agreeing, half infantilising their 'romantic' hopes, as he would tell me later. The words of the Kurdish village guards – from their political beliefs to their hopes for oil and Kurdish autonomy – embodied a series of contradictions. As paramilitary agents of the state, I expected Kenan and his family members to be AKP supporters, yet they shared the same political sentiments as those held by many of the Kurdish people, who were critical of both the state and its violent policies in the Kurdish south-east, including the village guard system. Oil's absent presence destabilised existing loyalties, or, even more, it exposed the already-unsettled and conflicted nature of these loyalties. In the story I narrate below, it further conjured up convoluted legacies of the violence that the collapse of the Ottoman Empire and the subsequent building of the Turkish nation-state had engendered.

Figure 8.4 Village guards Mahmut and Kenan on the mountains of Eruh, 2017. Photo by Zeynep Oguz.

Genocidal Pasts, Haunting Presents

In September 2015, Sıtkı Tayar, a middle-aged Kurdish man, a former construction worker from the province of Van, sent a letter to the exploration department of the Turkish Petroleum District Management in Batman. This was the fourth letter he had written to the company in the past two years, and he was already frustrated with the lack of attention he had received from the 'authorities'. In his letter, he declared that he was a 'gold, copper, and oil expert from Van' and that he had found 'the source of the oil' in Batman's famous Raman fields. 'And it's right under Van.' He added that he had contacted many officials and petroleum geologists about this, but none seemed to be interested in 'this crucial information that will elevate the country'. He concluded with a biting comment: 'Maybe these officials were not with the best of intentions.'

Yusuf told me that Tayar had been walking and exploring the hills and plateaux of Van for almost two years. Almost every other day, he would leave his house, drive for about an hour and then walk for hours, looking for signs, picking up rocks, sniffing them for traces of sulphur, photographing and sketching anything he found significant. His primary targets were oil seeps – or leakages, as he liked to call them. He had first noticed a thin trail of black liquid seeping into the ground while driving to Tatvan, a town on the south-west shore of Lake Van near the small district of Gevaş during a visit to a distant cousin. Realising it was petroleum, he followed the trail southwards, as it appeared here and disappeared there along his path. A month later, he borrowed the hand drill of another distant cousin, who was a part-time 'treasure hunter'. He drilled in areas wherever oil seeps accumulated and where they seemed to escape and disappear.

I asked Yusuf what his cousin, the 'treasure hunter', was doing exactly. 'You know, looking for gold or other valuables left behind.' 'Left behind?' I asked. 'You know … Eh, during the deportation', Murat Bey murmured. Then, it dawned on me. Tayar and his cousin were those semi-professional people called 'treasure hunters', which meant rummagers, grave diggers, gold hunters, who searched for the artifacts left behind by the victims of the Armenian Genocide of 1915, which the state still refuses to recognise as a genocide, and which is rarely dealt with in direct ways in

Turkey, including in Turkey's Kurdish south-east. This silence around or disavowal[14] of the genocide was especially significant in this region, since the atrocity took place in the very same place, the historic homeland of Armenians, and with the collaboration of the Kurdish people, although there were many cases of Kurds refusing to collaborate, and, in some cases, Armenian victims were protected by some Kurdish tribes. Yet this did not change the extent to which the Kurdish population was complicit in the atrocities and how, in the following years, property and capital that had once belonged to Armenians had been appropriated by the Kurdish population in the region.[15]

Equipped with expensive metal detectors, treasure maps of lands and the underground, protective or curse-breaking spells, myths about previous treasure hunters and hidden wealth, and an extensive network of material tools, know-how and sociability, these 'treasure hunters' had looked for Armenian wealth hidden in the subsoil, in Armenian ruins, graveyards, backyards and caves for decades now. Tayar's search for oil and treasure in Turkey's Kurdistan conjured up repressed histories of violence. The search for oil and the search for Armenian treasures were conjoined here, unearthing the genocidal act that erased Armenians from Eastern Turkey and enabled the emergence of the Turkish nation-state as a largely Muslim population. Seepages of oil that Tayar traced, and the subsoil that Tayar imagined to be abundant with petroleum, were entangled with the remains of Armenian genocide victims and survivors, objects that pointed to the negated histories of violence and Kurdish complicity with the Turkish imperial and nation-state projects.

A Belated Discovery

With a narrow 51 per cent approval for the transition to an executive presidential system, the results of the constitutional referendum on 16 April 2017 put an end to the parliamentary system Turkey has had for more than a century. The results paved the way for Recep Tayyip Erdoğan to become Turkey's first executive president, with few checks and balances, in the 2018 presidential elections. Turkish Petroleum heavily focused on offshore petroleum exploration per President Erdoğan's directive, and in August 2020 the state-owned petroleum exploration company announced the discovery of 'massive offshore gas and oilfields' in the Black Sea. Mainstream and state-funded media reported that the discoveries were harbingers of Turkey's economic wealth and resource independence to come. On their Twitter accounts, ministers celebrated the findings as signs of Turkey's 'inevitable progress', in line with the government's 2023 goals. The pro-government daily newspaper *Sabah* claimed that the discoveries were set to 'change the energy equation' and end Turkey's resource dependency.[16] At this moment, it seemed that petroleum in Turkey was no longer elusive, but finally decisively present and abundant – an abundance that the AKP government intentionally capitalised upon.

The moment did not last long. Many seemed to think that the announcements were just a ploy to detract attention from Turkey's real political and economic problems. Following popular reactions to social media from Chicago, I was struck by how many people seemed to think that these 'discoveries' in the Black Sea were actually fake. Anti-government news sites compiled lists of previous gas and oil discovery news in the past decade, which nobody had heard from afterwards. Others on Twitter mocked the news by sarcastically asking if there was an election on the horizon, or if a bitcoin mine discovery was next in line, for example.

In September 2020, I attended an online conference jointly organised by the Chamber of Geophysical Engineers of Turkey and the Chamber of Geological Engineers of Turkey on the recent offshore hydrocarbon discoveries in the Black Sea. The consensus in the virtual meeting room was that the government officials' initial remarks and local media coverage had clearly exaggerated the magnitude of the discoveries. Current reserve estimations provided by Turkish Petroleum were only preliminary, and thus inaccurate, remarked one retired geologist, who had formerly worked at Turkish Petroleum. He added that further drills and analysis should be made in order to get a clearer idea about reserve estimates. A geophysicist added that the cost and feasibility of these reserves are still not clear, as Turkey does not possess the technology to extract and transport offshore hydrocarbons.[17] The consensus was that, contrary to the claims of government officials and pro-government media, gas discoveries in the Black Sea fell short of making Turkey energy-independent. According to official sources, the total amount of hydrocarbon deposits recently discovered in the Black Sea in the last years amount to 540 billion cubic metres. If these numbers were correct, this could theoretically supply Turkey's gas demand for ten years. This was to say, only if production were commercially viable.

Finally, prospects for offshore production also seemed to be futile projects in a world facing climate emergency. An energy and climate consultant who had spent the last decade working for private firms brought up the planetary consequences of exploiting offshore petroleum fields in the Black Sea. If Turkey is to meet its climate targets, in the event that it ratifies the Paris Agreement,[18] he said, the hydrocarbon fields in the Black Sea should not, or simply could not, be burned. Thus, despite an initial appearance as a much-expected realisation of the hopes for oil, the offshore hydrocarbon prospects in the Black Sea went back to oscillating between presence and absence, this time along with modes of uncertainty regarding reserves estimates, technological capability and climate change mitigation. As of October 2021, the AKP government was still dedicated to tapping into the gas and oil fields in the Black Sea by the symbolic year 2023, as Turkish Petroleum continued to seek out partnerships with international firms for offshore drilling. Yet in both public and expert circles, petroleum's ability to reproduce national(ist) politics and state power seemed to be waning, as concerns for the planet and disillusionment with progressive narratives start to populate public life in Turkey.

Speculative Futures beyond Oil

Heavy and located in small deposits, oil has a material ontology in Turkey that does not fit into the well-known tales of abundant crude oil. Yet oil in Turkey has never ceased to haunt public life and politics, having oscillated between abundance and lack, visibility and invisibility, presence and absence. It is between these oscillations that speculations over oil have rendered repressed and negated pasts visible and brought into presence novel political pasts and futures. These speculations, which feed off the secrets and truths of an unequal world order whose contours have been drawn through the geographies of empire and oil, conjure up imperial pasts that are longed for, and these pasts hark back to aspirations to territorial expansion in the Middle East. They compose spectres of a future where that perceived loss of territory and power in the face of the global capitalist system might be regained in the near future. In doing so, they reproduce and consolidate the authoritarian, violent and populist regime of the AKP government.

The encounters around oil's speculative presence in Turkey also reveal the multiple and often conflicting loyalties of state actors, pointing to the non-monolithic and fragile character of the Turkish state. Oil, here, becomes a material medium through which the village guards I met with articulated their political imaginaries for a free Kurdistan. In other stories, such as those about treasure hunting, oil's material ontology between presence and absence, visibility and invisibility, renders a violent past marked by genocide and massacre visible. Oil operates as a convoluted mirror for the ongoing conflicts in Turkey, the after-lives of the violence of imperial collapse, the troubles of nation-state-making and coloniality, while further inflicting cracks on what it means to inhabit such a fault-ridden space.

Yet, if matter, and earthly matter specifically, is political, it is crucial to rethink the limits posed by the political worlds that oil brings into being. While oil's absent presence continues to animate political visions of autonomy and progress that have been dependent on the horizons limited by petro-modernity, in Turkey today, concerns for the planet are slowly starting to appear next to national ones. The question to be posed from our currently grim ecological predicament, then, is whether it is possible to link decolonial projects of political emancipation with ecological-environmental justice and care. Such a task requires actors to imagine emancipatory and anti-hegemonic futures that do not depend on the often-violent logics of petro-modernity and extractivism that have characterised not only Turkey, but everywhere in the past century. Facing a planetary ecological crisis, exploring the links between political worlds and the particular materials that are desired, explored and extracted as 'resources', is key to composing alternative political and environmental futures. A decolonised world cannot solely be about unsettling colonial and oppressive regimes, or even shifting to non-fossil-fuel sources of energy. It also requires a radically anti-extractivist reconfiguration of the world. Perhaps this is the key task of emancipatory projects today as the world moves beyond oil for habitable futures on the planet.

Notes

1. Hannah Appel, Arthur Mason and Michael Watts (eds), *Subterranean Estates: Life Worlds of Oil and Gas* (Ithaca: Cornell University Press, 2015). Fernando Coronil, *The Magical State. Nature, Money, and Modernity in Venezuela* (Chicago: University of Chicago Press, 1997).
2. Coronil, *The Magical State.*
3. Andrew Barry, 'The Political Geology of Area', *Political Geography* 57 (2016), 94–104. Elizabeth Grosz, Kathryn Yusoff and Nigel Clark, 'An Interview with Elizabeth Grosz: Geopower, Inhumanism and the Biopolitical', *Theory, Culture & Society*, 34, no. 2–3 (2017), 129–46.
4. Nigel Clark and Kathryn Yusoff, 'Geosocial Formations and the Anthropocene', *Theory, Culture & Society*, 34, no. 2–3 (2017), 3–23. Gisli Pálsson and Heather A. Swanson, 'Down to Earth: Geosocialities and Geopolitics', *Environmental Humanities* 8, no. 2 (2016), 149–71.
5. Grosz, Yusoff and Clark, 'An Interview with Elizabeth Grosz', 129–46.
6. Elizabeth A. Povinelli, *Geontologies: A Requiem to Late Liberalism* (Durham, NC: Duke University Press, 2016).
7. The massive purges that began shortly after the failed coup have resulted in hundreds of thousands of civil servants, teachers, judges, academics, security officials, and employees of state-run institutions being fired. As of July 2019, more than 150,000 have been forcefully dismissed and more than 96,000 arrested. https://turkeypurge.com/ (last accessed 29 January 2022).
8. MAPEG, 'Orta Dönemli Petrol ve Doğal Gaz Arz-Talep Projeksiyonu', 2017 (last accessed 15 February 2021).
9. Mandana Limbert, 'Reserves, Secrecy, and the Science of Oil Prognostication in Southern Arabia', in Appel, Mason and Watts (eds), *Subterranean Estates*, 340–52, particularly 347.
10. Mesut Yeğen, 'The Turkish State Discourse and the Exclusion of Kurdish Identity', *Middle Eastern Studies* 32, no. 2 (1996), 216–22, particularly 218.
11. These three Kurdish rebellions were the Sheik Said rebellion in Diyarbakir in 1925, the Ağrı rebellion in 1927–30 and the Dersim uprising in 1937.
12. Joost Jongerden, 'The Spatial (Re)Production of the Kurdish Issue: Multiple and Contradicting Trajectories', *Journal of Balkan and Near Eastern Studies* 13, no. 4 (2011), 375–88.
13. Şemsa Özar, Nesrin Uçarlar and Osman Aytar (trans. Sedef Çakmak), *From Past to Present: A Paramilitary Organization in Turkey: Village Guard System* (Istanbul: DİSA, 2013).

14. Alice von Bieberstein, 'Treasure/Fetish/Gift: Hunting for "Armenian Gold" in Post-Genocide Turkish Kurdistan', *Subjectivity* 10, no. 2 (2017), 170–89, particularly 174.
15. Tener Akçam and Umit Kurt, *The Spirit of the Laws: The Plunder of Wealth in the Armenian Genocide* (New York: Berghahn, 2015). Von Bieberstein, 'Treasure/Fetish/Gift', 174.
16. *Sabah*, 'Enerjide denklem değişiyor!', 19 October 2020.
17. http://Bianet.org. (last accessed 29 January 2022).
18. Turkey ratified the Paris Climate agreement in October 2021.

9

PETROPROTEIN DREAMS: HYDROCARBON BIOTECHNOLOGY AND MICROBIAL LIFE WORLDS IN THE MIDDLE EAST

Douglas Rogers

Roughly a third of the twenty-one experts in petroleum microbiology who gathered at the Kuwait Institute for Scientific Research (KISR) on 15–17 January 1979 were based in the Middle East. The remainder hailed from Western Europe, the USA and Japan, mostly from the research departments of some of the world's biggest oil and petrochemical companies, among them British Petroleum, Imperial Chemical Industries, Phillips Petroleum and the Mitsubishi Gas Chemical Company. The *Proceedings of the OAPEC Symposium on Petroprotein*, published the following year, suggest a distinctly torch-passing tone to the conference.[1] By 1979, the big international companies were almost all abandoning the petroprotein business, having judged their attempts to synthesise food supplements by fermenting yeast or bacteria on hydrocarbon feedstocks to have been scientifically and technically successful but impossible to make profitable. Emil Malick, President and CEO of Provesta Foods (a division of the US-based Phillips Petroleum), captured the state of this field succinctly in his presentation. Provesta's microbial protein supplement would not, his company had concluded, be able to compete with prices for soya or fishmeal as a protein source for the foreseeable future. But, he continued, reiterating a claim he had made in 1976 to the League of Arab States: 'nations in the Middle East and elsewhere that possess large oil and gas reserves … appear to be in the best strategic and economic position to make and sell [petroprotein] in world trade.'[2]

KISR's 1979 symposium thus contemplated the possibility that a combination of resources unique to the Middle East – abundant hydrocarbons and as-yet-undiscovered microbes – might yet realise the gathered scientists' dream of bringing to market a new, cheap, safe and efficient way to feed livestock and maybe, one day, humans. Scientists in the Biotechnology Department of the Food Resources Division at KISR had taken up this challenge in 1977, as one of a number of initiatives begun during a significant expansion of KISR led by General Director Dr Adnan Shihab-Eldin. Their work continued, without interruption and with significant success in meeting its targets, until 1990.

In the end, neither livestock nor humans in the Middle East ended up on a home-grown petroprotein diet. This phase in the global, decades-long quest to transform oil into food nevertheless affords some useful perspectives on the life worlds of Middle Eastern oil. Along with other chapters in this volume, it points beyond studies of oil that focus on production output levels, gyrating global prices or petro-state policies and towards an appreciation of the many other transformations and imaginations that have surrounded oil, especially those related to the refining, petrochemical and technoscientific sides of the industry. The story of petroprotein also enables a broader understanding of the 'life' part of 'life worlds' by expanding the term's default referent – human life – to include microbes. Yeast and bacteria have been transforming and making use of hydrocarbons for billions of years longer than humans, and the oil industry has sought to kill, grow, tame, engineer and otherwise manipulate them almost since its inception. It is not only humans, that is, whose lives are awash in hydrocarbons. Joining petroleum microbiologists in attending to the lives and proclivities of hydrocarbon-metabolising microbes can thus add some new dimensions to the ways in which we understand some the roles of more familiar players – states, corporations and international organisations – in shaping life worlds on earth.[3]

British Petroleum and the Discovery of Petroprotein

Once the near-exclusive province of cranks and hucksters, the idea that oil might be transformed into food through technoscientific wizardry began to look as though it might become reality in a British Petroleum laboratory in Lavera, France, outside of Marseille, in 1959. Alfred Champagnat, a microbiologist at BP's French subsidiary Société Française des Pétroles BP (SFPBP), was experimenting with ways to remove excess crude from refinery waste by treating it with hydrocarbon-metabolising bacteria and yeast. Champagnat's team realised that one of the strains of yeast with which they were experimenting produced protein in the process of breaking down paraffins – the wax of waxy crude oil. Could that protein, they wondered, be processed and used as a nutritional supplement?[4]

BP invested quickly, quietly and significantly in the basic science and technology of petroprotein. By 1963, when it announced its discovery to the world, the company was already opening a test plant at Lavera, next to Champagnat's laboratory, and a larger 'Protein Demonstration Unit' at its existing refining complex in Grangemouth, Scotland. The Grangemouth facility had an initial staff of thirty-one, ranging from microbiologists to chemical engineers to technical assistants. BP's Protein Exploitation Branch, charged with market research, contracting for field trials and licensing, began operations in 1964, and a new laboratory dedicated to basic petroleum microbiology, including but not limited to protein research, began operating in Epernon, France in 1965. In 1965, BP's annual expenditure on research dedicated to bringing microbially produced petroprotein to reality came to just under GBP 500,000 – more than the amount spent that year on other categories of research: Motor Fuels (around 227,000), Lubricants (413,000), hydrocracking (224,000) and other basic research (286,000).[5] As the company's proteins profile continued to grow in scale and complexity, BP's Board of Directors established a new corporate subsidiary, BP Proteins Limited, in 1970, to co-ordinate research and ongoing commercialisation efforts at an expanding array of sites.

BP's rapid expansion of basic microbiology research dedicated to protein supplements unfolded in the context of the broader shifts in which the company found itself in the decades between the 1950s and the 1970s. Since the late nineteenth century, oil had been steadily transforming the world of transportation, and with it the global economy. By the mid-twentieth century, a second revolution was under way that involved the petrochemical industry. Hydrocarbon-derived nitrogen-fixing fertilisers were contributing to the

Green Revolution, and fungicides promised more and better foods. Ever more grades and transformations of crude were possible in refineries, contributing to what Matthew Huber called the making of 'fractionated lives', the ways in which humans' answers to the question 'how am I going to live?' were increasingly tied to the proportions of different oil products flowing out of refineries and petrochemical plants.[6] This was the age of fascination with plastics, cosmetics, rubbers, asphalt, synthetic fibres and new medical equipment. The oil industry, Huber shows, became ever more closely tied to the post-war global bio-political order, especially in the high-consuming states of the West. The growing field of microbiology was a next logical step for BP and its competitors. With chemistry just having been conquered – 'Better Living through Chemicals', in the words of Dupont Chemical's 1935–82 tagline – manipulating the short but productive lives of hydrocarbon-metabolising microbes looked to be a new and exciting frontier. Indeed, this shift from chemistry to microbiology, and from agricultural to industrial protein, as an area of corporate growth animated much of BP's investment in the field. An internal planning document reflecting on ten years of work in this field was not modest about the company's achievements and possibilities:

> For the first time, a purely industrial process, relying neither on arable land nor on farm produce but nevertheless based on a biological *process*, yields a nutrient which can be incorporated into animal feed or food for human consumption ... Does this not indicate that a third era in man's history is now beginning?[7]

BP's microbiologists worked quickly, excited by the prospect of developing new ways of improving human lives by manipulating microbial ones. By the middle of the 1960s, they had developed two separate industrial biological processes for the production of petroprotein. Both processes relied on a strain of a yeast species – *Candida lipolytica* – that consumed the paraffin in petroleum as part of its regular metabolism. Both processes also drew on and modified fermentation technologies that up until that point had been used on an industrial scale in the pharmaceuticals sector (including the production of penicillin and other antibiotics) and in food and beverage industries, but were as yet unknown in the oil and petrochemical refining sectors. Dozens of careful experiments at BP's labs sought to pinpoint the precise conditions for optimal growth of *C. lipolytica*, while others aimed to discover new strains and species, usually isolated from oil-contaminated soil samples.

In what became known as the 'Lavera process', an inoculation of yeast was mixed with a gas oil feedstock (in petrochemical terms, a 'middle distillate' already produced by many oil refineries) and other nutrients, including mineral salts and ammonia to optimise acidity. In the process of fermentation, the yeast would consume around 10 per cent of the gas oil, leaving a mass of protein-rich yeast and a watery gas oil. Gas oil and water were then removed from the mixture through an array of centrifuges, detergents, solvents and other washes, and reused for other refining or fermentation processes. The remaining yeast was dried, freed of any harmful hydrocarbon residue, and made ready for testing as a protein supplement for animal fodder. In the 'Grangemouth process', yeast grew on a different form of hydrocarbon known as n-paraffins, which were not a typical refinery output and so required the co-construction of an n-paraffin plant. In the 'Grangemouth process', notably, all of the n-paraffin feedstock was consumed by the yeast in the fermentation process, meaning that the complicated post-fermentation steps at Lavera – the removal of residual gas oil – were not necessary.

In BP's economic modelling, the two processes were about equally costly and both produced essentially the same product: 'a cream-coloured, free flowing powder, odourless and tasteless, high in protein content and constant in quality, a notable feature being its large proportion of assimilable amino acids, particularly lysine'.[8] The 'Grangemouth process' came to be regarded as preferable by the company over the course of the 1960s, leading one BP manager to the conviction that protein from oil offered not only an ongoing opportunity for diversified revenue streams for the company, but the possibility that BP would, by the late 1980s, become 'one of the great food companies of the world'. Protein products, he suggested, could be sold to farmers at 'every BP service station in Europe'.[9] Although other oil and petrochemical companies were fast on BP's heels, the true competitors for BP Protein were other sources of protein supplements. It was indisputable that petroprotein could be produced faster than soybeans, but could it be produced more cheaply?

By the late 1960s, BP's production and testing had grown to the point where the company officially launched Toprina, a protein concentrate designed to be mixed into animal fodder by feed compounders. Toprina was cleared for use in most European markets, beginning with France in 1967. Although always cagey about the possibility of actually producing protein concentrate for human consumption, by the mid-1970s BP was quietly collaborating with Neville S. Scrimshaw of the Massachusetts Institute of Technology on early-stage human trials, employing MIT students as volunteers to eat petroprotein-infused cookies and meticulously record how their digestive systems reacted. Scrimshaw and his colleagues, at an international conference at MIT in 1969, suggested that neither the term 'petroprotein' nor the term 'microbial protein' was likely to win over consumers of animal, much less human, food. The class of protein concentrates derived from these processes was renamed as the more innocuous-sounding single-cell protein, or SCP.

These early decades of rapid growth for SCP research and productions were also, however, decades of pervasive anxiety for the global oil industry, from oil fields to boardrooms. The famous 'Seven Sisters' oil companies of the West were steadily losing control over their colonial-era concessions to nationalising movements the world over. In the Middle East, racialised labour unrest in the 'oil camps' was everywhere, as Robert Vitalis shows for the Saudi Arabian case with particular analytical insight.[10] The Abadan Crisis of the early 1950s, in which Iranian nationalists seized British Petroleum's main refinery in the country, shook the company to its core, and the six-month closing of the Suez Canal in 1956–7 exposed massive vulnerabilities in the emergent global oil supply chain. Petroprotein, rechristened SCP and marketed as Toprina, recruited microbes into BP's efforts to respond to challenges to the old oil order – from the Middle East and elsewhere – and fashion a new one.

The Race for Single-cell Protein in Iran

The boosters of SCP within BP, then, saw in their new product a potential route by which the company might diversify its income streams as it encountered challenge after challenge to its basic assumptions and operations. Income might be earned not just from the sale of new product (Toprina) in a new sector (food and nutrition), but also through intellectual property, by licensing the patents on the complex fermentation process that the company had vigorously pursued. Accordingly, BP Protein explored joint ventures and expansion projects in many directions, entering talks with the Soviet Union, Venezuela, Italy, and a number of Middle Eastern states. Plans developed most quickly and comprehensively in Italy, and in 1971 BP entered into a 50:50 collaboration with ANIC, the petrochemical division of the Italian state-owned oil company ENI, to build a 100,000 metric ton per year Toprina plant near ANIC's existing refinery in Sardinia. ENI's oil concession in Libya, which would provide a reliable supply of crude, made this an ideal site for the world's first massive SCP plant, slated to open in the mid-1970s and to supply Toprina to the French and Italian animal feed markets.

Iran was the first major Middle Eastern country to seriously and extensively consider partnering with a major Western company in the SCP business. BP's interest in Iran as a site for an SCP plant seems to have started as early as 1968, when Dr Bagher Mostafi of the Iranian National Petrochemical Company contacted BP representatives while visiting the UK.[11] Soon thereafter, a joint survey team composed of representatives from BP and Spillers Foods Limited, a British animal feed company that was collaborating on projects with BP Protein, spent three weeks in Iran evaluating the prospects for a joint venture that would construct and operate a feed mill in the country. Their report noted the growing affluence and purchasing power of Iranians, the increase in large animal farms – especially around Tehran – intended to supply livestock to the population, and, at least near the capital city, the adequacy of the feed mills and compounding operations for current demand levels. The team reported more promising possibilities for development in Iran's southern provinces, particularly if a joint venture could lead to export possibilities for the Iraqi and Kuwaiti markets and if the National Iranian Oil Company, which operated a major farm near Abadan, would sign on. The report ultimately took the position that markets in Iran were too uncertain for it to be sensible to proceed further.[12]

By the early 1970s, however, BP executives returned with somewhat more enthusiasm to the topic of 'assist[ing] Iran to develop the background technology which may eventually help in the establishment of protein and other microbiological/biochemical processes', including laboratory set-up and consultation with Iranian experts at the Iranian National Oil Company, as well as with university and government scientists.[13] Additional tours and surveys of Iranian feed producers followed, and a steady stream of delegations shuttled between the Iranian Government, the Iranian National Oil Company and British Petroleum. Delegation after delegation discussed possible sites, licensing terms and other details. The Iranian Protein Committee debated the terms on which protein supplements would be tested, whether livestock raised on those supplements would be acceptable to the Iranian palette, and which ministries would play what role in a joint venture or licensing agreement. Neither side moved ahead decisively.

The major difference between the desultory report of 1968 and the optimism of the early 1970s was that, around the world, a rush of evidence suggested that a wide variety of research projects dedicated to SCP seemed on the verge of becoming realities. Multiple United Nations development agencies had, since the 1950s, considered protein deficits to be one of the major impediments to human advancement in the Third World. Protein from oil – rapidly grown, easily transported and potentially very cheap and plentiful – appeared as a tantalising solution, a line of thinking that BP and others did much to encourage. With BP's Toprina approved for animal use and for sale in European markets, the BP–ANIC–ENI deal signed in 1971, and ever more scientists endorsing the promise of oil-into-food projects, SCP's potential seemed, finally, to be arriving. Indeed, BP delegations attending talks with their Iranian counterparts in the early 1970s found themselves time and again knocking down myths: that SCP as human food was already available; that SCP plants could be constructed everywhere and were easy to operate; that the product would be so cheap that it would immediately transform the lives of individual farmers and Iran's export profile.[14]

These persistent, overly optimistic 'nationalistic wishes'[15] from their Iranian counterparts aside, BP continued to develop plans for an SCP plant in Iran. In 1974, the diversification possibilities of SCP in the Middle East were clearly outlined in a BP internal memo, and just as clearly placed in the context of the revolutions in the oil world taking place at the time: 'At present major upheavals are taking place in the commercial and political relationships between the oil producing countries, the oil consuming countries and international oil companies who for so long have served as the main channel for oil distribution in the major part of the free world.' In this environment, the memo concluded, it 'would be useful to discuss … the opportunities which the present oil situation offers for the exploitation of BP's protein process in the Middle East'.[16]

By the 1970s, however, and despite its decades-long presence in the country, BP found that the competition for SCP joint ventures in Iran was fierce. The Shah had established the Iranian Protein Committee, composed of an array of government, scientific and educational officials, in the early 1970s, and charged it with improving the amount and quality of protein in the country's diet – in part by exploring joint ventures with foreign firms. By 1975, the US-based companies Amoco and Provesta had both pitched detailed plans for the construction of an SCP plant in Iran, at least one of them with the personal support of US Vice President Spiro Agnew.[17] A Japanese collaboration between Dal Nippon Inc. and Kanegafuchi, with support from the United Nations Industrial Development Agency, was said to be actively courting Iran as well.[18] But some of BP's biggest competition was closer to home – Imperial Chemical Industries (ICI).

ICI, the British chemical industry behemoth, had watched as BP developed its petroleum-based SCP processes through the 1960s, and had embarked on a parallel but microbiologically quite different track. The ICI process under development in the 1970s utilised methanol (a methane-derived alcohol) as a feedstock and *Methylophilus methylotrophus* bacteria as a growth agent. Methanol-metabolising bacteria, the company wagered, would beat out BP's oil-metabolising yeast in the protein supplements market, in part because ICI's supply of methane for methanol production came from nearby and reliable North Sea oil fields. By the mid-1970s, ICI was in the process of constructing a massive fermenter at its Billingham plant in northern England to produce its own SCP product, named Pruteen, for European markets. Even before final plans for Billingham were approved, however, ICI saw particularly strong possibilities in Iran, in part because BP's preferred n-paraffins process was not well suited to Iran's not-very-waxy crude. The n-paraffins for feeding BP's *Candida* yeast would have to be imported or a costly plant constructed to produce them. By contrast, ICI pointed out, Iran had abundant access to methane – much of it simply flared into the atmosphere and convertible into the methanol that ICI's bacteria preferred.[19]

The competition between BP and ICI over the Iranian SCP market nicely illustrates some of the larger forces at play as the world's oil companies turned their attention to the cultivation of microbes and the animal feed markets. The Iranian Protein Committee, in close collaboration with the Iranian National Oil Company and, ultimately, the Shah of Iran, was understood by all parties to be the deciding player in which microbial process, and which company, would move ahead. The Committee's decisions were to be undertaken with the health and modernisation of the Iranian nation in mind, as well as the fostering of Iranian agriculture and industry. Courting the Iranian Protein Committee were leading global companies which, in the face of challenges to their accustomed modes of accumulation and colonial-era governance, sought models of profit that diversified them away from a focus on the production and sale of hydrocarbons and the standard aims of the petrochemical industry of the mid-twentieth century. The key drivers of this competition were not just companies and states and the humans who peopled these institutions, but the actions, attributes and diets of species of bacteria and yeast isolated, examined, tended and deployed by all sides.

In this competition, BP's position in Iran was shaped by its long history in the country, attendant networks, and drawn-out efforts to retain a corner of the Iranian oil market throughout waves of nationalisation. It also had the advantage of being the clear global leader in SCP – it could claim significant experience in all manner of SCP development, extensive experiments and testing with its *Candida* yeast, and more advanced plans to move from pilot plants to full scale production. In 1972–3, in one indication of the company's priorities, it worked to install Alfred Champagnat, the original discoverer of oil-into-food technologies, as a temporary adviser to the National Iranian Oil Company.[20]

For its part, ICI did not have the benefit of BP's long presence in Iran, but it had developed its own influential connections – and microbiological experts – that were able to effectively lobby the Iranian Protein Committee. Chief among these was Sir Ernst Chain, a retired professor of microbiology at Imperial College London who had received the Nobel Prize for his contributions to the production of penicillin. Chain's expertise in pharmaceutical fermentation had given him close connections to both ICI and other international companies – like the Swedish firm Astra – and he began to take an active interest in Iran in 1974, aided by Iranian expatriate businessman and London notable David Alliance.[21]

Chain's Nobel Prize helped open doors in Iran, including to a private audience with the Shah to discuss the future of the country's biological industries.[22] He joined the Iranian Proteins Committee in an advisory capacity and the Biological Research Unit of the National Iranian Oil Company.[23] Chain addressed these bodies on multiple occasions, both insisting on his independent, dispassionate scientific view of the matter and rarely missing an opportunity to register his opinion that the ICI process for SCP animal feed was safer and cheaper for Iran, especially if it were undertaken in collaboration with the Swiss firm Nestlé (with whom Chain also worked extensively).[24] Chain considered BP's *Candida* yeast to be of inferior 'biological properties' as compared to ICI's *Pseudomonas* bacteria: the protein supplements grown by ICI were purer, had fewer uneven fatty acid chains (thought by some to be detrimental to digestion and health) and contained a more robust complement of amino acids (BP's product, by contrast, lacked high levels of the amino acid methionine, which had to be supplemented in the final product).[25] When ICI went ahead with its plans for a massive SCP fermenter in Billingham, Chain lost little time in recommending that Iran begin site exploration for a methanol plant and expand its existing negotiation with Nestlé to include the production of human food supplements. Indeed, Chain was several biological steps ahead of all parties in suggesting that ICI and Nestlé plants might be co-located to facilitate co-operation and that he would 'use his influence to get Nestle and ICI to exchange their strains [of *Methylomonas* bacteria] as a first step in the collaboration'.[26] Iran, he continued, would be able to play a world-leading role in the production of safe protein biomass for human consumption, a dream that had so far eluded everyone.

In the end, neither British Petroleum, its long-standing Iranian connections and their oil-fed *Candida* yeast nor Imperial Chemical Industries, Professor Sir Ernst Chain and their methanol-eating *Methylomonas* bacteria prevailed in Iran. BP was the first to pull out – not only of Iran but of the entire SCP business. The company's signature large-scale SCP project had been, since 1971, its joint venture with ANIC in Sardinia. By 1976, that new Italprotein plant stood idle, fully constructed but never commissioned, its opening blocked by Italian health authorities who refused to accept tests showing that Toprina was safe to produce or consume. An all-out lobbying and publicity campaign in Italy failed to change the situation, and the company decided that there were no further worthwhile prospects in SCP production.

BP Protein was, in fact, overwhelmed by a rapid succession of blows to its plans. On the production side, the mid-1970s spike in oil prices had rendered SCP far less lucrative than it once appeared to be. At the same time, prices for soya and fishmeal had stayed stubbornly low. Always built on expectations of small margins, SCP no longer looked feasible at all. On the consumption side, the product's chief intended market had always been the alleviation of the global protein deficit. But a new public health consensus was rapidly emerging, spurred on by a stunning and quickly influential 1974 article in *The Lancet* entitled 'The Great Protein Fiasco'.[27] Its author, Donald MacLaren, argued that public health and nutrition experts had wandered into a decades-long dead end by viewing all malnutrition around the world as a variant of protein deficiency originally diagnosed in Africa. The United Nations Protein Advisory Board – the chief international champion of SCP development – fought back for a little while but disbanded by 1977.[28] BP Proteins was liquidated in 1978, although the company retained a foothold in the animal feed sector through another subsidiary, BP Nutrition, into the mid-1990s.

ICI pressed on with the construction of its fermenter in Billingham, successfully launching Pruteen (the prices for North Sea methane that supported its methanol feedstock were not as affected by the OPEC embargo and global energy crisis). Chain continued to advise his Iranian contacts that this development augured well for the construction of an SCP plant in Iran, but by mid-1978 his repeated letters to his closest contact at the NIOC, Dr A. Badakhshan, were not receiving their customary speedy replies.[29] When Chain did re-establish a connection with Badakhshan later in the year, he first expressed his hope that 'this friendship will persevere through these difficult times and political upheavals that we have to face' and then acknowledged that the next meeting of the Iranian Protein Committee would likely be delayed.[30] That meeting never happened, for the Shah was swept from power in the Islamic Revolution of early 1979. Chain himself died only a few months later, in August 1979. The Iranian route to SCP production was over, and the industry's oldest and most experienced player, BP, had abandoned its protein subsidiary.

KISR and the Hunt for Thermo-tolerant Microbes

Iran had not been ICI's only promising candidate for an SCP plant in the Middle East. In the mid-1970s, the company had approached the Kuwaiti government and Kuwait National Oil Company, both of which took an active interest. Kuwait had at its disposal a unique – for the Middle East – set of possibilities for evaluating and pursuing a new source of protein. The Kuwait Institute of Scientific Research had been established in 1967 with funding from the Japanese Arabian Oil Company as part of its oil concession on the Saudi Arabia–Kuwait border. In 1973, an Amiri Decree provided KISR with a national charter and significantly expanded its mission to include research in both petroleum sciences and agriculture. KISR's Biotechnology Department, part of its Food Sciences Division, far outpaced anything else of the sort in the Middle East, eventually growing to fifteen to twenty dedicated researchers and an even larger support staff.[31]

Ibrahim Hamdan, recruited to KISR in 1974 after a fermentation-focused Ph.D. in Food and Agricultural Sciences at Ohio State University in the USA, recalled being asked to participate in a feasibility study of ICI's SCP proposals. He and his colleagues reported back to KISR leadership that the plans were not economically promising as they stood, but that the situation might look very different if, for example, the high costs of cooling the fermenters could be substantially reduced. The problem was well-known in the SCP community, and first identified at BP years earlier: the strains of yeast that the company had isolated and grown at its production facilities in Scotland and France were not well adapted for the much hotter Middle East. They grew at 32 degrees Celsius, which could be relatively easily maintained in refinery conditions in Europe but would require 'quite severe' refrigeration elsewhere – which meant significant infrastructure and expensive amounts of water for cooling.[32] At BP's Epernon microbiology facility in France, the search for thermotolerant yeast began in 1968 – the same year as the first BP/Spillers expedition to Iran to explore the Iranian feedstocks market. Already by 1971, BP's Grangemouth microbiology laboratory was screening soil samples shipped in from across Africa, the Far East and Australia that might grow comfortably at temperatures as high as 39–40 Celsius.[33] This was a cumbersome process, especially as the company's extensive library of *C. lypolitica* strains did not appear to offer any good thermotolerant options, and the most promising yeasts were either *C. tropicalis* or *C. blanki*, neither of which had undergone toxicological testing or economic modelling, processes that might take as much as two years.[34] ICI's bacterial process encountered similar hurdles, and its preferred microorganisms would also need massive cooling infrastructure to be successfully cultivated at scale in the Middle East.

Hamdan and his colleagues thought that they might succeed where BP and ICI had not: 'there should be a lot of organisms in the soil of Kuwait', he recalled suggesting to the KISR leadership, '[with so much] oil dripping here and there'. General Director Adnan Shihab-Eldin agreed, and the main direction of KISR petroleum biotechnology research from the late 1970s until 1990 was born. KISR would search for thermotolerant or thermophilic hydrocarbon-metabolising microbes that would produce a safe and efficient protein supplement, and they would eventually be grown in a full-scale SCP plant constructed by ICI or another major Western firm. This thrust fit well with KISR's overall goal as summed up by Hamdan: 'The philosophy of KISR was not to re-invent the wheel – it was mainly to adapt and develop existing technologies for the area, and in order to do that, we had a lot of connections with many institutions and organizations, in the US and in Europe.'

What this next generation in the development of petroprotein dreams required, in other words, was a significant realignment of the corporate, state, microbial and human configurations that had shaped the quest for SCP over the previous two decades. Transnational organisations again played a role. As ICI waited in the wings with its production plant plans and KISR scientists set about their laboratory work, the idea of a specifically Middle Eastern path to SCP received a substantial boost from the Organization of Arab Petroleum Exporting Countries (OAPEC). OAPEC saw in KISR's SCP research the possibility of transforming and modernising agriculture across the Arab world. In this, it followed – now at a regional level – the enthusiasm that the United Nations Industrial Development Organization and Protein Advisory Group had expressed for BPs original petroprotein plans more than a decade earlier.

OAPEC, founded in 1968 by Saudi Arabia, Libya and Kuwait, served as a forum for the exchange of expertise among Arab states whose main national source of income was petroleum. By 1972, Abu Dhabi, Algeria,

Bahrain, Iraq and Qatar had joined, with Egypt and Syria following by 1973 (following a relaxation of membership criteria requiring only petroleum to be an 'important' source of national income).[35] If the famous mission and purpose of the Organization of Petroleum Exporting Countries (OPEC) was largely directed outward, towards the domain of oil as an international commodity and source of monetary income, then OAPEC's mission was directed inward to the Arab world and largely concerned with regional oil sector development projects and infrastructure. Its earliest joint ventures give a good indication of the kinds of projects that OAPEC was pursuing: The Arab Maritime Petroleum Transport Company, the Arab Shipbuilding and Repair Yard Company, the Arab Petroleum Investment Company (a financial holding company that invested in members' oil-sector development projects) and the Arab Petroleum Services Company. Although it never became one of the organisation's main interests or goals, OAPEC's *News Bulletin* and *Annual Reports of the OAPEC Secretariat* describe a running series of meetings and conversations about the future of petroprotein in the second half of the 1970s, listing them among the lengthy series of conferences, committees, working groups and other initiatives dedicated to building a specifically Arab network of people, institutions and, indeed, microbes, that would populate the post-OPEC embargo life worlds of Middle Eastern oil.[36]

These plans were discussed with particular intensity in the early 1980s, as KISR researchers submitted proposals for OAPEC to assist with their research, especially by financing the constructing of fermentation equipment and facilitating ongoing connections with major oil and petrochemical companies around the world. Senior OAPEC officials meet with Amoco Chemicals specialists on the edges of the Seventh International Conference on Petrochemicals in Chicago in 1982, including a side trip to Amoco's plant in Hutchison, Minnesota (Amoco was the last US-based petroleum company harbouring protein dreams). The same delegation, along with KISR specialists, had met with representatives of ICI earlier that year. Plans were advanced enough that the investment wing of OAPEC, the Arab Petroleum Investments Corporation (APICORP), was reviewing plans for a 100,000 ton/year plant, to be established in one of its member states.[37] Summing up this flurry of activity since the 1979 symposium that it had sponsored at KISR, the OAPEC newsletter for December 1982 led with a front-cover editorial, 'SCP Production – An Instrument for Food Security in the Arab Region', that highlighted the many benefits that would come with a specifically Arab SCP industry:

> Production of single cell protein (SCP) to replace imports of fishmeal and soya beans offers a feasible option for meeting the increasing demand for animal feedstuff and raising the efficiency of livestock production. SCP production would also improve the Arab countries' use of their natural gas resources. Moreover, it would spearhead the introduction of the rapidly evolving field of biotechnology into the Arab world and would foster a scientific and technological environment, ultimately helping to expand and diversify the Arab industrial base.[38]

At the time of this editorial, OAPEC had good reason to be optimistic about the potential for SCP in the Arab oil-producing countries, for KISR was beginning to report some success. Ibrahim Hamdan had become the Principal Investigator of the SCP project at its inception, and former Shell SCP bioengineering expert Geoffrey Hamer had been brought on as a regular consultant. Ibrahim Banat, an assistant research scientist, did much of the microbiological bench work, gaining a Ph.D. in microbiology at the University of Essex along the way: 'At that time we were interested in thermophiles … We wanted thermotolerant or thermophilic organisms [that would grow at] a minimum of 40–45 degrees [Celsius].' According to the calculations of the KISR techno-economic department, 41 degrees Celsius was the minimum growth temperature that could cut cooling costs by a sufficient amount – as much as 16 per cent of the total costs of growing non-native microbes.[39] The SCP group quickly made choices that took account of these economic realities and narrowed their search. Following ICI's lead rather than BP's original discoveries, they chose bacteria over yeast because they were faster-growing and tended to have a more favourable protein content. As a substrate, Banat recalled, 'We chose methanol, which made it a little bit difficult. Methanol of course is not only a hydrocarbon but it is also an alcohol that inhibits microorganisms, so to find a microorganism that grows in the presence of methanol, which is a toxic alcohol, wasn't an easy task.'[40]

'After one or two years, we found a lot', recalled Hamdan. KISR researchers reported in international journals that they had discovered four novel, methanol-utilising, thermotolerant strains of *Methylophilus* bacteria that they considered to be good candidates for SCP production at a Middle Eastern facility. Each had been isolated from oil-contaminated soil in Kuwait. Samples of KISRI-5, KISRI-6.1, KISRI 512 and KISR-51112 were deposited in the National Collection of Industrial and Marine Bacteria in Scotland, where they received the patent numbers NCIB 12135, NCIB 12136, NCIB 12137 and NCIB 12138.[41] The group eventually received a United States Patent as well, #5,314,820, for 'Process and Microorganisms for Producing Single Cell Protein'.[42] Even as they wrote up their basic discoveries of new organisms, the KISR team was moving rapidly through the next phases of their work. KISR had built a 1.5 cubic metre capacity fermenter in 1979–80 to support work on the SCP project, and candidate strains of bacteria were quickly put to work producing enough protein supplement to begin nutrition and toxicology testing. The amino acid profiles of the resulting product, they reported, compared very favourably to the standards established by the United Nations for protein supplements for animal food, indicating that animal testing could begin. KISR researchers undertook this product testing in collaboration with the TNO Toxicology and Nutrition Institute in Zeist, the Netherlands. Trials with rats, following the basic protocols established by the proceeding two decades of research on SCP, proved promising, and longer-term studies in rats, as well as tests with poultry, calves and lambs, were scheduled to follow.[43]

As the 1980s came to a close, then, Hamdan and his team at KISR had essentially delivered on their late 1970s response to ICI's proposal of an SCP plant in Kuwait. They seemed to be on the verge of realising OAPEC's expansive vision of transforming agriculture in the region. By this time, however, ICI had joined BP in deciding to shut down its petroprotein operations for reasons of economic feasibility, and the last sales of its Billingham-produced Pruteen took place in the mid-1980s. There was one remaining international potential builder of an SCP plant, the German chemical company Hoechst, from which Kuwait had been buying protein supplements for some years and which had an existing relationship with the Kuwait National Oil Company. The proposal on the table, gaining traction over the course of several exchanges of delegations, was for Hoechst to build a major SCP plant in Kuwait's Shuaiba Industrial Complex, bringing together German fermentation technology and KISR's microbes.

'Negotiations were ongoing', Hamdan remembered, 'when everything exploded.' Iraq invaded Kuwait on 2 August 1990, when many of KISR's scientists were on their summer holidays. KISR and its labs were ransacked. 'They stole the fermenters. They stole the big bioreactors', recalled Ibrahim Banat, continuing, 'You lost your job, you lost the country you knew, the livelihood you had, the house you had, the money you had. Everything disappeared except for what you had published and what you had [submitted for publication].' Hamdan told me that the KISR fermenters were transported by truck to Iraq and stored at a microbiological research facility, where they were later destroyed by American bombs in the course of the US-led international campaign to destroy Iraq's biological weapons research and production facilities.

Most of the expatriate scientists who had staffed KISR's biotechnology department in the 1970s and 1980s never returned to work in Kuwait. Ibrahim Banat continued his petroleum microbiology research at the University of Ulster in Northern Ireland, and Ibrahim Hamdan went to work as a scientific collaborator at the United States Department of Agriculture (USDA) in Maryland and from there to a position as the Senior Regional Officer for Agro-Industries & Technology at the Food and Agriculture Organization (FAO) of the United Nations. He later he served as the Executive Secretary for the Association of Agricultural Research Institutions in the Near East & North Africa (AARINENA). Biotechnology at post-liberation KISR continued, but was no longer concerned with single cell protein. The department was headed by Nader al-Awadi, who had joined the SCP team in 1979 as a fermentation engineer, working mostly with KISR's experimental fermenter, and eventually completing his Ph.D. in Zurich on thermophilic bacteria. Under al-Awadi's guidance, petroleum microbiology at KISR turned to a new and urgent purpose: bioremediation, the microbe-assisted cleansing of oil spills left behind after the Iraqi army destroyed hundreds of oil wells on its retreat north. Of the SCP era at KISR, al-Awadi recalled, 'The knowledge, and expertise, and whatever know-how we had supported us in further work. We gained a lot from it.'

The plans and projects for synthesising protein supplements from hydrocarbons that migrated from the international oil companies to the Middle East in the 1970s and 1980s were punctuated by some of the most dramatic episodes of those decades: the rise of OPEC helped spark and, later, up-end BP's original petroprotein plans; the Iranian Revolution brought the BP-ICI competition in Iran to a crashing halt; and Iraq's invasion of Kuwait, along with the subsequent war, abruptly ended the most promising SCP research programme in the world at the time. Shifting the focus away from these high-profile events – and their often-debated links to the geopolitics and international political economy of oil – allows other dimensions of Middle Eastern life worlds to come into sharper focus. The history of petroprotein dreams shows that, in these decades, the actions of some of the largest players on the world stage were significantly concerned with and highly responsive to the life worlds of some of the smallest.

Notes

I thank Nelida Fuccaro, Mandana Limbert, and others at the NYU Abu Dhabi workshop 'Lifeworlds of Middle Eastern Oil' for encouraging me to step (far) outside my usual comfort zone in writing this chapter. I am especially grateful to the staffs at the BP Archive at Warwick University and the Wellcome Collection in London for permission to consult and cite material in their collections, and to Drs Ibrahim Hamdan, Ibrahim Banat and Nader Al-Awadi for sharing recollections of their work at the Kuwait Institute for Scientific Research in the 1970s and 1980s via Skype interview. Dr Hamdan and Dr Adnan Shihab-Eldin also sent generous written answers to some questions. A version of this paper was drafted and discussed at the 'Energy Materiality: Infrastructure, Spatiality and Power' Study Group at the Hanse-Wissenschaftskolleg Institute for Advanced Study in Delmenhorst, Germany, convened by Margarita Balmaceda.

1. *Proceedings of OAPEC Symposium on Petroprotein.* Kuwait, Organization of Arab Petroleum Producing Countries, 1980. Other symposia followed, including a second OAPEC-sponsored symposium in Algeria in 1983. See also Daham I. Alani and Murry Moo-Young, *Perspectives in Biotechnology and Microbiology: Lectures and Papers from the First Arab Gulf Conference on Biotechnology and Applied Microbiology* (New York: Elsevier, 1986).

2. Emil A. Malick, 'Status of SCP in the USA and Provesta's Technologies', in *Proceedings of OAPEC Symposium on Petroprotein* (Kuwait: Organization of Arab Petroleum Producing Countries, 1980), 151.
3. For a sense of the spread of topics in petroleum microbiology that is roughly contemporaneous with the era of petroprotein research, see Peter Hepple (ed.), *Microbiology: Proceedings of a Conference Held in London, 19 and 20 September 1967* (London: Institute of Petroleum, 1968). This chapter takes some inspiration from recent studies that also foreground the role of microbes, among them Stefan Helmreich, *Alien Ocean: Anthropological Voyages on Microbial Seas* (Berkeley: University of California Press, 2009); Anna Lowenhaupt Tsing, *The Mushroom at the End of the World: On the Possibility of Life in Capitalist Ruins* (Princeton: Princeton University Press, 2015).
4. The basic outlines of the discovery and development of petroprotein at British Petroleum are related in James Bamberg's corporate history, *British Petroleum and Global Oil, 1950–1975* (Cambridge: Cambridge University Press, 2000), 424–44.
5. BP ARC54417: 'Protein (France, Grangemouth, Nigeria, and Japan)'.
6. Matthew Huber, *Lifeblood: Oil, Freedom, and the Forces of Capital* (Minneapolis: University of Minnesota Press, 2013).
7. BP ARC47739: 'Petroleum Microbiology: Reflections on Future Orientations and Recommendations, May 1969', emphasis in original. In this report's historical framing, the first two stages were hunting/gathering and agricultural/livestock raising.
8. BP ARC117739: 'BP Proteins – Background Notes', 25 August 1971, 3.
9. BP ARC80934: 'The Future of Protein', 13 January 1969, 1.
10. Robert Vitalis, *America's Kingdom: Mythmaking on the Saudi Oil Frontier* (New York: Verso, 2009).
11. BP ARC38791: 'BP Proteins – IRAN'. BP also made some progress in plans for an SCP plant in Saudi Arabia, where BP engineers worked with international contractors, including Bechtel and Foster Wheeler, to draw up designs for a 100,000 ton per year SCP plant in the growing petrochemical complex at Al Jubail – part of Saudi Arabia's effort to 'reach national self-sufficiency in protein'. Although the new Saudi Toprina plant was planned to become operational in the early 1980s, it was never built.
12. BP ARC113219: 'Report on B.P./Spillers Survey of the Iranian Feed Market, 17 October–9 November 1968'.
13. BP ARC113219: Llewelyn to Fidler, 30 November 1971.

14. BP ARC113219: 'Interest in Protein from Hydrocarbons', May 1972. On United Nations interest in SCP, see, for instance, Josh Ruxin, 'The United Nations Protein Advisory Group', in David F. Smith and Jim Phillips (eds), *Food, Science, and Regulation in the Twentieth Century: International and Comparative Perspectives* (London and New York: Routledge, 2016), 151–66.
15. BP ARC113219: Llewelyn to Fidler, 30 November 1971.
16. BP ARC121514: 'Proteins in the Middle East', 9 January 1974.
17. BP ARC113219: 'Proteins – Iran – Position as at May 1976' and 'Proteins – Iran (Notes on a visit to Iran, June, 1975)'.
18. BP ARC38791: Watts to Llewelyn, 9 October 1972.
19. On ICI's internationalisation in these years, see I. M. Clarke, 'The Changing International Division of Labour within ICI', in Michael Taylor and Nigel Thrift (eds), *The Geography of Multinationals: Studies in the Spatial Development and Economic Consequences of Multinational Corporations* (New York: St Martin's Press, 1982), 90–116.
20. BP ARC38791: 'Proposed Arrangement for M. Champagnat to Act as a Consultant to N.I.O.C. on Petroleum Microbiology, 16 October 1972'.
21. The Wellcome Collection, Chain Papers, F. 272 Kermanshanchi to Chain, 28 April 1975; Chain to Alliance, 20 April 1974; Chain to Alliance 12 June 1974. See also Ronald W. Clark, *The Life of Ernst Chain: Penicillin and Beyond* (New York: St Martin's Press, 1985), 197.
22. The Wellcome Collection, Chain Papers F. 275, Chain to Gaguik, 24 October 1975.
23. The Wellcome Collection, Chain Papers F. 276, Badakshan to Chain, 23 December 1975.
24. The Wellcome Collection, Chain Papers F. 276, Chain to Badakshan, 13 April 1976.
25. The Wellcome Collection, Chain Papers F. 274, 'Single Cell Proteins: An Address given to the Iranian Protein Committee', 4 October 1975.
26. The Wellcome Collection, Chain Papers F. 278, 'Protein Committee meeting of November 22, 1976. Minutes of Professor Sir Ernst Chain's comments'.
27. Donald McClaren, 'The Great Protein Fiasco', *The Lancet*, 13 July 1974, 93–6.
28. See Ruxin, 'The United Nations Protein Advisory Group'.
29. For example, The Wellcome Collection, Chain Papers F. 281, Chain to Badakshan, 14 June 1978.
30. Ibid., Chain to Badakshan, 8 November 1978.

31. Although SCP was the Biotechnology Department's marquee project in the 1970s and 1980s, there was simultaneous research in other areas of hydrocarbon biotechnology, especially biosurfactants. Other Middle Eastern universities and research centres were home to individual scientists or small groups that participated in SCP or other biotechnological research projects, but none had anything near the staffing, connections or success of KISR.
32. BP ARC113206: 'Iran – Proteins/Nutrition', 1972–4.
33. BP ARC80578: 'Project Report #1 – Grangemouth, 1971'. See also BP ARC 113206 'Iran – Proteins/Nutrition', 1972–4.
34. BP ARC19475: 'Selection of Thermotolerant Organism for N-Paraffins Fermentation'. BP Proteins Grangemouth Division, 29 June 1972.
35. Pan-Arab organisations were nothing new, of course, and indeed OAPEC took over and expanded some of the mission that had been carried out by the Arab League's Department of Oil since the 1950s. For a detailed accounting of the founding and early years of OAPEC, see George Tomeh, 'OAPEC: Its Growing Role in Arab and Oil Affairs', *The Journal of Energy and Development* 3, no. 1 (January 1977); Mary Ann Tétreault, *The Organization of Arab Petroleum Exporting Countries: History, Policies, and Prospects* (Westport, CT: Greenwood Press, 1981).
36. See for instance 'Petroprotein Committee to Meet in March' 3, no. 1 (1977), 5; 'Petroproteins', *OAPEC News Bulletin* 6, no. 7 (1980), 11; 'Meeting of the Synthetic Protein Expert Group', *OAPEC News Bulletin* 7, no. 2 (1981), 2; *Secretary General's Second Annual Report*, OAPEC (1975), 66; *Secretary General's Fourth Annual Report*, OAPEC (1977), 98–9.
37. OAPEC *News Bulletin*, May 1982, 3.
38. Ibid., cover page.
39. KISR modellers had projected that plant cooling costs for non-thermotolerant SCP production would approach 16 per cent of plant operating expenses in an arid zone, compared to around 9 per cent in a temperate zone. See A. Prokop et al., 'Bacterial SCP from Methanol in Kuwait: Product Recovery and Composition', *Biotechnology and Bioengineering* 26, no. 9 (1984), 1,085.
40. On the process behind these decisions, see, for instance, Prokop et al., 'Bacterial SCP from Methanol' and G. Hamer and Ibrahim Y. Hamdan, 'The Transfer of Single Cell Protein Technology to the Petroleum-Exporting Arab States', *MIRCEN Journal* 1, no. 1 (1985), 23–32.
41. Ibrahim M. Banat, Mustafa Murad and Ibrahim Y. Hamdan, 'A Novel Thermotolerant Methyltrophic *Bacillus* Sp. and Its Potential for Use in Single-cell Protein Production', *World Journal of Microbiology and Biotechnology* 8,

no. 3 (1992), 290–5 and Ibrahim M. Banat, Nader Al-Awadhi and Ibrahim Y. Hamdan, 'Physiological Characteristics of Four Methylotrophic Bacteria and Their Potential Use in Single-Cell Protein Production', *MIRCEN Journal* 5, no. 2 (1989), 149–59.

42. United States Patent Office, 'Process and Microorganisms for Producing Single Cell Protein', No. 5,314,820, Inventors Ibrahim Y. Hamdan; Amin S. ElNawawy; Ibrahim M. Banat; Nader Al-Awadhi.
43. On the fermentation engineering and nutritional phases of work at KISR, see A. S. Abu-Ruwaida, Ibrahim M. Banat and Ibrahim Y. Hamdan, 'Chemostat Optimization of Biomass Production of a Mixed Bacterial Culture Utilizing Methanol', *Applied Microbiology and Biotechnology* 32, no. 5 (1990), 550–5; Nader M. Al-Awadhi, M. A. Razzaque, D. Jonker, Ibrahim M. Banat and Ibrahim Y. Hamdan, 'Nutritional and Toxicological Evaluation of Single-Cell Protein Produced from *Bacillus* Sp. KISRI-TM1A in Rats', *Journal of Food Quality* 18, no. 6 (1995), 495–509.

10

MAKING OIL MEN: EXPERTISE, DISCIPLINE AND SUBJECTIVITY IN THE ANGLO-IRANIAN OIL COMPANY'S TRAINING SCHEMES

Mattin Biglari

On 22 March 1951, on the first day of the Persian calendar year, and just two days after the Iranian parliament had passed the oil nationalisation bill, employees of the Anglo-Iranian Oil Company (AIOC) began a strike in the port of Bandar Mahshahr in South West Iran. Over the next few days and weeks, thousands more joined in the strike across the company's areas of operations in the oil-rich province of Khuzestan, especially in the refinery town of Abadan. Lasting until 27 April, the strike succeeded in halting the flow of oil in the Abadan refinery – the largest refinery in the world at the time – and signalled AIOC's imminent expulsion at the hands of popular resource nationalism.

Yet participation in the strike was not confined only to full-time employees of the oil company. In the words of the onlooking US embassy, the 'outstanding characteristic' of the strike was that it was 'created, and long sustained, by fewer than 400 students'.[1] These were students at AIOC's training centres in Abadan, especially the Abadan Technical Institute. The company had built such centres to produce skilled, loyal employees, yet these trainees used such spaces to mobilise against the company, performing sit-ins and holding secret meetings there to co-ordinate with workers in the refinery. Such was the fear among military officials about the Abadan Technical Institute and student hostel being 'hubs of the strike' that the military eventually blockaded both to prevent further intrigue.[2] Not deterred, students continued to be at the forefront of resistance against the police during the strike, consequently suffering bloodshed and fatalities. It is surprising, therefore, that their main demand should seem so trivial: that pass marks in exams were unfairly high.

This episode offers a window on understanding the purpose and impact of training in the history and politics of the global oil industry. AIOC envisaged training as not only producing productive, docile employees, but also as helping it to meet its political obligations to the Iranian government in 'Iranianisation', increasing the number of Iranians in senior staff positions. Thus, exploring AIOC's training schemes in the preceding years better illuminates how an oil company negotiated local entanglement in the desire to ultimately remain disentangled and removed from national or local politics. Scholars have already shown how oil companies have strived towards such aims through strategies of racial segregation, welfare paternalism and urban planning.[3] However, much less attention has been paid to how training functioned similarly as a form of labour management.[4] AIOC's training schemes in Abadan are particularly pertinent to such a study, in that they were the most extensive of all oil companies' at the time. Indeed, the Abadan Technical Institute, founded in 1939, was the first training institute directly run by an oil company to offer students the opportunity to attain a university-level degree. As it would happen, it would be the last: learning the lessons from the strike, AIOC discontinued the degree and other oil companies avoided introducing similar schemes.

Understanding how AIOC's very existence could be challenged by its own trainees, then, also invites a consideration of how the global oil assemblage could be modified in the 'grip of worldly encounter', challenging traditional narratives of oil and resource nationalism focused on elites and pioneering personalities.[5] During the strike, trainees not only accelerated the process of oil nationalisation by shutting down operations of a foreign oil company, but they also put forward a vision of what nationalisation should entail concretely on the ground, disrupting and altering the global oil industry 'from below' much like oil workers were able to.[6]

This chapter explores how training was constitutive of subjectivity at the intersection of the oil industry's various scales. It examines how training shaped subjects' identities and their notions of gender, class and nationality, complementing other works that have highlighted how oil has shaped quotidian experiences on the micro scale, especially in the Middle East.[7] In addition, it points to how AIOC deployed disciplinary practices of labour management and training schemes borrowed from wider corporate capitalism and the British empire. Above all, it focuses on the production of oil expertise, analysing how transnational flows of knowledge, actors and infrastructures coalesced in Abadan to result in ambivalent, contradictory conceptions of expertise. Thus, it engages with scholarship that shows how oil companies have historically built political arrangements into their technical and scientific components.[8] By paying attention to such themes, this chapter highlights the historical significance of training in the oil industry's power and its contestation on local, national and global scales.

It argues that AIOC's training schemes reflected a bifurcated system of knowledge production corresponding to a division between intellectual and manual labour. This privileged abstract, disembodied knowledge as the legitimate basis of oil expertise, marginalising embodied or 'tacit knowledge' as the domain of labour. In turn, this division instituted a class division between managers and workers, respectively, and designated their opposing relationships to oil operations. In this framing, expertise seemed to manage an objective, self-regulating technical system (the refinery), which was emptied of embodied knowledge and human labour (in a sense, black boxed). Nevertheless, the actual running of operations meant that human labour was integral to the running of the refinery, such that management sought to shape human subjectivity through its training schemes. This paradox produced tensions in company training schemes (especially in the Abadan Technical Institute), on the one hand creating aspirations in Iranian employees to gain expert status (defined against manual labour), and on the other, blocking those aspirations through disciplinary mechanisms aimed at producing employees as docile subjects. As a result, training schemes inadvertently produced anti-colonial subjectivities among trainees, culminating in their central involvement in expelling the oil company.

Standardisation and the Abadan Refinery in the Global Oil Industry

To understand the nature and significance of AIOC's training schemes it is necessary to first outline how the company assembled its operations in Iran and the implications this process had for the position of labour. Since oil had been struck at Masjed-e Soleyman in 1908, the Anglo-Persian Oil Company (APOC; AIOC from 1935) had gradually built its operations in Khuzestan to deal with the local specificities of Iranian crude to produce standardised products for an abstract, global market. A key node in this conversion was the Abadan refinery, which was built in 1912; although in its early years the refinery predominantly produced fuel oil – mainly for the British Royal Navy – by the 1920s the company worked on redirecting the refinery's purpose towards producing gasoline, in line with rising car consumption. This could not be done by simply applying established knowledge, instead requiring the company to create controlled conditions to understand the local properties of Iranian crude, helped by collaboration with local Bakhtiari khans to protect oil operations and secure property rights.[9] Subsequently, the company found that the particularly heavy nature of Iranian crude necessitated finding new refining processes that would enable its conversion into higher yields of lighter products, such as gasoline.

To do this, in the 1920s and 1930s the company made efforts to acquire and utilise scientific knowledge about the latest refining processes in the global oil industry. The company actively co-operated with other oil corporations to introduce new, standardised refining processes to Abadan, such as cracking, reflecting the standardisation and centralisation of scientific knowledge in large corporations during the inter-war period.[10] This allowed for the transnational flow of actors to Abadan, especially from the USA and the UK, operating within a domain of standardised knowledge and practices – a 'technological zone' – that was supposed to enable a seamless flow of oil from the wellhead to the consumer without being beholden to any locality or specificities of place.[11] As such, the refinery was built to privilege abstract, theoretical knowledge, thereby subordinating the importance of situated, embodied forms of knowledge.[12] By the early 1940s, the company's engineers and chemists were even confident that the local particularities of Iranian crude had been augmented to produce such high yields of gasoline that they now started to speak of refining processes as representing science's mastery over 'Nature'.[13]

Indeed, by the time the company was facing growing nationalist opposition in Iran in the late 1940s, it projected its expertise as a preformed set of objective, scientific knowledge necessary for the production of the country's 'black gold'.[14] In its extensive public relations machinery, the refinery figured as an important marker of this expertise.[15] It was represented as a complex, scientific and esoteric domain – a black box where only the inputs and outputs were known – to be contrasted with local society and its supposedly traditional, simple practices of using oil.[16] In various images, the refinery – along with refineries elsewhere – was depicted as a self-regulating, objective system devoid of human subjectivity and operating independently from local conditions; such images often omitted humans altogether while transposing the refinery onto a flat, featureless land (Figure 10.1). Not only did this dismiss the importance of local factors, including workers inside the refinery, but it also emphasised the modular nature of the refinery as a mobile, self-contained assemblage capable of repetition from one location to another – in effect, a visual representation of the technological zone.[17] In this framing, expertise appeared as an abstract set of knowledge managing the refinery from without

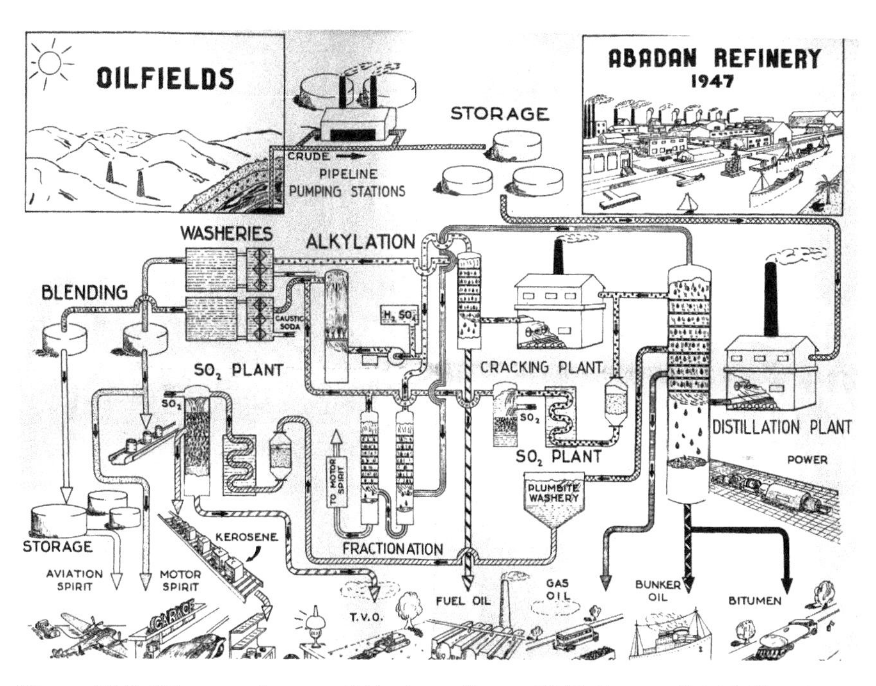

Figure 10.1 Diagramatic map of Abadan refinery, 1947. Source: British Petroleum ARC144669_003. © BP Archive.

rather than imbricated or emergent within as a situated, embodied form of knowledge.[18]

Controlling the 'Human Factor'

Yet in reality, throughout these decades AIOC management was consistently frustrated that local factors were undermining productivity in the Abadan refinery, especially the 'human factor' in operations: oil workers. Despite representations of the refinery, workers' manual labour was crucial in the functioning of the refinery, especially in construction and maintenance, so much so that there were over 30,000 workers in the refinery in 1949, with the biggest concentration being in its workshops.[19] Over the years, managers frequently complained about reduced output stemming from high rates of absenteeism, turnover and acts of everyday resistance. Above all, workers had the ability to affect production altogether through industrial action: strikes in 1920, 1922 and 1929 all caused disruption at the refinery, while in July 1946 a general strike across Khuzestan involving most of the company's workforce threatened to bring a complete halt to oil operations.[20]

As such, although the company represented the refinery as a black box devoid of human subjectivity, in practice it aimed not to remove human subjectivity but instead shape this through social engineering. From the 1920s, the company embarked on a series of interventions into Abadan town to control the lives of workers beyond the workplace, such as urban planning, segregation and the provision of leisure facilities.[21] Within the refinery itself, by the mid-1930s management turned to the time-and-motion studies of Taylorism in an attempt to regulate the daily rhythms of workers, maximising productivity by reducing superfluous movement, while also further aligning operations with the technological zone by standardising the oil worker.[22] This aimed to produce workers as docile subjects and mere extensions of the technical system, thereby reinforcing a class division between mental and manual labour.

Hence, despite the company obfuscating embodied knowledge in its performance of expertise, it realised the importance of tacit knowledge – or skill – in ensuring that labour conformed to the designs of the refinery, so that reality matched the plan. Until the mid-1920s, APOC preferred to meet its need for skilled labour by recruitment rather than training, in line with the practices of other oil companies on the world's oil frontier. In particular, the company relied on skilled Indian workers throughout the 1920s and 1930s rather than training Iranians, thereby maintaining a racialised division of labour in which 'unskilled' labour positions were filled exclusively by natives, in line with common practice in the global oil industry at the time.[23]

It was only due to pressure from the Iranian government that APOC began to change its training policy. To be sure, the company provided small-scale training to address a labour shortage caused by World War I and the discharge of unruly Indian workers involved in 1920 and 1922 strikes. However, the new concession agreement of April 1933 put pressure on the company to adhere to a policy of 'Iranianisation', which would entail the gradual replacement of foreign personnel by Iranians, including technical positions. Work began towards this in Abadan to prepare Iranians to take up skilled labour positions in the refinery. In particular, the company provided apprenticeships in its Artisan Training Shop (ATS), opened in 1933 and soon hosting over 450 apprentices each year.

An important part of this early training effort was to send a certain number of Iranian students to the UK, either to train in trade schools or to study at Birmingham University. In theory, the scheme opened the very first avenue available for Iranians to become educated with specialist knowledge about oil. In total, 62 students had been sent to study in the UK by 1950.[24] Some of these eventually took up management positions with AIOC in Abadan or went on to become some of the leaders of the Iranian oil industry in the post-nationalisation period, earning them the nickname the 'Birminghamers' (*birminghāmi-hā*).

The Abadan Technical Institute: Training Oil Experts

Soon the company built a new flagship training centre that seemed to offer the opportunity for Iranians to train as oil experts in Abadan as well. As a result of its negotiations with the Iranian government on 'Iranianisation' in the mid-1930s, the company built a new technical school in Abadan, which was completed in 1939 and established as the Abadan Technical Institute. The institute's expressed purpose was to train Iranian youths so that the company could 'obtain recruits of a high standard of skill and technical education'.[25] Within ten years, the institute hosted 1,161 students, illustrating how the company's training schemes in Abadan were among the most extensive of all oil companies in the world at the time.[26] While other companies did already provide vocational training to employees, such as Creole Petroleum in Venezuela, and some even sponsored more theoretical learning at local schools, such as its parent company Standard Oil New Jersey, the Abadan Technical Institute was unique for the time in its size and facilities.[27]

The design of the building indicated that its purpose diverged from the company's existing training centres. It was designed by the architect James Mollison Wilson, who had been integral to the company's new housing projects of the 1930s aimed at socially engineering workers.[28] The building was completed at an initial cost of £100,000 and comprised administrative, educational and recreational sections such as various sports grounds.[29] Significantly, it had multiple laboratories, which were increasingly common in technical schools in the UK and had become emblematic 'of the emancipation of mechanical engineering from the enduring primacy of apprenticeship'.[30] It was at this time that some UK universities, including Birmingham, were removing their workshops altogether and halting practical training on their engineering courses as the theoretical aspects of engineering education became more prominent; it was now widely believed that a workshop had 'no place in a modern university'.[31] Abadan's central clock tower resembled colonial boarding schools in the British Empire, such as Mayo College in India, serving as a constant reminder of temporal modernity and a means of instilling time discipline as well as a 'sense of Britishness'.[32] Indeed, some students even referred to it as 'little Big Ben'.[33] All this suggested the institute was to make professional employees or 'oil men', in contrast to the focus of existing training in Abadan on producing skilled labourers.

Yet early on, the emphasis of the institute's course was evidently more on vocational training. For instance, when Reza Shah visited the institute in 1940, he expressed satisfaction that he could see students in boiler suits doing 'practical things' and ordered the company to 'make young Iranians do physical exercises and make them strong. I do not want my people to sit down and read books.'[34] This comment should be understood within the wider social engineering programme of Reza Shah and nationalist elites in the inter-war period, which was fixated on producing uniform, idealised male bodies as a basis for national development.[35] From the Iranian government's perspective,

Figure 10.2 The clocktower of the Abadan Technical Institute in 1939.
Source: Iranian Petroleum Museum.

prominence should be given to vocational training over theoretical learning: the Technical Institute was to be yet another vocational school to add to the country's *honarestans* established in various cities during the 1930s as a means of training skilled labour for industry.[36]

Soon, however, the institute also offered opportunities for more theoretical learning. With the outbreak of World War II, there was increased demand for British skilled labour to serve in the military, and so a four-year course was introduced for Iranians that ended in a license or B.Sc. in Petroleum Technology, ratified by the Iranian Ministry of Education.[37] In theory, the Iranians studying this course had the chance to become first-grade technical workers, potentially spearheading Iranianisation of the oil industry.

Nevertheless, in reality, there was ongoing ambiguity over the institute's purpose. By 1944, there were two separate courses at the institute: an apprenticeship ending in a certificate (equivalent to the diploma that was awarded on completion of secondary school education) and a traineeship that ended in a B.Sc. Both courses offered 'engineering', but with divergent outcomes.

The final year of the apprenticeship allowed students to specialise in mechanical, electrical or civil engineering, yet most classwork was still in broad subjects like Persian, mathematics, physics and chemistry. In contrast, the B.Sc. course allowed students to specialise in engineering directly related to oil: for example, in the third and fourth year of its petroleum technology pathway, classwork in the institute each week included three hours for chemical decomposition, nine for physics and chemistry, six for chemical engineering and three for oil extraction and refining.[38]

Such divergence reflected the breadth of engineering as a discipline, especially in Britain, where it could be considered both a 'manual' and 'professional' occupation. Engineers came from a variety of educational backgrounds, while apprenticeships and junior technical schools could serve as avenues to engineering as much as, if not more than, formal technical education at graduate level.[39] For example, a 1942 article in the popular industry journal, *Engineering*, stated that there was general agreement that engineers who had learned on the job could obtain advancement equal to that of those who had attended a university or technical college.[40] Similarly, although engineering was offered as a practical subject in Iran's *honarestans*, at this time the country's new middle class – especially in Tehran – increasingly viewed engineering as a 'modern profession', imbuing the title *mohandes* with much cultural capital (which is even more the case today).[41]

Meanwhile, for apprentices at the ATS, most training went beyond the classroom and took place in the workshop, where they learnt multiple manual tasks through repetitive rote learning. For example, apprentices might learn 'chipping, sawing and filing' for three weeks, 'saw, drill, grinding' for three weeks, 'simple work orders' for five weeks and 'standard fitting' for three months. Even the classwork they did have was closely connected to their work in the workshop: 'engineering knowledge', for example, actually meant learning the 'shapes, names and uses' of various tools or the '[d]angers of loose clothing, long hair, moving parts, etc.', while a science class could be spent learning how to reduce friction through lubricants.[42]

The recruitment into training schemes also reinforced a division between mental and manual learning between the Technical Institute and ATS, respectively. For the Technical Institute's courses, there was a rigorous selection process to select the best students from across the country, especially in urban centres where Iran's nascent middle class was concentrated. Notices inviting applications were put out in national newspapers along with prospectuses, and a special selection committee was sent out to various towns to conduct interviews and oversee an examination that all candidates were required to take. The criterion for selection was the need for 'sturdy, intelligent, reliable, and adaptable youths in large quantities for the various levels of intake of the Company's Training Schemes'.[43]

In contrast, recruits to the ATS were selected by assessment in its workshop in Abadan, where there could be greater examination of physicality and manual dexterity in action. One of the prerequisites for candidates to the scheme was that they needed to 'have a reasonable physique', while it was stipulated that trade trainees training on the job needed to be at least five feet tall.[44] Company representatives prioritised candidates' possession of manual skills, which they essentialised on the basis of regional crafts: for example, boys from Isfahan were preferred because the town was said to be 'peopled by craftsmen in metal work of all kinds' who would 'take readily to mechanical work', whereas those from Kerman were undesirable as this was supposedly a 'town of carpet weavers'.[45]

Figure 10.3 Students of the Abadan Technical Institute, 1940s. Source: Iranian Petroleum Museum.

Despite this bifurcation, all training primarily aimed to prepare students for work in the refinery. Even for students at the Technical Institute, more time was spent gaining practical experience in the refinery than learning in the classroom: for the apprenticeship, students had to work four and a half days each week in the refinery, while technical trainees had to work forty-two hours part-time in the refinery alongside their three-month period of classwork, followed by another six months of full-time work in the refinery.[46] All training, then, was chiefly concerned with controlling the 'human factor' entering into the technological zone in the same way as other local factors had been, aligning local subjectivities with the standardised practices of the global oil industry. Moreover, given that the refinery had been constructed and represented to privilege disembodied, abstract knowledge rather than embodied, tacit knowledge, training reproduced class divisions between management and workers in the refinery.[47]

Yet controlling workers necessitated more than just teaching the skills and knowledge required for future work in the refinery. As summarised in one company report in 1946, the Technical Institute had to 'give much more out than the knowledge of mathematical formulas and getting degrees or diplomas'.[48] Training also required the shaping of students' overall character in everyday life beyond the classroom.

Discipline beyond the Classroom

The chief means of shaping character beyond studies was the Bahmanshir hostel. From initially accommodating thirty boys when it opened in 1932, it had grown to host 232 by 1945, including both apprentices from the ATS and students from the Technical Institute.[49] It was designed to insulate students from Abadan Town, which many managers considered produced poverty, malnourishment, ill-discipline, inefficiency and criminality. In a memo of 1943, sharing much with the language of the famous Beveridge Report of the same year detailing the state of poverty in Britain, Abadan was characterised as playing host to the 'giant evils of disease, ignorance and squalor', while the apprentice hostels were distinguished in allowing for the 'improvement in physique and health of those resident there'.[50] Similarly, the General Manager described the 'higher purpose' of the hostel as being 'to give apprentices a training of a general nature' so that the apprentice would become a 'good citizen', besides being a competent worker.[51]

In this way, the hostel was to link both spaces of social reproduction – the home and the school – in much the same manner as the British public school. It removed the apprentice from the supposedly femininised sphere of social reproduction in the town and placed him in a masculine domain of 'public life, requiring professional detachment and career ambition'.[52] Indeed, training had only been offered to small numbers of women in clerical work since 1935, preserving the refinery's workshops and processing plants as spaces of idealised masculinity.[53] Meanwhile, as elsewhere on the world oil frontier, the company attached importance (though it was rarely explicitly stated) to women's reproductive labour in ensuring the functioning of oil operations and the productivity of workers.[54]

The concentration of students in the same space allowed for surveillance and the inculcation of discipline in apprentices. It enabled the company to control the health and hygiene of boys, where meals could be regulated and weekly cleanliness inspections conducted. The shared experience could build a collective and enthusiastic drive towards discipline among students, which was further cultivated through group trips, weekend camps and dramatic productions.[55] Just as in the British boarding school, it was this sociality that was supposed to mould the boy into having the personality traits of a desirable employee. As one PR pamphlet of 1947 proudly proclaimed, the hostel's 'corporate life has done much to develop character'.[56]

Character training was exemplified no better than in the emphasis on sports. The Bahmanshir hostel had extensive sports facilities, possessing three football pitches, four tennis courts, one basketball court, a hall for indoor sports like badminton and table tennis, and a large asphalted ground for physical training, as well as its own swimming pool.[57] There was a sports programme in which all boys were expected to take part, symptomatic of a 'games ethic' in British public schools that was promoted as a 'potent means of developing leadership qualities of courage, decisiveness and self-confidence'.[58] Many former students attest that the company's head of training in Abadan, C. L. Hawker, was especially interested in sport for students and he could 'normally be found around the sports grounds', even personally leading a run with students each morning.[59]

Figure 10.4 Apprentices' football team, 1941. Source: Iranian Petroleum Museum.

Yet sport could also run counter to the company's plans. According to many memoirs of former students, sport was one of their favourite aspects of hostel life; indeed, sport became ingrained in local culture to the extent that Abadan became popularly considered a 'city of sport'.[60] But in taking on a life of its own, sport could be an arena for the formation of collective identities outside the company's control. For example, on one occasion during the half-time interval in a football match between the hostel team and a team of British soldiers, Hawker ordered students in the watching crowd to stop their one-sided support because this was 'unsporting' and instead cheer both teams, making reference to the Christmas truce football matches played between British and German soldiers during World War I.[61] As was the case in many colonial boarding schools in the British empire, although the corporate environment had helped engender discipline and masculinity in students, this did not necessarily entail obedience to British authorities: in bringing together boys from various parts of Iran and different socio-economic backgrounds into a shared living environment, it could also help create fraternal bonds necessary for the formation of national consciousness, defined against British management.[62]

Barriers to Expertise in Training

By the late 1940s, students were increasingly dissatisfied about being blocked from pathways to progression. Although evidence for students' involvement in the 1946 strike is scarce in company documentation, police blockaded the hostel to prevent refinery workers meeting with students there. This was in part due to the fact that students had been sending letters of complaint to newspapers in Tehran at the time. In one such letter sent to the Iranian newspaper *Bahar* in March 1946, students from the Abadan Technical Institute made several complaints, including: poor material conditions; being assigned work in the refinery beneath their training; that the Technical Institute had only produced twenty technicians and engineers thus far; and that tuition was only carried out in English. Even the students sent to the UK to study were 'not admitted to Cambridge or Oxford but to Birmingham University where all the natives of British colonies are enrolled'. The students demanded that the school be placed under the authority of Iranian Ministry of Education and that studies be conducted in Persian instead.[63] Implicitly, then, students understood their status within a wider context of British colonial education policy, but appealed to Iran's independence in demanding better treatment than that afforded to colonial subjects.

Even after the clampdown during the strike, students continued to voice their dissatisfaction, complaining especially of the barrier to the highest levels of company training. Some of this dissatisfaction was channelled through the underground work of the communist Tudeh Party, which reportedly became more popular among students after the strike.[64] Party organs revealed students' grievances over being expelled from the Technical Institute for failing exams that had unfairly high pass marks and subsequently being treated like 'coolies' and 'simple workers' when working in the refinery.[65] These perceptions reflected the reality that the pass rates for courses in the Technical Institute were extremely low. In fact, out of a total of 125 students who could have attained a B.Sc. degree at the Technical Institute by 1947, only twenty-three had done so.[66]

In response, company managers eventually decided to repurpose training. Figures responsible for training, such as Hawker, had concluded that pass rates were so low because the quality of recruits was poor, owing to limited primary education, malnourishment and a lack of 'serious character training' schemes such as boy scouting.[67] Therefore, the company and the government would have to start 'from scratch', co-operating in 'far-reaching social measures if the teeming populations' of the region were 'ever to get sturdy and self-reliant'.[68] Hence, the company agreed to assist the Iranian government in building more primary schools and focusing its training efforts on more elementary levels, preferring this to refashioning advanced courses such as those in the Technical Institute. Indeed, Hawker even suggested discontinuing the B.Sc. course altogether, claiming that the Technical Institute lacked the resources to develop 'the mental and cultural standards necessary to a degree man' and that there was 'no known case of an industry (let alone a foreign firm)' running a full four-year-long university course, especially 'on a desert island with none of the cultural amenities of a large city such as Rangoon'.[69] Although the B.Sc. courses were maintained, this was only done to appease public opinion, especially as complaints about the slow rate of Iranianisation were gathering momentum.[70]

Consequently, company training reverted to primarily producing skilled workers across all levels. The company deflected responsibility for producing Iranian oil experts through higher education onto the Iranian government (and in addition Birmingham University), helping establish Tehran University's Engineering Faculty by providing laboratory equipment worth £150,000 and three full-time British lecturers, a laboratory supervisor and a training shop supervisor.[71] The implication was that students currently studying advanced courses in the Technical Institute were being kept on only symbolically. In this way, the company was following common practice in the global oil industry. The petroleum companies at the ILO meetings concerning education and training of local nationals up to 1951 had put forward a common policy: that although they should provide some opportunities for training, ultimately it was the responsibility of national governments to provide basic and technical education.[72]

The problem for the company was that it had already produced a layer of students in the Technical Institute with aspirations to becoming oil experts themselves. In May 1950 there were reports of high rates of absenteeism among artisan apprentices and trainees, who were complaining that their privileges were being taken away from them and that they were being treated as 'simple labourers'. Such 'resistance to discipline', managers noted, 'did not augur well for their potentialities as eventual foremen'. In response, each trainee was issued with a brass check to display at his locker to show that he was present.[73] Although the trainees were reported to have eventually accepted the new system, defiance continued throughout 1950 and early 1951: students were said to be engaging in various acts of everyday resistance such as not attending sports, disrespecting teachers and boycotting hostel food.[74] Such discontent had the potential to manifest in political activism, especially in a context of growing resource nationalism.

Oil Nationalisation and the 1951 General Strike

By late 1950 and 1951, students in AIOC's training schemes were increasingly turning towards the Iranian oil nationalisation movement. In Tehran, the campaign to nationalise AIOC's operations had been gathering momentum – coalescing in the National Front and around its most prominent politician, Mohammad Mosaddeq – and it was also now gathering support in Abadan.[75] According to Majid Javaherizadeh, who was an apprentice in the ATS at the time, pro-nationalisation sentiments were late in gathering significant support: although there was 'talk here and there' about it from the few people who had a radio and who read the national newspapers, many 'did not know what nationalisation meant and did not pay attention to the subject'. Gradually, however, 'talk about nationalisation of oil got more common', and student protests broke out in Abadan.[76] Similarly, a former student in the Technical Institute, Qobad Fakhimi, recalls how most students there supported nationalisation, such that the feeling for nationalisation was nowhere stronger than in Abadan and that it was the 'heart' of the country during the nationalisation struggle.[77]

Indeed, it was no coincidence that students in the Technical Institute decided to go on strike in Abadan just when Parliament had passed the oil nationalisation bill. Although oil workers were already on strike in other parts of Khuzestan over allowances, these students were important in starting the strike in Abadan, holding demonstrations in the grounds of the Technical

Institute and meetings of up to 2,000 students at the hostel, to which they invited refinery workers.[78] Despite a police blockade of the Technical Institute and hostel, students were central in organising the strike up until its conclusion on 27 April, publicly identifying with the oil nationalisation movement in many protests and calling for the expulsion of AIOC.

On the surface, students' main reason for going on strike was for lower pass marks in their examinations so that they could access more senior positions in the company. But students linked this demand to wider discrimination, considering inordinately high pass marks to maintain inequalities between British management and Iranians in everyday life. On 17 March, before the strike broke out in Khuzestan, technical and commercial apprentices wrote a letter to management with several demands: repeat years to be allowed for all apprentices at every stage of training in the Technical Institute; maintenance of a 'correct' attitude by teachers towards pupils; formation of a students' union; promotion to non-graded staff status on completion of the second year (eleventh-class standard), whether or not students passed the examination; provision of staff buses and access to staff clubs and swimming pools, from which most Iranians were segregated; and the formation of a special hospital ward for students.[79]

Meanwhile, apprentices in the ATS expressed opposition to the repetitive rote learning and barriers to more theoretical knowledge. On 4 April they created an artisan apprentice organisation and issued a manifesto calling for various reforms, including: new academic courses every year after passing examinations; to be trained under 'experienced and qualified instructors'; for no job of a 'non-technical nature' to be given to them; for apprentices who scored a mark of 50 out of 100 to be promoted; that every apprentice be given the right to sit for the higher courses of the Technical Institute; provision of staff bus passes and accommodation upon completion of studies; and that every apprentice should be able to live in the apprentice hostel for five years.[80]

Ultimately, therefore, students' involvement in the strike challenged the fundamental purpose of company training. As summarised in a proclamation by Technical Institute trainees: '[t]he Oil Company brought each one of us to Abadan from one corner of Iran with fallacious promises that we would have a pleasant and comfortable life in Abadan; but immediately on arrival in

Abadan we found that not only those promises did not materialise, but every day we were disgraced and aspersed'.[81] These promises created aspirations for boys to become oil men by acquiring the knowledge necessary for attaining expert status. The timing of the strike suggests that students saw nationalisation as an opportunity via which they might finally realise such aspirations after years of systematically being blocked from doing so, linking the figure of the oil expert to anti-colonial nationalism.[82]

At the same time, students shared management's attitude towards abstract, disembodied knowledge as constituting the legitimate basis of oil expertise. Despite linking up with refinery workers, they disavowed manual work and demanded that the content of advanced courses should not be changed, only access to them. They explicitly did not want to be 'simple workers'. Despite subversively using the spaces designed to produce compliant employees against the company, they still identified with the masculine, corporate and scientific world of its managers.

In the wake of oil nationalisation, then, AIOC's training schemes resulted in the opposite outcome to what the company had intended. The company aimed for training to help disentangle itself from local and national politics by appeasing nationalist aspirations and controlling the human factor of operations, enabling it to operate in a transnational technological zone to ensure the frictionless flow of oil. Yet training had actually drawn the company further into locality, such that its global operations could be challenged by a few hundred students struggling over the seemingly parochial issue of examination pass rates. However, it produced a layer of individuals who aspired to be oil experts, as had been defined in the company's knowledge production system. Thus, although AIOC was expelled, the foundations were set for its expertise to be reproduced and for operations to remain in the technological zone of the global oil industry in the years ahead.[83]

Notes

1. 'The Strike in the AIOC Oil Concession Area', 17 May 1951, p. 1, despatch 835, box 5498, RG 59, The National Archives of the United States, College Park, MD (hereafter NARA).
2. Nosratallah Bakturtāsh, *Chand Yādemān Az San'at Melli Shodan-e Naft Dar Ābādān va Qeireh* (Tehran: Enteshārāt-e ārtemis, 1396/2017), 113.

3. For example, see Robert Vitalis, *America's Kingdom: Mythmaking on the Saudi Oil Frontier* (New York: Verso, 2009); Miguel Tinker Salas, *The Enduring Legacy: Oil, Culture and Society in Venezuela* (Durham, NC: Duke University Press, 2009); Elisabetta Bini, 'From Colony to Oil Producer: US Oil Companies and the Reshaping of Labor Relations in Libya during the Cold War', *Labor History* 60, no. 1 (2 January 2019), 44–56; and Hannah Appel, *The Licit Life of Capitalism: U.S. Oil in Equatorial Guinea* (Durham, NC: Duke University Press, 2019).
4. The role of training has been neglected in major histories of oil such as Daniel Yergin, *The Prize: The Epic Quest for Oil, Money, and Power* (New York: Simon & Schuster, 1991). Vitalis covers Aramco's training in *America's Kingdom*, while Beasley examines training sponsored by oil service companies in Houston from the 1950s to 1970s in Betsy A. Beasley, 'Service Learning: Oil, International Education, and Texas's Corporate Cold War', *Diplomatic History* 42, no. 2 (1 April 2018), 177–203. The only studies that examine AIOC's training schemes in detail are BP's company histories, Ronald W. Ferrier, *The History of the British Petroleum Company: Vol. 1, The Developing Years 1901–1932* (Cambridge: Cambridge University Press, 1982) and James H. Bamberg, *The History of The British Petroleum Company: Vol. 2, The Anglo-Iranian Years, 1928–1954* (Cambridge: Cambridge University Press, 1994), in addition to Michael E. Dobe, 'A Long Slow Tutelage in Western Ways of Work: Industrial Education and the Containment of Nationalism in Anglo-Iranian and Aramco, 1923–1963' (Ph.D. diss., Rutgers, State University of New Jersey, 2008). Dobe argues that AIOC training ultimately served to block Iranianisation by mirroring the 'Hampton–Tuskegee model' of racialised vocational education adopted by Aramco and ultimately preventing Iranians from accessing senior jobs, although he suggests that the company's training policies were slightly more progressive in the opportunities offered.
5. Anna Lowenhaupt Tsing, *Friction: An Ethnography of Global Connection* (Princeton: Princeton University Press, 2005), 1.
6. In recent years there has been burgeoning scholarship on the agency of oil workers; for example, see Touraj Atabaki, Elisabetta Bini and Kaveh Ehsani (eds), *Working for Oil: Comparative Social Histories of Labor in the Global Oil Industry* (Cham: Springer, 2018); and Elisabetta Bini and Francesco Petrini, 'Labor Politics in the Oil Industry: New Historical Perspectives', *Labor History* 60, no. 1 (2019), 1–7. For the case of Iran, more specifically, see Touraj Atabaki, 'Writing the Social History of Labor in the Iranian Oil Industry', *International*

Labor and Working-Class History 84 (2013), 154–8; Kaveh Ehsani, 'The Social History of Labor in the Iranian Oil Industry: The Built Environment and the Making of the Industrial Working Class (1908–1941)' (Ph.D. diss., Leiden University, 2015); Maral Jefroudi, 'Revisiting "the Long Night" of Iranian Workers: Labor Activism in the Iranian Oil Industry in the 1960s', *International Labor and Working-Class History* 84 (2013), 176–94; and Peyman Jafari, 'Reasons to Revolt: Iranian Oil Workers in the 1970s', *International Labor and Working-Class History* 84 (2013), 195–217.

7. This theme is particularly rich in scholarship on oil urbanism in the Middle East; for an introduction see Nelida Fuccaro, 'Introduction: Histories of Oil and Urban Modernity in the Middle East', Special Issue in *Comparative Studies of South Asia, Africa and the Middle East* 33, no. 1 (2013), 1–6. Other notable works include Mandana Limbert, *In the Time of Oil: Piety, Memory, and Social Life in an Omani Town* (Stanford: Stanford University Press, 2010); Farah al-Nakib, *Kuwait Transformed: A History of Oil and Urban Life* (Stanford: Stanford University Press, 2016); and Arbella Bet-Shlimon, *City of Black Gold: Oil, Ethnicity, and the Making of Modern Kirkuk* (Stanford: Stanford University Press, 2019). On Abadan itself see Kaveh Ehsani, 'Social Engineering and the Contradictions of Modernization in Khuzestan's Company Towns: A Look at Abadan and Masjed-Soleyman', *International Review of Social History* 48, no. 3 (December 2003), 361–99; and Rasmus Elling, 'On Lines and Fences: Labour, Community and Violence in an Oil City', in Ulrike Freitag et al. (eds), *Urban Violence in the Middle East: Changing Cityscapes in the Transition from Empire to Nation State* (New York: Berghahn, 2015), 197–221.
8. This has already been demonstrated in detail in the case of AIOC in Katayoun Shafiee, *Machineries of Oil: An Infrastructural History of BP in Iran* (Cambridge, MA: MIT Press, 2018). Of course, the most influential such study concerning the global oil industry is Timothy Mitchell, *Carbon Democracy: Political Power in the Age of Oil* (London: Verso, 2011). Other important works include Geoffrey C. Bowker, *Science on the Run: Information Management and Industrial Geophysics at Schlumberger, 1920–1940* (Cambridge, MA: MIT Press, 1994); Matthew T. Huber, *Lifeblood: Oil, Freedom, and the Forces of Capital* (Minneapolis: University of Minnesota Press, 2013); Andrew Barry, *Material Politics: Disputes along the Pipeline* (Chichester: Wiley Blackwell, 2013); and Hannah Appel, Arthur Mason and Michael Watts (eds), *Subterranean Estates: Life Worlds of Oil and Gas* (Ithaca: Cornell University Press, 2015).
9. Shafiee, *Machineries of Oil*, 70.

10. David F. Noble, *America by Design: Science, Technology, and the Rise of Corporate Capitalism* (New York: Knopf, 1977).
11. The concept of 'technological zones' was developed by Andrew Barry in Andrew Barry, *Political Machines: Governing a Technological Society* (London: Athlone Press, 2001); and, more specifically to the oil industry, in idem, 'Technological Zones', *European Journal of Social Theory* 9, no. 2 (May 2006), 239–53; Shafiee, *Machineries of Oil*; and Appel, *The Licit Life of Capitalism*.
12. James Scott similarly argues that developmental projects in the twentieth century saw the dominance of abstract, dislocated knowledge (what he calls *techne*) over more localised, embodied knowledge (*metis*); see James C. Scott, *Seeing Like a State: How Certain Schemes to Improve the Human Condition Have Failed* (New Haven: Yale University Press, 1998).
13. 'Chemistry and the Petroleum Industry – II', *The Naft Magazine*, vol. 19, no. 3 (October–December 1943). In effect, this was the active production of 'Nature' for profit, as argued in Jason W. Moore, *Capitalism in the Web of Life: Ecology and the Accumulation of Capital* (London: Verso, 2015).
14. This was the underpinning logic of developmental projects more generally in the wider context of the time, as argued in Timothy Mitchell, *Rule of Experts: Egypt, Techno-Politics, Modernity* (Berkeley: University of California Press, 2002).
15. For an analysis of AIOC's PR strategy at this time, with a focus on film, see Mona Damluji, 'The Oil City in Focus: The Cinematic Spaces of Abadan in the Anglo-Iranian Oil Company's *Persian Story*', *Comparative Studies of South Asia, Africa and the Middle East* 33, no. 1 (January 2013), 75–88.
16. On black boxing see Bruno Latour, *Science in Action: How to Follow Scientists and Engineers through Society* (Cambridge, MA: Harvard University Press, 1987).
17. In her study of offshore oil rigs, Hannah Appel emphasises modularity as a key characteristic of the way in which oil companies are able to operate seamlessly across the transnational technological zone of the global oil industry, disentangled from local and national politics; see Appel, *The Licit Life of Capitalism*.
18. On the resituating of knowledge (into centralised centres of calculation) as a basis for expertise see Mitchell, *Rule of Experts*, especially chapter 3.
19. K. J. Hird, 'Labour Developments – Anglo-Iranian Oil Co.' (2 February 1949), 2, LAB 13/519, The National Archives of the United Kingdom, London (hereafter NAUK).
20. On the 1929 strike see Stephanie Cronin, 'Popular Politics, the New State and the Birth of the Iranian Working Class: The 1929 Abadan Oil Refinery Strike', *Middle Eastern Studies* 46, no. 5 (September 2010), 699–732; on the 1946 strike

see Touraj Atabaki, 'Chronicles of a Calamitous Strike Foretold: Abadan, July 1946', in Karl Heinz Roth (ed.), *On the Road to Global Labour History* (Leiden: Brill, 2017), 93–128.

21. See Ehsani, 'Social Engineering'.
22. On Taylorism see Charles S. Maier, 'Between Taylorism and Technocracy: European Ideologies and the Vision of Industrial Productivity in the 1920s', *Journal of Contemporary History* 5, no. 2 (1970), 27–61; and Judith Merkle Riley, *Management and Ideology: The Legacy of the International Scientific Management Movement* (Berkeley: University of California Press, 1980).
23. On the company's recruitment of Indian labour see Touraj Atabaki, 'Far from Home, but at Home: Indian Migrant Workers in the Iranian Oil Industry', *Studies in History* 31, no. 1 (February 2015), 85–114. On the racialised division of labour in the oil industry in Venezuela see Tinker Salas, *The Enduring Legacy*; and, on Saudi Arabia, Vitalis, *America's Kingdom*. The racialised division of labour drew on prevailing stereotypes about which 'races' suited certain types of work best, as illuminated in Michael Adas, *Machines as the Measure of Men: Science, Technology, and Ideologies of Western Dominance* (Ithaca: Cornell University Press, 1989).
24. BP ARC142640: 'AIOC Students Trained in the UK, 1927–1950'.
25. BP ARC54206: Memo, 'Plan of Article 16 III', 1, 10 November 1947.
26. BP ARC142640: 'Education and Training in Iran, 1928–1951', Appendix 9.
27. On Creole's training see 'Venezuela', *The Lamp* 26, no. 6 (December 1944) Box 2.207/D163, Exxon Mobil Historical Collection, Briscoe Center for American History, University of Texas, Austin (hereafter ExMo); on Standard New Jersey see 'Refinery Technical Schools to Open', *The Esso Refiner*, September 1934, Box 2.207/D93B, ExMo.
28. See Mark Crinson, 'Abadan: Planning and Architecture under the Anglo-Iranian Oil Company', *Planning Perspectives* 12, no. 3 (January 1997), 341–59.
29. BP ARC129288-006: Pamphlet, 'Education and Training for the Oil Industry in Abadan', 1950, 12–15.
30. Anna Guagnini, 'Worlds Apart: Academic Instruction and Professional Qualifications in the Training of Mechanical Engineers in England, 1850–1914', in Robert Fox and Anna Guagnini (eds), *Education, Technology, and Industrial Performance in Europe, 1850–1939* (Cambridge: Cambridge University Press, 1993), 16–41. Quotation at p. 36.
31. NAUK: 'University Courses in Engineering: Problem of Practical Training', *Trade & Engineering*, May 1942, ED 46–292.

32. Sanjay Srivastava, *Constructing Post-Colonial India: National Character and the Doon School* (London: Routledge, 1998), 41.
33. Qobād Fakhimi, *Si sāl naft-e Iran: Az melli shodan-e naft tā enqelāb-e eslāmi* (Tehran: Mehrandish, 1387/2008), 50.
34. BP ARC142640: 'Education and Training in Iran, 1928–1951', 7. On Reza Shah's wider education policies see David Menashri, *Education and the Making of Modern Iran* (Ithaca: Cornell University Press, 1992); and Rudi Matthee, 'Transforming Dangerous Nomads into Useful Artisans, Technicians, Agriculturalists: Education in the Reza Shah Period', in Stephanie Cronin (ed.), *The Making of Modern Iran: State and Society under Riza Shah 1921–1941* (London: Routledge, 2003), 123–45.
35. See Sivan Balslev, *Iranian Masculinities: Gender and Sexuality in Late Qajar and Early Pahlavi Iran* (Cambridge: Cambridge University Press, 2019); and Wendy DeSouza, *Unveiling Men: Modern Masculinities in Twentieth-Century Iran* (Syracuse, NY: Syracuse University Press, 2019).
36. Reza Arasteh, *Education and Social Awakening in Iran* (Leiden: Brill, 1962), 43.
37. BP ARC96458: C. L. Hawker, 'Academics and Industry', 28 August 1946, 1–2.
38. Pamphlet, '*rahnema-ye amuzeshgah-e fani-ye naft-e Abadan*' (Guide to the Abadan Technical Institute), ordibehesht 1323 (May 1944), Iranian National Archives, Tehran (hereafter INA), document 240/18273.
39. Chris Smith and Peter Whalley, 'Engineers in Britain: A Study in Persistence', in Peter Meiksins, Chris Smith and Boel Berner (eds), *Engineering Labour: Technical Workers in Comparative Perspective* (London: Verso, 1996), 27–60.
40. M. W. Humphries, 'Post-Graduate Training for Engineers', *Engineering*, 8 May 1942, in NAUK: ED 46–292.
41. See Cyrus Schayegh, *Who Is Knowledgeable, Is Strong: Science, Class, and the Formation of Modern Iranian Society, 1900–1950* (Berkeley: University of California Press, 2009), 5.
42. BP ARC12306: Fields Training Department: Apprentice Training Shops, 'Programmes and Syllabuses for Office, Craft and Artisan Apprentices', 1951, 7–8.
43. BP ARC15918: 'The New General Plan: Report from Iran on Education and Training – 1947', 28.
44. BP ARC12306: 'Recruitment'.
45. BP ARC55806: 'The Apprentice Training Shop', 31 May 1947.
46. Pamphlet, '*rahnema-ye amuzeshgah-e fani-ye naft-e Abadan*', INA 240/18273.

47. On tacit knowledge see Michael Polanyi, *Personal Knowledge: Towards a Post-Critical Philosophy* (Chicago: University of Chicago Press, 1958); and, on its place in the workplace, see Tony Manwaring and Stephen Wood, 'The Ghost in the Labour Process', in David Knights, Hugh Willmott and David L. Collinson (eds), *Job Redesign: Critical Perspectives on the Labour Process* (Aldershot: Gower, 1985), 171–96.
48. BP ARC54206: General Manager's Annual Report for 1945, 27 February 1946, p. 21.
49. Ibid., p. 24.
50. BP ARC54206: Memorandum of October 1943, Relating to Part II of the General Plan of 1936.
51. BP ARC54206: General Manager's Annual Report for 1943, p. 4.
52. Patrick Joyce, *The State of Freedom: A Social History of the British State since 1800* (Cambridge: Cambridge University Press, 2013), 290.
53. This hyper-masculinity was perhaps symptomatic of the way in which the popularisation of engineering as a profession simultaneously hardened boundaries of race and gender, preserving 'technology' as an exclusively 'white, middle class, and male enterprise', as demonstrated in Ruth Oldenziel, *Making Technology Masculine: Men, Women and Modern Machines in America, 1870–1945* (Amsterdam: Amsterdam University Press, 1999), 53–4.
54. For example, see Myrna Santiago, 'Women of the Mexican Oil Fields: Class, Nationality, Economy, Culture, 1900–1938', *Journal of Women's History* 21, no. 1 (2009), 87–110; Elisabetta Bini, 'Building an Oil Empire: Labor and Gender Relations in American Company Towns in Libya, 1950s–1970s', in Touraj Atabaki, Elisabetta Bini and Kaveh Ehsani (eds), *Working for Oil: Comparative Social Histories of Labor in the Global Oil Industry* (Cham: Springer, 2018), 313–36; Nathan Citino, 'Suburbia and Modernization: Community Building and America's Post-World War II Encounter with the Middle East', *The Arab Studies Journal* 13/14, no. 2/1 (Fall 2005/Spring 2006), 39–64; and Benjamin Jones, 'Women on the Oil Frontier: Gender and Power in Aramco's Arabia', *Rice Historical Review* 2, Spring (2017), 55–69.
55. BP ARC54206: General Manager's Annual Report for 1945, 25.
56. BP ARC44269: Pamphlet, 'The Anglo-Iranian Oil Company in Iran', 1947, 15.
57. BP ARC129288-006: Pamphlet, 'Education and Training for the Oil Industry in Abadan', 1950, 8.
58. J. A. Mangan, *The Games Ethic and Imperialism: Aspects of the Diffusion of an Ideal* (Harmondsworth: Viking, 1986), 85.

59. Fakhimi, *Si sāl naft-e Iran*, 25; Majid Javāheri'zādeh, *Pālāyeshgāh-e Ābādān dar 80 Sāl-e Tārikh-e Irān 1908–1988* (The Abadan Refinery in 80 Years of the History of Iran 1908–1988) (Tehran: Nashr-e Shādegān, 1396/2017), 47.
60. Iraj Valizadeh provides a comprehensive history of sport in Abadan in his memoir in a chapter entitled 'Abadan, the City of Sport'; see Iraj Valizādeh, *Anglo va Bangolo dar Abadan* (Tehran: Simiā Honar, 1390/2011), 481–608.
61. Fakhimi, *Si sāl naft-e Iran*, 25.
62. This is an important theme in Heather Sharkey's study of Gordon College in Sudan; see Heather J. Sharkey, *Living with Colonialism: Nationalism and Culture in the Anglo-Egyptian Sudan* (Berkeley: University of California Press, 2003). On a similar case of youth and sports clubs facilitating the rise of national consciousness in Bahrain, see Nelida Fuccaro, 'Shaping the Urban Life of Oil in Bahrain: Consumerism, Leisure, and Public Communication in Manama and in the Oil Camps, 1932–1960s', *Comparative Studies of South Asia, Africa and the Middle East* 33, no. 1 (January 2013), 59–74. Quotation at 63.
63. NAUK: FO 371/52722 'Anglo-Iranian Oil Company's Technical School'.
64. Fakhimi, *Si sāl naft-e Iran*, 51–2. On the Tudeh's role in Khuzestan at this time see Ervand Abrahamian, 'The Strengths and Weaknesses of the Labor Movement in Iran, 1941–1953', in Michael E. Bonine and Nikki R. Keddie (eds), *Modern Iran: The Dialectics of Continuity and Change* (Albany: State University of New York Press, 1981), 211–32.
65. 'At the AIOC's Technical School', *Mardom*, 14 April 1947.
66. BP ARC15918: 'The New General Plan: Report from Iran on Education and Training – 1947', 20.
67. Ibid., 28.
68. Ibid., 29.
69. BP ARC96458: C. L. Hawker, 'Academics and Industry', 28 August 1946, 2–3.
70. See Dobe, 'A Long Slow Tutelage', 96–102.
71. BP ARC142640: 'Education and Training in Iran, 1928–1951', 11.
72. Ibid., 20.
73. BP ARC35198: General Manager's monthly reports: industrial relations, May 1950.
74. BP ARC12306: Report by A. P. David, 12 November 1951.
75. The most comprehensive studies of the National Front and the oil nationalisation movement are Mostafa Elm, *Oil, Power, and Principle: Iran's Oil Nationalization and Its Aftermath* (Syracuse, NY: Syracuse University Press, 1992); Homa Katouzian, *Musaddiq and the Struggle for Power in Iran* (London:

I. B. Tauris, 1999), 2nd edn; and Ervand Abrahamian, *The Coup: 1953, the CIA, and the Roots of Modern U.S.–Iranian Relations* (New York: The New Press, 2013).

76. Javāheri'zādeh, *Pālāyeshgāh-e Ābādān*, 48–51.
77. Fakhimi, *Si sāl naft-e Iran*, 43.
78. BP ARC68908: 'General Strike – Abadan, April 1951', 1–3.
79. BP ARC32535: S. K. Kazerooni, 'Refineries Industrial Relations Report March 1951', 9.
80. Ibid., 10.
81. BP ARC68908: 'List of the Proclamations and Pamphlets', 8–9.
82. As Dietrich shows, the fusion of these two would shape international oil politics in the years ahead; see Christopher R. W. Dietrich, *Oil Revolution: Anticolonial Elites, Sovereign Rights, and the Economic Culture of Decolonization* (Cambridge: Cambridge University Press, 2017).
83. In this respect, like decolonisation struggles elsewhere, Iranian oil nationalisation exhibited the reproduction of colonial epistemologies that concerned Fanon in his discussion of 'native elites'. See Frantz Fanon (trans. Constance Farrington), *The Wretched of the Earth* (London: Penguin, 2001 [1961]), 36.

NEW SPACES AND MOBILITIES

11

SHIFTING SOLIDARITIES: STRIKES, INDIAN LABOUR AND THE ARABIAN SEA OIL INDUSTRY, 1946–1953

Andrea Wright

During the first half of the twentieth century, the British government sought to maintain its dominance over oil production in Iran and the Arabic-speaking Gulf. Central to maintaining this position was curbing the influence of American, Russian and local political groups. The British attempted to realise this geostrategic goal through oil concessions and also through controlling the workforce at oil projects in the region. One avenue used to achieve a stable workforce was a policy that specified that British subjects, including persons from British India, should staff oil company projects, as British subjects were seen to be more sympathetic to the British government.[1] The staffing of oil projects by Indians was possible given the already strong presence of Indians in the Gulf, and the skills Indian workers had developed at other British oil projects, such as Burmah Oil Company; and also because large numbers of Indian labourers could be efficiently mobilised through the refashioning of the system used to move Indian indentured labour throughout the British empire in the nineteenth century.[2] As a result, Indians had worked in the Gulf's oil industry since oil was first discovered at Masjid-i-Sulaiman in South West Iran in 1908 and at the Awali oilfield in Bahrain in 1932.

While British companies and the British colonial government preferentially sought to hire Indians for oil projects, Indian workers were not always sympathetic to British imperial projects. For example, in 1920, 3,000 Indians went on strike at Abadan, Iran, and were soon joined by Iranian workers.[3] In Bahrain, less than six years after oil was discovered at Awali, the Indian workers at Bahrain Petroleum Company (BAPCO) went on a series of disruptive strikes in co-ordination with Bahrainis and other Arab workers.[4] At these strikes and other moments of collective action in the first half of the twentieth century, workers in the Gulf's oil industry often formed alliances across ethnic, national and linguistic divides. However, by the mid-twentieth century, worker solidarities increasingly segmented along national lines. The reasons for this shift from class- to nation-based solidarity include workers' experiences, postcolonial nationalism and corporate management practices.

Labour is often invisible in examinations of the oil industry, except in moments of 'spectacular collective action'.[5] As a result, the impacts of labour movements on corporate practices and state governance are often overlooked. In order to attend to the complex dynamics between labour, oil and governance, this chapter de-centres the nation-state as the focus of analysis. De-centring the nation-state requires historians to critically examine our archives, as oil company managers and British colonial officials often interpreted nationalism as the motivation behind worker agitations. It also requires a new approach to labour history, as, historically, much labour history has relied on the nation-state as the 'main analytic or expository frame'.[6] In considering the actions of migrant workers in the oilfields, we find that nationalism influenced worker action, but we are also reminded that nation-states are only one factor informing labour politics.

This chapter looks at how workers expressed solidarity and formed alliances to navigate their working and living conditions in the 1940s and 1950s Gulf. We see changing worker solidarities when we compare work stoppages in Iran, complaints of human rights violations in Bahrain and hunger strikes in Aden. At each of these sites, I consider the claims made by workers, with whom workers formed alliances and how workers mobilised. This focus brings to light how worker solidarities do not always follow the state or imperial logics. Rather, within the context of India's independence in 1947, workers both interrogated and reinforced borders as they agitated for better working conditions. During this same period, managers increasingly segregated workers by nationality, applied a racialised hierarchy rooted in British colonial imaginings of civilisational progress and American Jim Crow policies, and naturalised inequality through the application of economic models. Exploring migrant worker actions in these contexts demonstrates the need to look beyond the analytic of the nation-state in order to understand the multiple actors that shaped historic entanglements between labour and the nation-state.

Wildcat Strikes at Abadan, Iran, 1946

Beginning in March 1946, the employees at the Anglo-Iranian Oil Company (AIOC) participated in a series of strikes without the approval of union officials, so-called 'wildcat' strikes. British and Iranian officials believed that the Tudeh Party of Iran was the driving force in these strikes. This communist party had much influence in 1940s and early 1950s Iran and was able to mobilise tens of thousands of demonstrators in Iran's capital, Tehran, and on the streets of Abadan, where the AIOC oil refinery was located. This was the largest refinery in the world at the time and the site where most Indians working in Iran were stationed.[7] While Iranians officials understood the Tudeh Party to be a movement by workers to agitate for higher wages and better living conditions, British officials based in India and the Gulf saw it as symptomatic of Russia's growing influence in the Gulf and harmful to British companies' businesses, especially the oil industry. This view of the British was reinforced when, also in 1946, the Iranian government, in response to pressure from Tudeh Party leaders, gave the USSR concessions to the northern Iranian oilfields.

The mass agitations against working conditions were not surprising. Manucher Farmanfarmaian, a high-ranking Iranian official,[8] visited labour camps near Abadan and was appalled by the conditions in which workers lived. He wrote: 'Wages were 50 cents a day. There was no vacation pay, no sick leave, no disability compensation. The workers lived in a shanty town called Kaghazabad, or Paper City, without running water or electricity ...' He contrasted Kaghazabad with the British management's accommodations, which included air conditioning, swimming pools and tennis courts.[9] Iranian officials were not the only ones to notice this stark disjuncture between the treatment of British workers and others, and British managers employed in Iran also described the poor working conditions. One British supervisor working for AIOC in Iran wrote: 'The vast majority of our labour force are still living in scandalous accommodation in urine-tainted squalor – in little *manzels* [houses].' These houses, the supervisor explained, were 'without sanitation, water supply or lighting, and with either too much ventilation or no ventilation at all'. Not only was housing for labourers poorly made, he wrote, but '[o]vercrowding is rampant and up to 10 persons sleep on the damp floor of a 10 foot room, huddled together to keep warm in the winter or to keep out of the sun in the summer afternoons'.[10] Indeed, living conditions were so poor that AIOC managers cautioned against re-creating similar living conditions.[11]

The wildcat strikes at AIOC's oilfields and refinery culminated in a large general strike in July 1946, during which fifty employees were killed and 170 injured. Scholars who discuss these strikes often consider the strikes' effects and, in particular, the resulting relations between the Iranian government and Iranian citizens.[12] However, the labourers who participated in these activities were not limited to Iranians and a diverse group of workers participated in collective action against AIOC. One scholar attentive to this diversity is Ervand Abrahamian. In his discussion of the Tudeh Party's actions in Iran, Abrahamian explores the high percentage of minorities within the Tudeh Party and the Tudeh Party's promise of citizenship and secularism.[13] His focus on the multi-generational residents of Iran who supported the Tudeh Party points to the diversity of workers participating in the strikes, sheds light on how solidarities were formed across groups, and the complexity of political development in Iran. In addition to the groups explored by Abrahamian, temporary migrant labourers also participated in the Tudeh Party's political movements.

Attentiveness to the temporary workers on oil projects in Iran and their role in collective action expands the implications of the July 1946 strikes by de-centring the nation in narratives about labour, and, instead, shows a movement towards the construction of a global working class. One place in which the importance of temporary workers arises is in the case of Indian labourers at the refinery. At Abadan, four days before the general strike, five Muslim men from the Punjab, in British India, were forced to resign from their positions at the Abadan refinery because they had joined the Tudeh Party. Despite their resignation, they continued to live in the camps near Abadan.

Figure 11.1 Demonstrations at Abadan Refinery, 1951. British Petroleum Archives, J. H. Bamberg papers.

In the coming days, these five men helped organise a coalition of hundreds of Indians who, in solidarity with Arabic-speaking and Farsi-speaking labourers, went on a strike that influenced AIOC and imperial policies in the coming years. A few days after these July strikes, the company, recognising the activities of these men, paid for the five Tudeh Party labour organisers from the Punjab to fly back to Karachi as their 'continued presence on the oilfields was considered so undesirable'.[14]

Some aspects of the July 1946 coalition deserve attention. First, the coalition was class-based. Indian managers did not participate in the strike and remained 'loyal' to the company.[15] While Indian managers lived in similar conditions as the striking workers, their managerial position influenced their actions and oriented them towards a set of politics that was strictly tethered neither to place of origin nor native language. The coalition for the 14 July strike gained momentum from shared ideological (Tudeh) and class sympathies and formed multi-lingual and multi-ethnic alliances. Despite this broad coalition of actors, the five men who were considered the Indian organisers of the strike all came from a few districts in the Punjab near Lahore. These men shared a natal village and a religion, and spoke the same language. This points to the second interesting characteristic of the strike: while workers formed broad coalitions, local affective ties also influenced worker actions. The activities of these Indian workers were concerning for some British colonial officials, and Lieutenant-Colonel Thomas, a British military intelligence officer based in Karachi, travelled extensively throughout the Punjab in order to learn more about the workers who participated in strikes at Abadan and other oil projects in the Gulf. In 1946, Thomas spent the majority of the year locating the men who had been dismissed from their jobs following these strikes, interviewing workers on leave from the Gulf, and searching for the organisers of the strikes at their homes in northern India.[16]

Thomas's research and the shifting policies at AIOC elucidate the strength of class-based solidarities for political mobilisation; the lasting, powerful effects of affective local ties in transnational movements; and the importance of inter-group solidarity. For example, in order to mitigate the effects of these strikes, the management at AIOC attempted to break apart this inter-group solidarity. One strategy AIOC used was to encourage Arabic-speaking employees to form their own union that operated, and negotiated, separately from the Farsi-speaking employee union. The hope of this policy was that it would decrease the Tudeh Party's influence among non-Iranian employees. The policy may have been effective to some extent, as strikes increasingly segmented along linguistic and native lines.

For Indian labourers, local networks intersected with international political movements and both informed workers' attempts to renegotiate their labour and living conditions. These mobilisations were not always supported by other groups working at Abadan. On 4 August 1947, Indians working there again tried to strike in the hope of improving their working conditions. Organisers of this strike were able to get between four and five hundred Indians to participate, but they were unable to bring Iranians or Arab workers into the strike. The result was that the Indians returned to work on 6 August without any changes in their labour conditions. As in the earlier July strikes, the suspected organisers of the strike were dismissed.[17] This second, smaller strike indicates the importance of solidarity among workers for their actions to be effective. The cohesiveness of this smaller strike also points to the power of a regional or national identity to mobilise workers from all over the Indian subcontinent.

Despite the dismissal of workers perceived to be organisers of labour agitations, strikes continued in the Abadan refinery and at other oil worksites in the Gulf, with varying impacts. In Iran, the government's response to these strikes was mixed from workers' perspectives. The first minimum wage law came into effect and day rate workers (or unskilled workers) were converted into a more stable form of employment.[18] However, the government also implemented marshal law and began the regulation of unions.[19] In 1946, not only did workers go on strike in Iran, but workers at Kuwait Oil in Kuwait, a subsidiary of AIOC and Gulf Oil (an American company), also carried out a series of strikes that were influenced by the Tudeh Party. These strikes utilised multi-ethnic, multi-linguistic solidarities and in many ways appear similar to strikes at AIOC in Iran. Also, as with the strikes in Iran, the Indian organisers of the Kuwait strikes were all from the same area of India and shared a linguistic background. Kuwait Oil, too, deported these workers before they could cause further disturbances.[20] British agents, such as Thomas, were troubled by the return of these workers to India and feared the consequences of 'Tudeh propaganda' in the subcontinent. While many British officials in the Gulf were concerned with maintaining Britain's access to oil, Thomas was also worried about the rise of communist sentiments among workers in India. He believed the communist ideas that spurred the strikes came to workers based in Iran and Kuwait via the Soviet Union, and he feared such ideas would spread throughout India. Thomas reported that many of the participants in the strikes were 'fervent admirers' of the Tudeh Party. He wrote repeatedly to warn his superiors that the workers would continue to disseminate information on communism in India.[21] The dangers the British saw in communist movements certainly grew from the British understanding that such movements were connected to USSR.

Communism, from Thomas's perspective, was dangerous because of geopolitical tensions among imperial powers. For workers, communism's strength derived from its ability to effectively mobilise workers through an inclusive logic of membership based on class identity. Such a movement could be contrasted with nationalist movements, which often limited participation on the basis of citizenship. While strikes along national lines limited the numbers of workers usually involved, following India's independence in 1947 Indian workers were able to appeal directly to the Indian government to protect their rights.

Skilled Workers in Bahrain, 1948

In the late 1940s, Indians increasingly moved to the Arabic-speaking Gulf for work, but their working and living conditions did not improve. Workers, from managers to labourers, ran into obstacles when they tried to change company policies. In 1948, the Bahrain Petroleum Company (BAPCO), which employed one thousand Indians,[22] fired eight Indians employed as chemists. According to the company, the chemists were dismissed from their position because they failed to follow orders. The chemists, on the other hand, disputed the company's reason for their dismissal and argued they had been dismissed after they had complained about the racism with which they had been treated by company managers. Seeing their dismissal as both a breach of contract and a violation of their rights, the chemists wrote to the Indian government and Indian newspapers. In their letters, the chemists argued that their dismissal violated their fundamental right of association because they believed the real reason they had been fired was that they had co-ordinated among themselves when they had filed their complaints with management. Because of this violation of the right to association, the chemists asked the Indian government to intercede on their behalf with the company.

The chemists insisted upon the Indian government's intervention on the basis of their citizenship, the premise of universal rights and the principle of the equality of Indians on a global stage. In their letters, the chemists also connected BAPCO's violation of their rights with the racism with which they said Indians were treated by the company's management. In one news article, BAPCO managers were accused of treating Indians at the company with 'racial hatred'. The author also compared these experiences with Indian experiences in South Africa – only with the caveat that Bahrain was much worse than South Africa.[23] In another news article, the chemists again argued that they experienced discrimination because of racism and bias against Indians. 'The Americans are careful not to over-step the boundary as far as the sentiments of the Arabs are concerned. But an Indian remains a slave of the Company, segregated, discriminated, and snubbed on account of his colour.'[24] Arguments such as these highlight that the chemists were not seeking to form class-based, inter-group solidarities. Rather, they saw their position, as Indian citizens, to be uniquely different from that of other employees of the company. In addition, the chemists, who were university-educated, were offended that they were so poorly treated in spite of their education, their expertise, and India's status as a newly-independent state.

The poor working conditions and the racism with which the chemists were treated were not out of the ordinary; a common American managerial tactic in both North America and the Gulf was to divide the labour force at oil projects on the basis of race, nation and ethnicity.[25] As a result, there were multiple common complaints by all non-white workers, and Arabic-speaking and Farsi-speaking workers, for example, also complained of living in deplorable conditions and being subjected to racist treatment. Despite this common complaint, the segregation of workers by nationality was often effective at preventing workers from forming the kind of political alliances that resulted in large-scale strikes. In this context, the chemists developed a new strategy: they asserted solidarity based on their profession and their citizenship in the new Indian nation. Differently from Indian participation in the strikes at Abadan, the workers in this case were appealing to a country outside of the country in which they were working. At Abadan, workers co-ordinated across ethnic, linguistic and religious divisions to organise large strikes and directed many of their complaints to the company. In contrast, the chemists appealed to the Indian government as citizens and justified the government's intervention through a discourse of rights.

The Indian chemists cited these rights in their written petitions to the Indian government, arguing for their 'fundamental Right of Association and of speech recognised even under the U.N.O. [United Nations Organization] charter', and the chemists used this argument to critique their dismissal by BAPCO. By complaining of human rights violations to the Indian government, the chemists pointed to the importance of a strong state to support citizens' rights and the obligations of the Indian government to its citizens. In order for these claims to be made upon the state, India's independence from colonial rule was essential. This was because, according to the UN, human rights were given to most peoples, but not to colonised subjects, who were understood to be protected by their colonial governments.[26] Rights were ensured through states recognising citizens, acting within their territories, and negotiating with other states and inter-state actors.[27] Thus, the chemists' demand for rights could only be successful when appealing to an independent state that also acted on behalf of its citizenry.

The chemists' appeal to the Indian government to protect their human rights mobilised a then-current political discourse that was being developed and propagated in the United Nations (UN). In the late 1940s, the same period as the United Nations was focused on ratifying the Declaration of Human Rights, such a strategy was not uncommon; for example, Pakistani workers in Saudi Arabia made similar claims.[28] For the chemists, appealing to the Indian government on the basis of human rights had two implications. First, it made the implicit argument that ensuring the recognition of such rights was one of the duties a liberal democratic state owed to its citizenry. Second, it banked on the notion that the Indian state would be particularly invested in protecting these rights given the country's own anti-colonial struggles. The Indian state, for example, had encouraged the International Labor Organization, even before India's independence, to develop a Declaration of Human Rights in 1944 as a method of critiquing colonial rule.[29] After India's independence from British colonial rule in 1947, India's leaders, including the country's first prime minister, Jawaharlal Nehru, saw the internationalism of the United Nations as holding promise for critiquing imperialism and forming international communities. Simultaneously, Indian leaders were also wary of the potential the United Nations had for reasserting imperial power.[30]

In response to the chemists' letters and the ensuing press coverage, the Indian government launched an investigation into the conditions of workers in the Gulf and, in particular, Bahrain.[31] One Indian official, based in Baghdad, visited Bahrain, Kuwait and the Trucial Coast in order to ascertain the position of Indians in the oilfields. In his letters to the Indian government in New Delhi, he indicated that workers took oil jobs because there were no jobs in India and were thus compelled to sign contracts that were unfavourable to them. In his opinion, however, the lack of jobs in India did not mitigate the conditions workers faced. Indian government officials understood the end of colonialism in India to be a strong factor in mobilising Indian citizens and motivating them to challenge poor treatment. The official wrote: 'As soon as they reach these places, they find themselves in an inferior position, and being discriminated against. This could be endured in the past, but the change over in India makes these young men feel humiliated.'[32] Here, the official indicated that India's new independence altered how workers understood their position internationally and that India's independence was central, as they made claims based on their rights.

Not only did independence inform the chemists' complaints, but class was also a factor, and other skilled workers at BAPCO appealed directly to the Indian government in the hopes of improving their working and living conditions. Some skilled workers, who were employed as clerks, reported that they had been terminated because of their participation in collective action. These workers claimed they had been fired because they, along with three hundred other Indian employees, had collectively written a petition requesting medical facilities near their camps, provisions that were guaranteed in their contract. In speaking with the Indian government, these workers hoped that the government would act to enforce their contract with BAPCO and contest their unjust dismissal. Other skilled workers complained to the Indian government that they were treated with racism. These complaints, like the complaints of the chemists and those seeking medical facilities, came from well-educated, professional employees.

Skilled workers made up one-third of the Indian workforce at BAPCO, and the other two-thirds of the Indian employees working at BAPCO were daily wage labourers. Unskilled workers at BAPCO were also unhappy with their working conditions, but this only came to light after the Indian government arranged interviews with twelve unskilled workers and enquired about their working conditions.[33] These unskilled workers' complaints included being injured on the job, not being paid a full salary upon termination and being terminated without due cause. While many of these workers filed complaints with the company, they did not attempt to contact the Indian government to negotiate on their behalf or enforce their contracts. Thus, while these unskilled employees were unhappy with their working conditions, they did not directly petition the Indian government for assistance. In the wake of independence, migrants with higher educations and income levels than the majority of Indians in the Gulf rhetorically situated themselves within a cosmopolitan and international workforce. Through situating themselves in such a workforce, skilled workers made claims upon the nation-state to ensure their rights. In making these claims, skilled migrants influenced the Indian government's emigration policies and the government's understanding of its obligation to citizens living outside of the state.

The form of labour mobilisation that relied upon international human rights discourse and the citizenship differed markedly from the strikes in Iran just a few years earlier. These differences may indicate a shifting sense of both political action and national obligation for workers. As industry practices began to shift, companies encouraged recruiters to hire workers from diverse areas of India. In addition, the Indian government increasingly began to negotiate with oil companies on behalf of workers. These changes indicate shifting tactics used by workers to protest oil company practices and the role of a newly independent Indian state in advocating for worker rights. While petitions to the Indian government were predominantly from skilled workers in the late 1940s, by the early 1950s, Indians from diverse backgrounds and with varying skill and education levels began to increasingly call upon the Indian government to mediate with oil companies concerning their working and living conditions.

Indian Hunger Strike at Aden Refinery, 1953

With the oil crisis in Iran from 1951 to 1953 and the related nationalisation of the refinery at Abadan, AOIC directed its refinery investments elsewhere and, in particular, to building a new refinery in Aden. In 1952, British Petroleum (BP)[34] hired over six hundred Indians to help build this new refinery. These workers included a couple of Indian doctors hired exclusively to attend to workers from Asia, as well as a few clerks and a large number of cooks.[35] At the Aden refinery construction project, workers were employed on a contract for eighteen months.[36] They worked most Sundays and reported putting in at least ten hours a day.[37] While the company claimed ten-hour work-days were stipulated in the contract, workers reported that these hours were too long and that they had not been informed of these hours before arriving in Aden.[38] Furthermore, workers were confused by the payment system for overtime hours. The company claimed that workers would be paid overtime upon completion of their contracts, but workers were unaware of this condition.[39]

Their camp was also twenty miles from the site of the refinery. As a result, workers commuted a long distance every day for work, and they were not

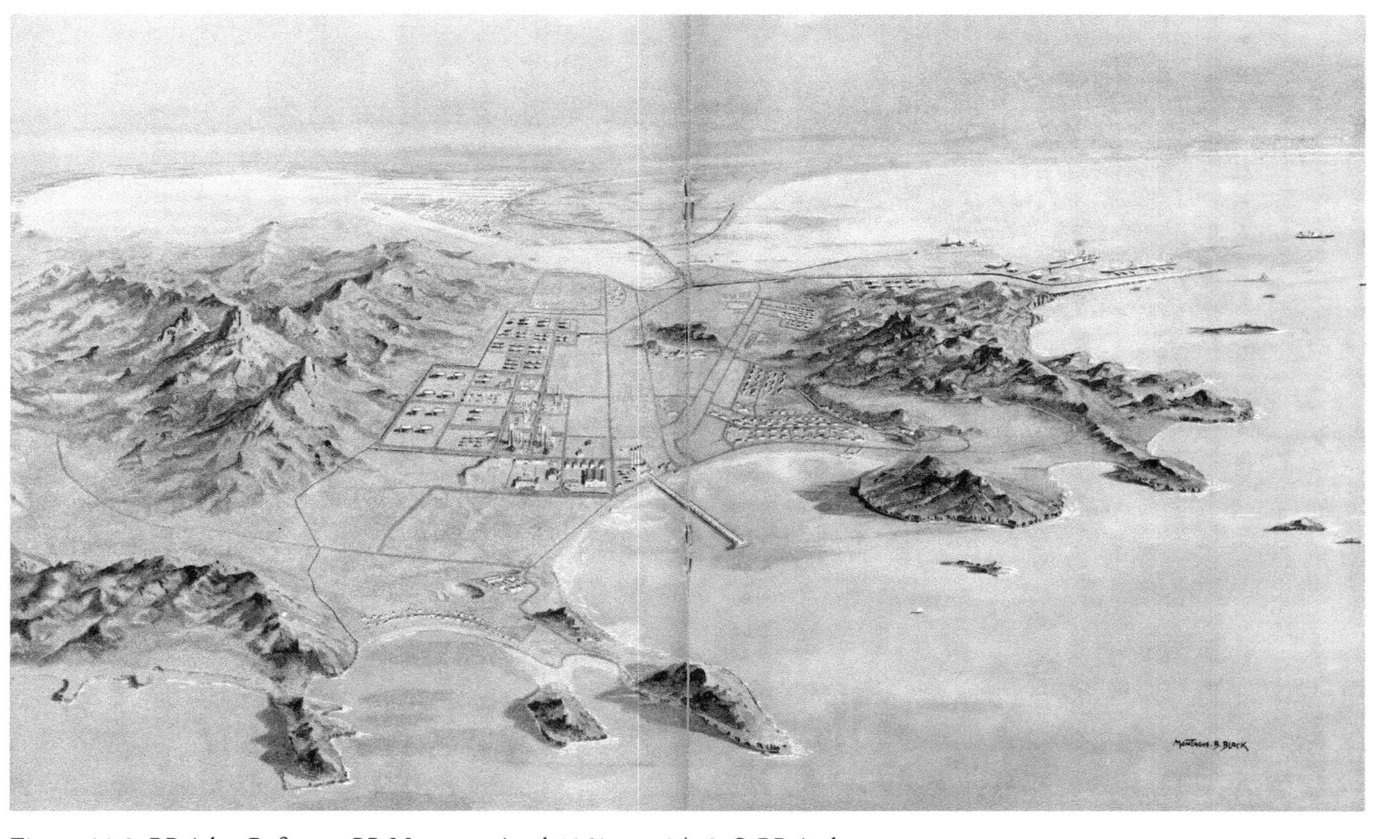

Figure 11.2 BP Aden Refinery. *BP Magazine*, April 1953, pp. 24–5. © BP Archive.

paid for this time. The isolation of the camp also impacted how workers could advocate for changing conditions. First, the distance of the camp from the city meant that it was hard for workers to contact the Indian government and ask the government to negotiate with the company on their behalf.[40] A second way the camp's isolation impacted workers was that it made it challenging for them to form solidarities with Arabic-speaking and Farsi-speaking workers through regular interactions and shared experiences. Indeed, the isolation helped foster discord between groups, as I discuss in greater detail below. Separate work camps were not new at this time, but camps were becoming increasingly strict in their segregation practices. In comparison, at Abadan in 1946, Indians also lived in 'enclaves', and the British understood Indians to be an intermediary group that operated as a 'buffer zone' between Iranians and themselves.[41] Scholars, however, show that these spatial separations, particularly in cities, exist far more strongly in coloniser imaginaries than in practice – they are largely discursively constructed within an Orientalist framework, and in practice are frequently transgressed as workers move through colonial spaces.[42] Such movements facilitated affiliations across national lines, whereas isolation reinforced separation between groups.

In 1953, 350 Indian employees at the Aden Refinery construction project went on a forty-eight-hour hunger strike.[43] Before beginning their hunger strike, the workers had conveyed their grievances to the Commissioner of the Government of India in Aden. In their communications, they told the Commissioner that they were working ten-hour days and not getting overtime. In addition, they reported that if an employee resigned, he was forced to pay for his own return passage. Paying for this passage was an economic hardship and had the effect of 'virtually keep[ing] them in bondage for the period of their contract'.[44] When the Commissioner failed to instigate changes on their behalf, Indian workers began their hunger strike and sent a telegram to the Indian ambassador in Egypt.[45]

During this strike workers asked for improvements to their housing and an end to the racial discrimination with which they were treated by managers.[46] Workers lived in makeshift tents of plywood, and these were extremely hot as a result of having no fans.[47] Not only were the tents hot, but they were also located near to the latrines, which smelled bad and were of poor quality. One visiting Indian official described them as 'insufferable'.[48] In addition to their miserable living quarters, workers were upset that the company charged them a significant portion of their earnings for food that was of poor quality, 'unpalatable to Indian taste', and did not allow for Indians to follow dietary restrictions.[49] Even a European personnel supervisor at the refinery agreed that the food Indians were served was not good. As a result of bad food, this supervisor observed, many workers lived on bread and jam and could not perform hard labour.[50] The shared experiences of migration and dealing with difficult working conditions and unresponsive employers thus served to bring Indian workers together and foster a sense of solidarity.

The poor quality of the food may have been one reason workers chose to hold a hunger strike, but this form of worker action also connects with India's independence. Building upon the popularity of trade unions in India after World War I, strikes in India were a preferred method of agitating against working conditions and promoting solidarity amongst workers. In 1951, there were 120 registered strikes and hundreds of additional strikes in the Indian bidi industry alone.[51] Hunger strikes, in contrast, were infrequent in India and uncommon in the Gulf. Facing restrictions on labour unions and formal strikes, the hunger strike at Aden gave workers a means to address their working conditions. This method of protest, popularised by a leader of India's independence movement, M. K. Gandhi, was especially poignant given the history of India's colonisation and the contemporary discrimination workers faced. By appealing to the Indian government, workers defined themselves as a community of Indians and, through their strike, situated their actions within the country's own nationalist movement.

Evoking India's struggle for independence and equality also drew attention to the outrage Indians felt when facing racial discrimination. Workers argued that discriminatory practices at the refinery meant that they were forced to work long hours, were segregated from white employees and were treated disrespectfully. One worker said that he often worked twelve or fourteen hours a day and was given no credit for overtime. He described spending all day in the hot sun and, upon asking for a transfer, being refused. When he quit, he said that the European worker who replaced him was given a 'fat' salary and assistants.[52] Americans and Europeans worked shorter hours for higher wages, and their bar was always open, whereas the Indian bar was open only part of the day. Such discrepancies in treatment were compounded by verbal abuse workers faced. Multiple workers wrote that the American manager in charge of the Indian employees' mess hall reportedly abused their nationality in the most 'insulting of terms'. This verbal abuse was compounded by the fact that Indians were barred from leisure spaces. The cinema, for example, was off-limits to Indian workers and designated 'European only' – something the Indian workers said was in breach of their contract – and Indians did not receive equal access to the cricket field.[53] Overall, Indian employees argued that the discrimination they faced was based on skin colour.[54]

In response to worker complaints of racial discrimination, managers explained that the policy of treating workers differently on the basis of their nationality was a result of global economic inequalities. One manager argued that the company was 'in no way responsible for the different standards of living and comparative wage scales throughout the world'.[55] Through such an argument, the managers claimed that the differences in treatment and pay of employees were based on the nationality of the employee, even as the management simultaneously refused to deal with employees through national employee unions. This argument was common in the oil industry in the Gulf. One oil company claimed that in Saudi Arabia, 'racial discrimination is unknown in the Company's operation', and that all employees 'recruited from countries located east of the Atlantic' were treated equally.[56] Such claims by managers used economic models to naturalise differences between American management and workers hired from the Middle East or India. The international oil management practices applied in Bahrain and Aden, as well as in other parts of the Gulf, such as Saudi Arabia, used racial discrimination as a method both to control the workforce and then to excuse such practices, by pointing to global inequalities that just existed naturally. This reasoning obscured how managers used segregation as a technique to minimise worker coalitions among nationalities and, therefore, reduce both the number of strikes and their impact on oil production. Such practices were informed by Jim Crow laws in the American South that segregated Black Americans. As Robert Vitalis demonstrates in his history of the USA's involvement with oil production in Saudi Arabia, key methods developed by American oil companies to curtail labour strikes included the racial segregation of workers and discrepancies in pay based on race.[57] The use of racial segregation as a management technique was not unique to the oil industry in the Middle East, and racial labour hierarchies were also mobilised by managers in, for example, the oil industry in Mexico and factories in India.[58]

Despite experiencing inequalities, Indians were not above marshalling discriminatory discourse in their own favour, when necessary. Indian workers complained about being forced to live with Arabs,[59] whom the Indians described as having a 'lower standard of living' and not having 'clean habits'.[60] Indeed, it was deemed particularly outrageous that they, Indian workers, were not even treated as well as the Arab employees. They complained of being paid less than workers coming from the Levant and the same amount as local workers with fewer skills and less experience. One Indian highlighted the situation by pointing out that the Adenese 'orderly who serves the office with tea and has little or no qualifications to be called literate draws as much pay as us ... the Indian is getting a very raw deal'.[61] Through these arguments of inequality, Indian workers defined themselves as a discrete community by emphasising their differences from European and Arabic-speaking employees.

The national solidarity that workers exhibited at Aden made the strike uniquely different from other attempts by Indian workers to strike in the late 1940s and early 1950s Gulf. Earlier strikes were often ineffective not only because they did not include workers of other nationalities but also due to lack of unity among the Indian employees. In these earlier strikes, the gap between Indian employees was most pronounced between senior and junior employees, at least according to one Indian government observer, who argued that senior workers were loath to join collective action. Government officials attributed this lack of unity to a generational difference – namely, that senior workers were more submissive, whereas junior workers felt more strongly about their rights and were more emboldened by the new independence of India.[62] In addition, the demographics of the Aden strike were different from those applying to the strikes at Abadan or the complaints of the chemists in Bahrain. Notably, at Aden striking workers were all Indians and did not form alliances with workers from other countries, as was the case in Abadan and Bahrain. Distinct from the 1946 strike at Abadan in which skilled Indian workers did not participate, or the filing of complaints by skilled workers at BAPCO that unskilled workers did not participate, the hunger strike at Aden mobilised workers at all levels, from day labourers to skilled employees. The fact that it was not just skilled, educated Indians making claims on the Indian government based on citizenship, but also workers in unskilled positions and with less formal education, suggests that an expanding notion of citizenship was forming and gaining popularity among workers abroad. This broadening definition of citizenship cut across regional differences existing within India.

Through appealing to national identity, the striking Indian employees drew the attention of both the company and Indian government officials, with mixed results. After the strike, there were still no fans in workers' sleeping areas; cold water became available to Indians; the food quality continued to deteriorate, but beef was served less often; the number of hours worked remained the same; and workers eventually moved into a new camp.[63] The Indian government's intervention also did not meet workers' expectations. While workers hoped the government would exert some pressure on the oil company to improve their conditions, this was not enacted. At the onset of the strike, the Indian government briefly stopped all Indians from emigrating to work at the Aden refinery project, but this halt lasted for only a brief period. When emigration reopened, the Indian workers already in Aden became fearful they would be fired.[64] This was because the company used a 'spy system' to control workers, and many workers feared this system would be used to uncover the leaders of the strike.[65] With the reopening of emigration, workers knew that they could be fired and easily replaced by other Indians hoping to work in the Gulf.

Conclusion: Citizenship, Worker Rights and Oil Companies

Strikes, protests and complaints lodged by workers demonstrate the importance of the oil complex as a site where international carbon managerial practices and localised politics met.[66] In the Middle East, the oil industry has relied, and continues to rely, on labourers from all over the world, and many of these workers are from former colonies. At the conjunctures of intensifying oil production and the emergence of postcolonial states, workers developed and negotiated their rights within the contexts of class and citizenship. In these moments, they developed solidarities that invoked scales of the local, national and transnational. Looking at worker action before and after India's independence illustrates how worker action shifted from class-based to state-based solidarities. In the wake of India's independence, Indian workers felt they were due dignity, and this belief served to bolster their claims and foster alliances between Indians. In this context, workers, when they attempted to improve their working and living conditions, selectively invoked the postcolonial nation-state.

Nationalism did not exist within a vacuum, and the increasing formation of worker solidarities by nationality was further reinforced by oil companies' managerial practices. These practices segregated workers by nationality and housed workers in remote areas. They also gave different nationalities different benefits and pay, thereby spurring conflict between nationalities and decreasing the likelihood of workers forming class-based solidarities that had the potential to halt oil production. By the mid-twentieth century, materials, ideas and people were all moving through circuits that cross-cut empires and nations, solidifying the salience of those political entities in some instances, challenging them in others. This examination of political claims points to how nationalism was constructed outside of the nation and challenges the boundaries of organisations, nations and empires.

Notes

1. Ian J. Seccombe and Richard I. Lawless, 'Foreign Worker Dependence in the Gulf, and the International Oil Companies: 1910–50', *International Migration Review* 20, no. 3 (August 1986); Hassan Mohammad Abdulla Saleh, 'Labor, Nationalism and Imperialism in Eastern Arabia: Britain, the Shaikhs and the Gulf Oil Workers in Bahrain, Kuwait and Qatar, 1932–1956' (Ph.D. diss., University of Michigan, 1991).
2. Thomas A. B. Corley, *A History of the Burmah Oil Company, 1886–1924* (London: Heinemann, 1983); Andrea Wright, 'From Slaves to Contract Workers: Genealogies of Consent and Security in Indian Labor Migration', *Journal of World History* 31, no. 2 (2020); Andrea Wright, 'Imperial Labor: Strikes, Security, and the Depoliticization of Oil Production', in Neilesh Bose (ed.), *South Asian Migrations: A Global History: Labor, Law, and Wayward Lives* (New York: Bloomsbury, 2020).
3. Touraj Atabaki, 'Indian Migrant Workers in the Iranian Oil Industry, 1908–1951', in Touraj Atabaki, Elisbetta Bini and Kaveh Ehsani (eds), *Working for Oil: Comparative Histories of Labor in the Global Oil Industry* (London: Palgrave Macmillan, 2018), 208.
4. After Bahrainis, British Indians were the largest nationality working at BAPCO, and the 352 British Indians at BAPCO made up 13 per cent of BAPCO's

workforce and outnumbered all Americans, Canadians and British workers combined. Jane Kinninmont, 'Bahrain', in Christopher Davidson (ed.), *Power and Politics in the Persian Gulf Monarchies* (New York: Columbia University Press, 2011), 35; 'Annual Report of the Bahrain Petroleum Co. Ltd 1940', National Archives of India, External Affairs, Near East Branch, 1941. F 360-N/41 (hereafter NAI).

5. Kaveh Ehsani, 'Disappearing the Workers: How Labor in the Oil Complex Has Been Made Invisible', in Atabaki, Bini and Ehsani (eds), *Working for Oil*, 18, 25.
6. Geoff Eley, 'Review of *Transnational Labour History: Explorations* by Marcel van der Linden', *Labor: Studies in Working-Class History of the Americas* 3, no. 1 (2006), 164–6; Geoff Eley, *A Crooked Line: From Cultural History to the History of Society* (Ann Arbor: University of Michigan Press, 2005).
7. In 1910, the Anglo-Persian Oil Company, renamed AIOC when Persia changed its name to Iran, began recruiting large numbers of Indians to work in Persia. By 1946, AIOC employed the most Indians in the Gulf and approximately 2,560 Indians worked for the company. Seccombe and Lawless, 'Foreign Worker Dependence in the Gulf', 563.
8. Farmanfarmaian was influential in the development of Iran's oil industry. He worked in the military, then in 1943–9 was in the Ministry of Finance; in 1949–58 was Director General of Petroleum, Concessions and Mines; in 1958 was Director of Sales for National Iranian Oil Company; and later was Iran's first ambassador to Venezuela.
9. Manucher Farmanfarmaian and Roxane Farmanfarmaian, *Blood and Oil: Inside the Shah's Iran* (New York: Modern Library, 1999), 184–5.
10. BP ARC118823: 'Lessons of 1946: An Essay on the Personnel Problems of the Oil Industry in South Iran' by Donald MacNeill, 1949. In addition, a British parliamentary delegation characterised the living conditions of labourers as '"a penal settlement in the desert" with accommodation "little better than pig-styes [*sic*]".' Timothy Mitchell, *Carbon Democracy: Political Power in the Age of Oil* (London: Verso, 2011), 107.
11. BP ARC14706: 'The Kuwait Report, 1949'.
12. Mitchell, *Carbon Democracy*, 107; Habib Ladjevardi, *Labor Unions and Autocracy in Iran* (Syracuse, NY: Syracuse University Press, 1985).
13. Ervand Abrahamian, *Iran Between Two Revolutions* (Princeton: Princeton University Press, 1982).
14. NAI: MEA, Middle East Branch [ME], 1946. 10-(91)-ME/46. Thomas' letters to General Headquarters, 10 July 1946.

15. NAI: MEA, ME, 1946. 10-(91)-ME/46. Note by Sd. H. C. Beaumont, 5 September 1946.
16. NAI: MEA, ME, 1946. 10-(91)-ME/46. Thomas's letters.
17. NAI, MEA, ME, 1946. 10-(91)-ME/46. Thomas to General Headquarters, 24 August 1946.
18. Farmanfarmaian and Farmanfarmaian, *Blood and Oil*, 186.
19. Mitchell, *Carbon Democracy*, 107.
20. NAI: MEA, ME, 1946. 10-(91)-ME/46. Thomas letter to General Headquarters, 31August 1946.
21. NAI: MEA, ME, 1946. 10-(91)-ME/46. Thomas to General Staff Branch, 10 July 1946.
22. This is four times more than the 352 Indians working at the BAPCO in 1940. NAI: MEA, Near East Branch, 1941. 360-N/41. 'Annual Report of the Bahrain Petroleum Co. Ltd. for the year 1940'.
23. Indians in South Africa faced racism, discrimination and exclusion from social spaces. Parvathi Raman, 'Being Indian the South African Way', in Annie Coombes (ed.), *Rethinking Settler Colonialism: History and Memory in Australia, Canada, Aotearoa New Zealand and South Africa* (Manchester: Manchester University Press, 2006), 193–208.
24. NAI, MEA, Emigration. F. 22-8/48-Emi. 'Skilled workers engaged by the Bahrein [*sic*] Petroleum Co., Bahrein'.
25. For example, Vitalis outlines in detail the work done to maintain colour lines at the oil projects in Dhahran, Saudi Arabia. See Robert Vitalis, *America's Kingdom: Mythmaking on the Saudi Oil Frontier* (New York: Verso, 2009), 98–105.
26. Lydia H. Liu, 'Shadows of Universalism: The Untold Story of Human Rights around 1948', *Critical Inquiry* 40 (Summer 2014).
27. Hannah Arendt, *The Origins of Totalitarianism* (New York: Meridian, 1958).
28. This problem was also being discussed in the Pakistani press in the 1940s and early 1950s. See Vitalis, *America's Kingdom*, 103.
29. Daniel R. Maul, 'The International Labour Organization and the Globalization of Human Rights, 1944–1970', in Stefan-Ludwig Hoffman (ed.), *Human Rights in the Twentieth Century* (New York: Cambridge University Press, 2011), 301–20.
30. In 1955, Nehru and Indonesia's president, Sukarno, organised the Bandung Conference, a meeting of African and Asian states. This meeting is seen as a key event in the formation of the Non-Aligned Movement in 1961. This movement created an agreement between former colonies that respected each state's

sovereignty and attempted not to take sides during the Cold War. Member states needed to agree to the principles which were established at the Bandung Conference, including 'respect for fundamental human rights and for the purposes and principles of the Charter of the United Nations'. 'History and Evolution of Non-Aligned Movement' (New Delhi: Government of India, 22 August 2012); Manu Bhagavan, 'A New Hope: India, the United Nations and the Making of the Universal Declaration of Human Rights', *Modern Asian Studies* 44, no. 2 (2010), 311–47; Vijay Prashad, *The Darker Nations: A People's History of the Third World* (New York: The New Press, 2007).

31. The Indian ambassador to Iran argued that the condition of workers in Bahrain was much worse than in Iran. NAI: MEA, Emigration, 1948. 22–8/48. Extract from note on 'Bahrain'. Original in AWT. Branch F. No. 3(9)-AWT/50.
32. NAI: MEA, Emigration, 1948. F 22-8/48-Emi. 'Skilled workers engaged by the Bahrain Petroleum Company'.
33. Worker complaints against BAPCO continued into the 1950s. In 1953 in Bahrain, the main problem was housing. In particular, landlords had a lot of power over workers because housing was not supplied by the company and Indians could not buy land in Bahrain. In addition, foreign labourers had little job security. Indian labourers were hired on a daily contract, and this did not provide them with the same stability in income or job security as monthly wage employment. NAI: MEA, Emigration, 1953. F6-6/53-Emi. 'Oil Companies in the Persian-Gulf and Mid-East'.
34. BP was a subsidiary of AIOC. In 1954, AIOC renamed itself British Petroleum (BP).
35. In addition, the company was hiring 8,000 local labourers, and it was thought by Indian government representatives that only a few Indians would be kept at the refinery once construction was finished. NAI: MEA, Emigration. 6637/52-Emi. Express Letter to All Passport Issuing Authorities in India, 24 November 1952. NAI: MEA, Emigration, 1953. F.23-9/52-Emi. 'Aden – Recruitment of 600 skilled workers from India'.
36. NAI: MEA, Emigration, 1953. F.23-9/52-Emi. 'Aden – Recruitment of 600 skilled workers from India'.
37. NAI: MEA, Emigration, 1953. F.23-9/52-Emi. Letter to Ghatge from Thadani, 19 March 1953.
38. NAI: MEA, Emigration, 1953. F.23-9/52-Emi. 'Aden – Recruitment of 600 skilled workers from India'.

39. NAI: MEA, Emigration, 1953. F.23-9/52-Emi. Letter to Thadani from Ghatge, 26 March 1953 and letter to Ghatge, 26 March 1953.
40. NAI: MEA, Emigration, 1953. F.23-9/52-Emi. 'Aden – Recruitment of 600 skilled workers from India'.
41. Atabaki, 'Indian Migrant Workers', 191, 204.
42. William Glover, *Making Lahore Modern: Constructing and Imagining a Colonial City* (Minneapolis: University of Minnesota Press, 2008).
43. NAI: MEA, Emigration, 1953. F.23-9/52-Emi. Telegram to Embassy of India, Cairo.
44. NAI: MEA, Emigration. 11(5)CA/53. Letter to the Secretary, MEA, from Commissioner, 22 May 1953.
45. NAI: MEA, Emigration, 1953. F.23-9/52-Emi. Telegram to Embassy of India, Cairo.
46. NAI: MEA, Emigration, 1953. G/16307. Letter to POE, Bombay, from BP, 10 June 1953.
47. NAI: MEA, Emigration, 1953. 2954/53-Emi. Letter to the Secretary, MEA, from Thadani, 10 June 1953.
48. NAI: MEA, Emigration, 1953. F.23-9/52-Emi. Extract from Monthly Report No. 5 of 1953. NAI: MEA, Emigration, 1953. 2954/53-Emi. Letter to the Secretary, MEA, from Thadani, 10 June 1953.
49. NAI: MEA, Emigration, 1953. F.23-9/52-Emi. Aden – Recruitment of 600 skilled workers from India.
50. NAI: MEA, Emigration, 1953. F.23-9/52-Emi. Extract from Monthly Report No. 5 of 1953.
51. Rina Agarwala, 'From Work to Welfare: A New Class Movement in India', *Critical Asian Studies* 38, no. 4 (December 2006), 430; Dipesh Chakrabarty, 'Conditions for Knowledge of Working-Class Conditions: Employers, Government and the Jute Workers of Calcutta, 1890–1940', in Ranajit Guha (ed.), *Selected Subaltern Studies* (New York: Oxford University Press, 1988), 179–230.
52. NAI, MEA, Emigration, 1953. F.23-9/52-Emi. Extract from Monthly Report No. 5 of 1953; Letter to the Trade Commissioner, Aden, from Bechtel and Wimpey, 20 May 1953.
53. NAI: MEA, Emigration, 1953. 2954/53-Emi: Letter to the Secretary, MEA, from Thadani, 10 June 1953. NAI: MEA, Emigration, 1953. F.23-9/52-Emi. Letter to Indian Trade Commission from Indian Employees Committee, 4 July 1953. NAI: MEA, Emigration, 1953. G/16307 Letter to POE, Bombay, from BP, 10 June 1953.

54. NAI: MEA, Emigration, 1953. 2954/53-Emi. Letter to the Trade Commissioner, Aden, from Bechtel and Wimpey, 20 May 1953.
55. NAI: MEA, Emigration, 1953. F.6-6/53-Emi. Report Dated 30 July 1953 from Messrs. Middle East Bechtel Corporation and George Wimpey & Co., Ltd., Aden.
56. NAI: MEA, Emigration, 1953. F.6-6/53-Emi Letter to POE, Bombay from E. E. Evans, Recruiting Agent, Arabian American Oil Co, 30 March 1954.
57. Vitalis, *America's Kingdom.*
58. Ranajit DasGupta, *Labour and Working Class in Eastern India* (Calcutta: K. P. Bagchi & Co., 1940); Myrna I. Santiago, *The Ecology of Oil: Environment, Labor, and the Mexican Revolution, 1900–1938* (Cambridge: Cambridge University Press, 2006).
59. NAI: MEA, Emigration, 1953. G/16307. Letter to POE, Bombay, from BP, 10 June 1953.
60. NAI: MEA, Emigration, 1953. 2954/53-Emi. Extract from Monthly Report No. 4 of 1953.
61. NAI: MEA, Emigration, 1953. F.23-9/52-Emi. Letter to Indian Trade Commission from Indian Employees Committee, 4 July 1953.
62. NAI: MEA, Emigration. F. 22-8/48-Emi. 'Skilled workers engaged by the Bahrein [*sic*] Petroleum Co., Bahrein'.
63. NAI: MEA, Emigration, 1953. 3496/53-Emi. Letter to Das Gupta from Ghatge, 1 August 1953. NAI: MEA, Emigration, 1953. F.6-6/53-Emi Report Dated 30 July 1953 from Messrs. Middle East Bechtel Corporation and George Wimpey & Co., Ltd., Aden.
64. NAI: MEA, Emigration, 1953. 27321/53-Emi. Telegram to Foreign Office, New Delhi, 8 July 1953.
65. NAI: MEA, Emigration, 1953. 2954/53-Emi.Extract from Monthly Report No. 4 of 1953. NAI: MEA, Emigration, 1953. 2779/53-Emi. Letter to Secretary, MEA, from Commissioner, 27 June 1953.
66. Michael Watts, 'Righteous Oil? Human Rights, the Oil Complex and Corporate Social Responsibility', *Annual Review of Environment and Resources*, 30 (2005), 373–407.

12

GENDER, DOMESTICITY AND SPEED: AMERICAN PETRO-MODERNITY

Nathan J. Citino

American engineer Larry Barnes was working on a Wednesday at Abqaiq, one of the main Arabian American Oil Company (Aramco) installations in Saudi Arabia, when he heard about 'some kind of trouble' taking place seventy kilometres away in Dhahran. Barnes listened on the phone as an Aramco secretary 'shriek[ed], "They are burning Dhahran! They are setting cars and houses on fire!"' She was referring to Aramco's Arab employees, many of whom were protesting US support for Israel. This was no ordinary Wednesday, but 7 June 1967, the third day of what some would later call the 'Six Day War' and a day that came to be known as 'Rock Wednesday' among the thousands of Aramco employees and their families who lived in the Saudi Eastern Province.[1] According to the Central Intelligence Agency (CIA), 'Aramco was hit by a mob which broke in the gates, looting and overturning cars', forcing the company to shut down its refinery and loading facilities. Demonstrators then surrounded the US consulate before advancing on the family quarters serving the US Military Training Mission at the Dhahran Airbase. 'Despite the fact that women and children were inside', the CIA reported, 'homes were entered and almost everything inside destroyed.'[2]

Barnes immediately became concerned about his own family. Like other Americans, Barnes, his wife Marion and their three children resided in Dhahran's suburban Senior Staff Community, once an all-white company town that Aramco was slowly integrating as it promoted Saudi Arabs in response to government demands and previous strikes. According to his written account, Barnes thought in that moment of a Belgian friend's terrifying experience in the Congo, which had witnessed its own 'native uprising'. Ominously, this Belgian claimed that relations between locals and westerners there had seemed at least as friendly as they were in Saudi Arabia. Barnes telephoned Marion, who was 'semi-hysterical'. He could 'hear glass shattering as rocks came through the windows'. Barnes ordered his wife to load his .38 revolver and lock herself in the bedroom closet with their three-year-old daughter. Meanwhile, Barnes had his American foreman gas up a new 'V8 Chevy sedan'. Barnes put his 'foot flat down on the gas pedal' and raced at top speed to Dhahran. 'I don't know what a V8 Chevy will do flat out', he wrote, 'but I do know that I have the Abqaiq-to-Dhahran speed record.' Although Dhahran was not burning, the secretary had accurately reported that Arab demonstrators were targeting the residences and vehicles that symbolised Aramco's presence in Saudi Arabia. They smashed windows and protested in front of Americans' homes, including that of Aramco president Tom Barger. As Barnes' American neighbour whisked Marion and her daughter to safety, an Arab youth tried to set the neighbour's car 'afire by dropping lighted matches into the fuel tank'. An 'old Bedu gardener decided to even up some old scores', Barnes wrote, by attacking the windshields of passing cars with a piece of wood. Barnes' Saudi Arabian neighbour was absent that Wednesday and his next-door house was left untouched, raising Barnes' suspicions about Saudi Arabs' loyalty to Aramco. The next day, the Barneses and other American families left Saudi Arabia on an emergency flight home chartered by Aramco and the US government. 'What would you take with you in a situation like this?', Barnes asked rhetorically. He and Marion took 'our silver flatware, some vases from Hong Kong' and 'Marion's crystal Christmas trees'. At the last minute, they rescued their Dachshund, Penny. These were the trappings of the domestic consumer lifestyle that the Barneses and many other Americans cultivated while residing in one of the global oil industry's most productive sites.[3]

As the consortium of American companies that developed Saudi oil resources, Aramco went to great lengths to evoke an exceptional US–Saudi partnership that was successfully modernising the kingdom. This chapter argues that Aramco's presence in fact constituted a neo-colonial order whose modernising rationale the company promoted in response to anti-colonialism and resource nationalism.[4] It illustrates the central role that gender played in Aramco's campaign and describes the social hierarchies and distinctions between public and private that characterised the US-led order. More than that, it transcends the dichotomy in US history between 'domestic' and 'foreign' by analysing American suburbanisation and the growing overseas presence in Saudi Arabia as parts of the global phenomenon known as 'petro-modernity'.[5] Petro-modernity fostered hierarchies between and within societies in the mid-twentieth century, a decolonising era when Western corporations and governments built neo-colonial economic relations with oil-producing states that enjoyed formal political independence.[6] It was characterised not only by technologies of speed that relied on hydrocarbons, such as the automobile and the airplane, but also by a gendered domesticity focused on consumerism. This complementarity between speed and domesticity could be observed in the car-centred suburbs of post-war America and in the oil camps of the Saudi Eastern Province, which became a centre of global hydrocarbon production as well as home to new communities built for oil workers and their families. Similar to other neo-colonial narratives, Barnes' account establishes a hierarchical relationship with the 'natives' while also mapping distinct gender roles onto public and private spheres.[7] But its emphasis on masculine dominance over speed, as the power to both enforce separate spheres and rapidly overcome distances (such as between Abqaiq's service centre and Dhahran's Senior Staff Community), distinguishes it as an artifact of *petro*-modernity. Aramco and Arabic sources can thus help us to appreciate how US relations with Saudi Arabia involved a reordering of societies and the remapping of space comparable to other neo-colonial settings.[8] Racial segregation of residential communities, in Saudi Arabia and the post-war American suburbs, constituted one example of petro-modernity's spatial dimensions. So did the construction of oil-transit infrastructures, such as the Trans-Arabian Pipeline (Tapline) built by an Aramco spin-off company. Tapline connected the Abqaiq oil fields to Sidon in Lebanon via a route that planners chose in order to circumvent Palestine and the Arab–Israeli conflict.[9] Ultimately, petro-modernity functioned as a distinctly mid-twentieth-century neo-colonial order that transformed landscapes along with social relations in both the USA and Saudi Arabia. As Barnes' account illustrates, petro-modernity was inherently unstable and therefore unsustainable, because subordinated, feminised subjects refused to respect its boundaries. Such transgressions occurred during the series of strikes against Aramco on the part of Saudi and other workers in 1945, 1953 and 1956, years prior to 'Rock Wednesday'. This was likewise the case when both American and Saudi women insisted on the right to drive automobiles, and when Arab oil states' embargo and price hikes threatened post-war American consumerism.

The American Studies scholar Stephanie LeMenager defines petro-modernity as 'modern life based in the cheap energy systems made possible by oil'. Twentieth-century modernity found expression in the built environment: 'the now ordinary U.S. landscape of highways, low-density suburbs, strip malls, fast food and gasoline service islands, and shopping centres ringed by parking lots or parking towers'. This 'auto-highway-sprawl complex' was a 'utopian project', she writes, created from what modernist architect Le Corbusier 'called the "rapture of power … and speed"'.[10] Historians such as Lizabeth Cohen have recognised the purportedly classless, but highly gendered characteristics of this 'landscape of mass consumption', where shopping centres were 'feminized public space'.[11] As Elaine Tyler May has argued, post-war gender roles were not a reversion to 'traditional' norms but emerged among white suburban families as responses to post-war affluence and cold-war fears.[12] Predicated on automobility, the decentralised communities described by LeMenager were the settings where many Americans formed ideas and practices around gender in the twentieth century.

But a focus on gender also provides a useful way of relating these post-war changes in the US 'domestic' sphere (in both senses) to an analysis of global American power. Many studies have analysed official US–Saudi relations, but petro-modernity depended on power relationships enforced at global and local levels, encompassing diplomacy, public investments and private contracts.[13] At the global level, petro-modernity included the appropriation of Middle Eastern oil to fuel European reconstruction; new corporate practices of net-back pricing that pegged the price of cheaply extracted Persian Gulf oil to that produced at greater expense along the US Gulf coast; a global network of military bases including the Dhahran Airbase; and state subsidies, in the forms of Export–Import Bank loans to finance infrastructure projects in Saudi Arabia and tax breaks granted to Aramco in the USA.[14] Aramco vice-president James Terry Duce combined the global and domestic aspects of petro-modernity when he lobbied Congress in 1948 for the scarce steel needed to construct Tapline. With abundant Middle Eastern oil, he claimed, the Marshall Plan could fuel Western European reconstruction, while more people in the USA could 'put oil burners in their homes' and 'the more general use can be made of petroleum products'. Without oil from the Middle East, warned Defense Secretary James Forrestal, Americans would 'have to return to a four-cylinder motorcar rather than a six or eight, within a period of about 10 years'.[15]

In Saudi Arabia, petro-modernity fostered power disparities at the micro level between segregated communities of Americans and non-Americans and within individual households, in gender and family relations. Anti-colonial criticism of Aramco was later published in Arabic by 'Abdullah al-Tariqi, the former Aramco employee who became Saudi oil minister and a member of Aramco's board. Saudi Arabia's subordinate relationship to Aramco and the USA was reflected in al-Tariqi's personal experiences as a student and Aramco employee. He had faced racial discrimination while at the University of Texas and then harassment when he and his American wife sought to integrate the Senior Staff Community in Dhahran. As *Time* magazine reported, 'American matrons took his wife aside and reproved her for marrying an Arab'.[16] Tariqi would go on to describe Tapline and Aramco as instruments of American power. He characterised oil companies as a 'means of neo-colonialism [*wasila al-isti'mar al-jadida*]' in the era of decolonisation.[17]

The type of neo-colonial order imposed by the US presence in Saudi Arabia may have been distinct, but it was not exceptional. In *Empire's Tracks*, Manu Karuka explains how the transcontinental railroad, built across North America by the 'war-finance nexus' of the US Army and private corporations, had destroyed Native American women's gendered 'modes of relationship'.[18] Europeans practised railroad imperialism and built the Suez Canal in Egypt at the same time as they sought to reform Egyptians' gender roles. 'Modern technology is at the heart of most projects for reforming "the developing world"', writes On Barak. For British empire-builders in the nineteenth century, this meant 'the development of the steam engine, making trains, steamers, and trams into emblematic driving forces of liberal modernity'.[19] In a similar vein, William Eddy, the US intelligence officer and diplomat who worked for Aramco, described flying to Hofuf and driving thirty miles into the desert to inspect the construction of the Dammam–Riyadh railroad. Beyond the Americans' air-conditioned camp, he wrote, 'Saudis were spiking down the rails and laying rails at the rate of 4,000 feet a day'. Eddy's wife Mary received a 'special invitation' from the Eastern Province governor Saʿud bin Jiluwi to visit Hofuf: 'by far the most Arab city I have ever seen', she wrote, 'and the most beautiful'. She noted the presence of 'women clothed in black' who 'hid their faces' from the Americans. To complement this Orientalist imagery, Mrs Eddy wrote that the Emir's palace had 'installed modern bathrooms, electric fans, [and] telephones'. She 'even saw a vacuum cleaner in the corner of the bedroom' and, on one of the Emir's 'heavy plush chairs', a 'tag which said "Sears Roebuck"'.[20]

Accounts of the 'Age of Speed' during the era of European imperialism in the Middle East explain how new technologies met existing social divisions. Yoav Di-Capua describes aviation as a project by which Egypt's upper class promoted its authority over the *effendiyya*, or middle class. According to Di-Capua, aviation reflected the class divisions that emerged following Egypt's 1919 revolution and subsequent pseudo-independence under the British: 'the upper class purchased and piloted the airplanes, and the middle-class *effendiyya* contributed to literary magazines and stormed the police barriers to congratulate the landing heroes.'[21] Kristin Monroe similarly argues that elite women's access to automobility in France's mandate of Lebanon transcended the gender restrictions that characterised middle-class respectability. 'Behind the wheel', she writes, 'upper-class women not only became more visible, they asserted agency and an authority over technology and public space that defined gender norms, in Lebanon and beyond, of the time.'[22]

In other words, the quintessentially modern experience of speed was not universal but depended on one's social position. As Enda Duffy argues, speed 'should be thought of as political'. It was first of all 'a politics of access: this newly intense experience was offered to citizens based on their ability to pay, on their gender, proximity to centres of production, consumption, and power'.[23] These restrictions applied to women who demanded the privilege of driving in the USA. If men experienced the car as 'an assertion of masculine identity, predicated on power, control, and freedom', writes Katherine J.

Parkin, then women experienced automobility through a patriarchal culture that stereotyped them as 'terrible drivers, harping passengers, and naïve mechanics'.[24] Americans scrutinised women's 'physiognomy' to assess their capacity to drive, assumed that women drivers required 'gender-specific instruction', and agonised over the dangers, especially the sexual peril, awaiting women who used the car to escape domestic spaces and the surveillance of their families.[25] A law judge memorably told sociologists Robert and Helen Lynd for *Middletown*, their study of 1920s Muncie, Indiana, that the auto served as a '"house of prostitution on wheels"'.[26] Although car advertisements promised women freedom, many married women in post-World War II America seldom drove 'except to shop' or 'chauffeur their children', wrote feminist Betty Friedan.[27] This history of automobility and gender in the USA shares striking similarities with later opposition to women driving in Saudi Arabia. Opponents cited supposed '"damage to women's ovaries"' and '"loss of shyness"', or raised objections such as 'the need for establishing special women's lanes on roads' and 'special training for traffic police to deal with women'.[28] Petro-modernity promised freedom through speed, but that opportunity had never been accessible to US women on an equal basis with men.

Segregated communities and new gender relations were features of petro-modernity in both the USA and Saudi Arabia. Many historians have studied the racial segregation of US suburbs through redlining, racial covenants and violence. The interstate highway system exacerbated racial inequalities by cleaving African-American and Hispanic neighbourhoods and by maintaining white commuters' physical and economic distance from urban spaces.[29] Toby Craig Jones notes the marginalisation of established Eastern Province towns in Saudi Arabia such as al-Hasa and Qatif, 'that sat on the edges of the new oil metropolis' of Dhahran, Dammam, Ras Tanura and Abqaiq. Aramco facilitated social distance between Arabs and Americans by providing buses that ferried Arab workers between their homes in towns and villages and the company's work sites (Figure 12.1). The older, largely Shi'i communities experienced impoverishment and under-development, especially of roads, water infrastructure and housing, along what Jones calls the 'Black Gold Coast'. This phrase comes from the title of Muhammad Said al-Muslim's history *Sahil al-dhahab al-aswad*, which offers a history of al-Hasa before it became incorporated into Saudi Arabia's 'Eastern Province' and descriptions of the '*mudun mustahdatha* [modern towns]' built by Aramco.[30] Petro-modernity brought foreign gender relations together with changes to the built environment. In Abdelrahman Munif's fictionalised account *Cities of Salt*, the Americans' arrival at the Black Gold Coast is marked by astonishment on the part of Arabs constructing the new oil towns at immodestly dressed women mingling freely with men. 'Where else in the world were there women like these', wonders the narrator, 'who resembled both milk and figs in their tanned whiteness?' Particularly scandalous were the music and dancing brought by the Americans. New gender relations were personified by the 'Billy Goat', a bearded American 'King Solomon' surrounded by several 'Queens of Sheba'.[31]

Petro-modernity therefore offers an alternative way of reading the numerous accounts of the Saudi oil industry written by Americans. Rather than

Figure 12.1 Aramco workers at Dhahran main gate waiting for transportation to their home communities, 1955. Photo by R. Lee. William E. Mulligan Papers, Booth Family Center for Special Collections, Georgetown University. Courtesy of Saudi Aramco.

exhibiting Aramco's modernisation of Saudi Arabia, these accounts provide evidence of how American power regulated physical distances and enforced gendered social hierarchies. Indeed, American descriptions of Aramco's enterprise feature formulaic images of speed and power on the one hand, and those of domesticity and consumption on the other. In the mid-twentieth century, Americans introduced oil-powered technology into Saudi Arabia to facilitate the production and consumption of petroleum. Early exploration by Barger, later Aramco's president, was conducted in 'a four-wheel-drive Ford station wagon'.[32] American geologists 'followed old camel trails and crisscrossed the desert many times over in Ford V-8 touring cars and a Fairchild monoplane'.[33] President Franklin D. Roosevelt gifted Saudi Arabia's founder, King 'Abd al-'Aziz ibn Sa'ud, with a DC-3 Dakota airplane, and Aramco built his successor, King Sa'ud, a 'custom-designed Chrysler hunting car' that was a 'a rich, dark blue'.[34] Descriptions of Americans arriving in Dhahran typically begin with the perspective of looking down onto the desert and gas flares from Aramco's planes, the 'Camel' and the 'Gazelle'.[35] The company cited its construction of the railroad and the steep rise in the number of motor vehicles in the kingdom as tangible measures of Saudi modernisation. By 1959, boasted the company's *Handbook*, 'the Aramco transportation fleet included 1,334 automobiles and light trucks, 419 medium and heavy trucks and truck-tractors, and 358 trailers'. Vehicles were modified with huge tyres so that the timeless desert sands would not impede twentieth-century progress (Figure 12.2).[36] Before Aramco, writes journalist Thomas Lippman, Saudi Arabia lacked any 'highways, airline, or cross-country telephone lines'. Saudi Arabia 'came late to the aviation age', when Trans-World Airlines (TWA) helped to establish Saudi Arabian Airlines.[37]

Speed was initially the privilege of Americans, while its strangeness to Saudis marked their backwardness. Aramco public relations officer Michael Cheney 'cross[ed] the desert in a midget car', a Hillman Minx that covered the 200 miles per day between pumping stations following Tapline's route from Lebanon to Dhahran. We were 'discouraged … from giving Saudis lifts in our cars', he wrote. 'Still unused to speed', they might suddenly jump out of the car 'while the vehicle was still traveling at forty miles an hour.'[38] In contrast to al-Tariqi, who was dismissed as oil minister by Crown Prince Faysal, Saudi Arabs who rose in the company ranks retrospectively

Figure 12.2 Aramco's Dodge Power Wagons in the Rub' al-Khali desert, 1955. Photo by R. Lee. William E. Mulligan Papers, Booth Family Center for Special Collections, Georgetown University. Courtesy of Saudi Aramco.

validated Aramco's modernising narrative. Saudi Aramco executive Nassir al-Ajmi remembered that, as a boy playing near the Ghawar oil field, the 'first time that I saw a vehicle was a frightening experience'.[39] Oil Minister Ali al-Naimi was nearly killed when he 'darted onto the tarmac' to meet a plane at the Dhahran Air Base and 'came dangerously close to the whirling propellers'. But al-Naimi could later compare his early memory of handling a camel that had slipped with 'pulling an oil company truck out of a hole after it had sunk up to its axles in the mud'.[40] In a cliché of Aramco literature, as a result of the Americans' modernising presence Saudi males gradually exchanged the camel for modern speed. Cheney wrote that 'tenth-century shepherds ... have jumped straight from a donkey onto a Kenworth truck', and the 'Chevrolet is replacing the dromedary as the favored means of transport'.[41] Lippman declares: 'TWA succeeded in implanting among its Saudi employees the understanding that time discipline is essential to running an airline'.[42] Kai Bird, whose father was the US consul, writes that by the early 1960s, 'a common sight in the roads around Dhahran was a Bedouin driving a small Chevy pickup with a large camel kneeling in the flatbed, evidence of Saudi adaptation to modern ways'.[43]

In these same accounts, domesticity and consumption provide a feminine complement to masculine power and speed. Entering the American community in Dhahran, Cheney encountered the 'carbon-copy of an American town' (Figure 12.3):

> Ahead, a broad, paved avenue curved away past neat blocks of houses set in lush gardens. Homes built of stone, mellow orange brick and white plaster, with broad windows and long verandas, rested in the shade of palm, acacia and feathery tamarisk. Smooth lawns lay between jasmine hedges and clumps of oleander heavy with bright pink blossoms. Arcs of spray from whirling sprinklers flashed in the sunlight.[44]

Fadia Basrawi, whose Arab family was among the first to break the colour barrier by settling in the Senior Staff Camp, noted that each of Aramco's communities of Dhahran, Ras Tanura and Abqaiq 'had its own library, golf course, clay tennis court, bowling alley, yacht club, horseback riding stables and Olympic-sized swimming pool'.[45] Lippman described Dhahran as '*Pleasantville*' while Bird compared it to a 'Dallas suburb' that featured 'an elementary and junior high school, a commissary, swimming pools, a movie theatre, a bowling alley and a baseball field'.[46] By 1955, writes Cheney, 'wives and children already outnumbered actual [male] employees'.[47] Anthropologist Solon P. Kimball explained that Aramco had decided 'to provide family housing' so that 'some greater stability of personnel might be achieved'.[48] Reducing costs associated with employee turnover required approximating America's suburban built environment and consumer standard of living. Houses 'had one to three bedrooms, usually a combination living and dining room, bath, pantry, and kitchen', and 'were equipped with modern plumbing, range, and refrigerators'.[49] Librarian Mary Elizabeth Hartzell, among the 300 'bachelorettes' who staffed Aramco's pink-collar service jobs, wrote to her mother describing the cornucopia in the company's dining hall: 'a big buffet with candles and a camel made of ice for a center piece … salads & meat & fish, cheese, delicious cake with real butter icing, peaches & pears,

Figure 12.3 Senior Staff Community and Aramco Administration Building, Dhahran, 1952. William E. Mulligan Papers, Booth Family Center for Special Collections, Georgetown University. Courtesy of Saudi Aramco.

roast turkey & dressing'. In the commissary, she could purchase 'celery, lettuce, tomatoes, grapes, bananas, pears, apples, lemons, oranges, potatoes, onions, cabbage, limes, avocadoes'.[50] According to Cheney, the company used two converted British bombers 'to shuttle over 300,000 pounds of fresh fruits and vegetables each month from Beirut to Dhahran'.[51] As centres of both oil production and domestic consumerism, Aramco's Senior Staff camps became showplaces of petro-modernity.

The separate communities that Aramco built for American and Arab workers were characterised by distinct but related gender hierarchies. 'The oil industry is a man's world', wrote Kimball, predicated on a nuclear family headed by a male breadwinner.[52] In Dhahran, writer and Aramco spouse Nora Johnson languished in a pink-and-green ranch house that for her evoked Englewood, New Jersey. Johnson and other wives 'not only cooked, but strained to outcook each other'. She had travelled halfway around the world only to confront 'the problem that has no name' in Aramco's 'phony, plastic Levittown'.[53] Working in the Saudi oil fields 'was the biggest thing that was going if you were a geologist or a petroleum engineer', noted Ellen Speers, who gave up a diplomatic career to follow her husband to Dhahran. She told Lippman: '[T]hey were on a high. And here was the wife at home.'[54] Aramco's pro-natalist policy assigned priority for scarce housing to married couples with children, so that when Cheney's nameless wife (formerly 'Aramco's most attractive secretary') gave birth to a daughter, it 'entitled us to larger quarters'.[55] Hartzell commented that the 'Mrs.' who hosted a sewing circle had a 'lovely six room house', in contrast to the portable housing Hartzell shared with other 'bachelorettes'. In a letter to her mother, Hartzell confessed that she felt envious when she learned about an acquaintance's engagement, but not out of desire for a husband: 'They will have a brand new house.'[56]

Although Aramco camps were racially segregated until Arabs such as al-Tariqi and Basrawi breached the colour line in the 1950s, a corresponding gender hierarchy existed in the separate communities built for Arab families. Harvard Ph.D. candidate Phebe Marr, sent to interview Arab employee wives about declining participation in Aramco's Home Ownership Loan Program, found many in the Arab company towns of Rahimah and Madinat al-Abqaiq who felt isolated from their extended families. One young Shiʿa wife told Marr that 'she missed her family and had no one to talk to all day'.[57] Another Aramco researcher found that for an Arab wife isolation could produce 'acute mental depression, constant nagging, and even physical illness, any of which is often reflected by a decrease in her husband's job efficiency and morale'.[58] A Saudi driver refused to move into company housing because, while he was at work, 'his wife might have to stay by herself and this would leave her conduct and his honor open to question'.[59] Marr's colleague Malcolm Quint noted that Saudi Arabs from the 'traditional' community of Qatif largely refused to move into company towns because the extended 'family bond and kinship ties are central to the Qatifi's way of life'.[60] Aramco's privileging of the nuclear household imposed constraints on both American and Arab women. Acknowledging the inadequacy of defining family in the Saudi context as a nuclear unit, Aramco's Industrial Relations department consulted experts in Islamic law to arrive at a functional, if circular definition of the Saudi family:

'The family of a Saudi employee of the Company consists of the employee himself and of those of his relatives to whom he is obligated by Shariʿah law to give full support.'[61] Aramco faced greater challenges in making Saudi females into wage labourers than it did with with men. While the company equated male employment with modernisation, in the case of women management claimed to respect existing gender roles. The company studied the possibility of training and hiring Saudi women for certain jobs but concluded, without any apparent irony, that in Saudi society 'the only socially sanctioned adult role for a woman ... is that of wife and mother, who is to be kept at home'.[62]

Petro-modernity, based on speed and domesticity, defined hierarchies of gender and power in the oil camps. American women complained about not being allowed to drive in the kingdom, and when they did it required cross-dressing across lines of race as well as gender. Bird recalls Helen Metz, wife of CIA officer and Aramco's liaison to King Saʿud Ronald Metz, disguising herself as an Arab male for the purpose of driving King Saʿud to the airport so that he could be airlifted to Dhahran for medical treatment. Bird's own mother 'drove all the way to Riyadh, wearing a kaffiyeh wrapped around her head so passing drivers might mistake her for a man'.[63] Even arriving by plane at Dhahran was an experience inflected by gender. Kimball repeats the story of '"bachelors" [who] greeted every incoming plane with new personnel from the United States for the purpose of scrutinizing the arriving female contingent'.[64] For an Arab woman like Basrawi, the initial descent into Dhahran required a quick trip to the airplane's lavatory to shed Western clothes. One by one, Saudi women 'disappeared into the bathroom and reappeared incognito enshrouded from head to toe in the voluminous black *abayas*'.[65] Only airplane crashes, such as a devastating accident at Dhahran that claimed many American and Arab lives in 1964, erased social distinctions.[66]

Americans living in segregated communities seldom encountered what Cheney called 'Arab womanhood at close range', but his Orientalist descriptions of these 'hooded black shadows' secluded behind a 'hideous black mask' called into question Aramco claims to respect local gender norms.[67] Hartzell's comparisons between Arab women as 'shapeless black figures' and her own 'bachelorette' status fixed her coordinates in the neo-colonial order as racially privileged but subordinate with respect to gender.[68] Johnson went further in using Arab women to protest her own captivity in the pink-and-green ranch house: 'I *was* the thief whose hand was cut off, the woman buried up to her chest and stoned.' Arab women seemed 'dreadfully like me', she wrote: 'silent, protected, infantile creatures'.[69] Several accounts feminise the Indian houseboys and Saudi gardeners employed by many Americans, whose gendered descriptions emphasise the male servants' confinement to the domestic sphere. Johnson writes that a Goanese houseboy and Saudi gardener 'came along with the house'. The latter 'spent most of his time dozing under the oleanders, or else praying on a small rug he rolled up when he was finished'. She wondered 'if he knew how tenderly protected he was'. He 'seemed very precious, like a rare artifact'.[70] Arab women rarely visited American communities, writes Lippman, but when they did 'they would search uninvited through drawers and closets, not out of rudeness but out of simple curiosity. The American had gadgets and garments that were outside their experience.' Jane Hart, a US diplomat's wife, would leave *Vogue* on the table for female Arab guests to examine, and when she visited the household of the Eastern Province governor Saʿud bin Jiluwi, Hart reported that women there took particular interest in her brassière.[71]

Only by interpreting Aramco as a petro-modernity-based, neo-colonial order is it possible to fully understand labour uprisings against the company. Cheney attributed strikes to the raised expectations produced by encounters with American consumerism. 'American movies, salesmen, teachers and diplomats have given the Arabs glimpses of a dazzling new world', he wrote. It included 'not only our glittering cars and gadgets but something of our standard of living and our freedoms'.[72] Yet Arab workers revolted against their place in Aramco's neo-colonial order. They organised strikes to protest segregated housing, failure to promote Arabs out of the lowest-paid jobs, and the company's refusal to recognise a union. In October 1953, their protests targeted Western vehicles, while many strikers remained in their home communities and away from the Aramco oil towns of Dhahran, Abqaiq and Ras Tanura. The strike therefore unfolded within the decentralised geography of segregated residences and worksites along the Black Gold Coast. The US consulate reported that on 16 October, a crowd of '500 to 2000 spilled over the road leading from Aramco's main gate' and 'suddenly began stoning passing cars. A U.S. Air Force bus was badly damaged.' Demonstrators also attacked a TWA vehicle. Strikers held 'mass meetings in nearby villages', and when

Aramco buses arrived to bring Saudis to work 'they were frequently met by an angry crowd and stones were often thrown'.[73] Arabic accounts of the 1953 strike compiled by journalist Sultan al-Jumayri corroborate that workers 'pelted American cars with rocks [*rami sayyarat al-Amrikan bil-hijara*]'.[74] The Saudi government moved around 2,000 troops into the Eastern Province, arresting strike leaders and screening workers both in their home communities and in the oil camps. Aramco made limited concessions that rejected workers' demands for the company to build them housing and instead offered them a 'loan plan for housing'.[75] After the 1956 strike, labour leader Nasir al-Sa'id excoriated King Sa'ud for refusing to accept workers' petitions. Al-Sa'id cited Nasiriyya Palace outside Riyadh, with its black and white slaves, cinema and electric lights, as a symbol of how Aramco had corrupted the Saudi government and ruling family. As an agent of American imperialism, wrote al-Sa'id, Aramco had 'occupied' the kingdom 'economically, politically, culturally, and socially'.[76]

In the years following 'Rock Wednesday', petro-modernity began to unravel. What had been a neo-colonial order built around the production and consumption of cheap oil sustained challenges to hierarchies at every level. Growing oil demand relative to supply gave members of the Organization of Petroleum Exporting Countries (OPEC) leverage to insist on higher prices and weakened the power of Western corporations. Saudi Arabia's government negotiated an agreement to nationalise Aramco, and the embargo imposed by Arab exporters on the USA during the 1973 Arab–Israeli War reinforced the longer-term trend towards higher prices.[77] The resulting 'oil crisis' heightened fears in the USA about shifting domestic power relations and threatened post-war gender norms. American women were increasingly entering the paid labour force at a time when lay-offs and inflation generated by higher fuel costs diminished men's earnings and household purchasing power.[78]

During the oil crisis years, Americans and Saudis demonstrated against the authorities within spaces created by petro-modernity. Truck drivers protesting high fuel costs parked their rigs on I-80 and brought the interstate highway system to a standstill across five states. This was just the beginning of trucker protests that paralysed the country and occasionally turned violent during the oil price hikes of 1973–4 and 1979. Congressional approval of a new 55-miles-per-hour speed limit further provoked these '"new highway guerillas"', who withheld delivery of food and consumer goods to protest the inability to support their families.[79] In Saudi Arabia, writes Jones, Shiʿa protests in 1979 against the persistent impoverishment and neglect of their communities in spite of staggering oil wealth were 'concentrated in Qatif and the villages and towns surrounding it'. A nineteen-year-old graduate of Aramco's Industrial Training Center, Hussein Mansur al-Qalaf, became the first demonstrator killed by the Saudi National Guard in a town north of Dammam.[80]

Yet the association between gender, domesticity and speed persisted in the US–Saudi encounter. Protests by Saudi women demanding the right to drive coincided with the presence of hundreds of thousands of American troops during the Gulf War.[81] Pascal Menoret has described the high-speed drifting of cars by young Saudi men as 'road revolt'. Their dangerous and antisocial experimentation with speed signifies a 'loss of future' and the failure of Saudi development plans to provide males from Bedouin backgrounds with education, housing and jobs.[82] Today's Riyadh, with its congested, two-kilometre-square superblocks and millions of daily car trips through a 'vast car-based suburbia', may represent the fullest realisation of American petro-modernity. For women prohibited from driving and unable to leave home except when chauffeured, the car is an 'instrument of control' by male family members. While cars may 'give the illusion of freedom and self-determination', argues Menoret, automobility creates 'extensive systems of oppression'.[83] Despite Americans' exceptionalist claims about promoting freedom and modernisation, studying the history of petro-modernity as a gendered, neo-colonial order reveals underlying hierarchies of power in the US–Saudi relationship.

Notes

1. Larry Barnes, *Looking Back Over My Shoulder* (n.p., 1979), 168–9.
2. 'Arab–Israeli Situation Report (as of 2:00pm EDT)', 7 June 1967, p. 3, Directorate of Intelligence, Central Intelligence Agency, CIA-RDP79T00826A002100010005-7, CREST, https://www.cia.gov/library/readingroom/ (last accessed 2 February 2020).
3. Barnes, *Looking Back Over My Shoulder*, 169–72. See also Kai Bird, *Crossing Mandelbaum Gate: Coming of Age Between the Arabs and Israelis, 1956–1978* (New York: Scribner, 2010), 209–10.

4. For the key study of Aramco that uses race as a critical analytical category see Robert Vitalis, *America's Kingdom: Mythmaking on the Saudi Oil Frontier* (New York: Verso, 2009). See also Rosie Bsheer, 'A Counter-Revolutionary State: Popular Movements and the Making of Saudi Arabia', *Past & Present* 238, no. 1 (February 2018), 233–77. On Aramco's place in the history of global capitalism see Laleh Khalili, *Sinews of War and Trade: Shipping and Capitalism in the Arabian Peninsula* (New York: Verso, 2020).
5. See Stephanie LeMenager, *Living Oil: Petroleum Culture in the American Century* (New York: Oxford University Press, 2014).
6. See Christopher R. W. Dietrich, *Oil Revolution: Anticolonial Elites, Sovereign Rights, and the Economic Culture of Decolonization* (Cambridge: Cambridge University Press, 2017); and Betsy A. Beasley, 'Service Learning: Oil, International Education, and Texas's Corporate Cold War', *Diplomatic History* 42, no. 2 (April 2018), 177–203.
7. See Ann Laura Stoler, 'Tense and Tender Ties: The Politics of Comparison in North American History and (Post) Colonial Studies', *Journal of American History* 88, no. 3 (December 2001), 829–65. See also Amy Kaplan, 'Manifest Domesticity', *American Literature* 70, no. 3 (September 1998), 581–606.
8. For example, see Eric Lewis Beverley, 'Colonial Urbanism and South Asian Cities', *Social History* 36, no. 4 (November 2011), 482–97; and Todd A. Henry, *Assimilating Seoul: Japanese Rule and the Politics of Public Space in Colonial Korea, 1910–1945* (Berkeley: University California Press, 2014). For other examples from the Gulf see Nelida Fuccaro, 'Introduction: Histories of Oil and Urban Modernity in the Middle East', Special Issue in *Comparative Studies of South Asia, Africa, and the Middle East* 33, no. 1 (2013), 1–6.
9. See Asher Kaufman, 'Between Permeable and Sealed Borders: The Trans-Arabian Pipeline and the Arab–Israeli Conflict', *International Journal of Middle East Studies* 46, no. 1 (February 2014), 95–116.
10. LeMenager, *Living Oil*, 67, 74.
11. Lizabeth Cohen, *A Consumers' Republic: The Politics of Mass Consumption in Postwar America* (New York: Knopf, 2003), 278–86.
12. Elaine Tyler May, *Homeward Bound: American Families in the Cold War Era* (New York: Basic Books, 1988).
13. For the best recent study of US–Saudi relations, see Victor McFarland, *Oil Powers: A History of the U.S.–Saudi Alliance* (New York: Columbia University Press, 2020).

14. See David S. Painter, *Oil and the American Century: The Political Economy of U.S. Foreign Oil Policy, 1941–1954* (Baltimore: Johns Hopkins University Press, 1986).
15. U.S. House of Representatives, Armed Services Committee, Subcommittee on Petroleum, 'Petroleum for National Defense', 80th Congress, 2nd Sess. (Washington: Government Printing Office, 1948), 10, 213.
16. Quoted in Vitalis, *America's Kingdom*, 136.
17. 'Abdullah al-Tariqi, 'Sharikat al-ihtikariyya al-'alamiyya wasila al-isti'mar al-jadida fi al-saytara wa al-istighlal [Parts 1 and 2]', in Walid Khadduri (ed.), *'Abdullah al-Tariqi: al-A'mal al-Kamila* (Beirut: Markaz dirasat al-wahda al-'Arabiyya, 1999), 320–5, 350–3.
18. Manu Karuka, *Empire's Tracks: Indigenous Nations, Chinese Workers, and the Transcontinental Railroad* (Berkeley: University of California Press, 2019).
19. On Barak, *On Time: Technology and Temporality in Modern Egypt* (Berkeley: University of California Press, 2013), 5. On gender see Lisa Pollard, *Nurturing the Nation: The Family Politics of Modernizing, Colonizing, and Liberating Egypt, 1805–1923* (Berkeley: University of California Press, 2005).
20. See William A. Eddy, 'Dear Family', 26 February 1950, folder 4, box 6, William A. Eddy Papers, Seeley G. Mudd Library, Princeton University, Princeton, NJ.
21. Yoav Di-Capua, 'Common Skies, Divided Horizons: Aviation, Class and Modernity in Early Twentieth Century Egypt', *Journal of Social History* 41, no. 4 (Summer 2008), 924.
22. Kristin V. Monroe, 'Automobility and Citizenship in Interwar Lebanon', *Comparative Studies of South Asia, Africa, and the Middle East* 34, no. 3 (2014), 526.
23. Enda Duffy, *The Speed Handbook: Velocity, Pleasure, Modernism* (Durham, NC: Duke University Press, 2009), 1, 7.
24. Katherine J. Parkin, *Women at the Wheel: A Century of Buying, Driving, and Fixing Cars* (Philadelphia: University of Pennsylvania Press, 2017), x, xiv.
25. Ibid., 67, 105. See also Beth L. Bailey, *From Front Porch to Back Seat: Courtship in Twentieth-Century America* (Baltimore: Johns Hopkins University Press, 1988).
26. Robert S. Lynd and Helen Merrell Lynd, *Middletown: A Study in Contemporary American Culture* (New York: Harcourt, Brace & Co., 1929), 114.
27. See Emily Rosenberg, 'Consuming Women: Images of Americanization in the "American Century"', *Diplomatic History* 23, no. 3 (Summer 1999), 479–97;

and Betty Friedan (ed. Kirsten Fermaclich and Lisa M. Fine), *The Feminine Mystique: A Norton Critical Edition* (New York: Norton, 2013), 11.

28. Madawi al-Rasheed, *Muted Modernists: The Struggle over Divine Politics in Saudi Arabia* (London: Hurst, 2015), 117; idem, *A Most Masculine State: Gender, Politics, and Religion in Saudi Arabia* (Cambridge: Cambridge University Press, 2013), 169.
29. See Cohen, *A Consumers' Republic*; Owen D. Gutfreund, *20th-Century Sprawl: Highways and the Reshaping of the American Landscape* (New York: Oxford University Press, 2004); Kevin M. Kruse, *White Flight: Atlanta and the Making of Modern Conservatism* (Princeton: Princeton University Press, 2005); Thomas J. Sugrue, *The Origins of the Urban Crisis: Race and Inequality in Postwar Detroit* (Princeton: Princeton University Press, 1996); and Keeanga-Yamahtta Taylor, *Race for Profit: How Banks and the Real Estate Industry Undermined Black Homeownership* (Chapel Hill: University of North Carolina Press, 2019).
30. Toby Craig Jones, *Desert Kingdom: How Oil and Water Forged Modern Saudi Arabia* (Cambridge, MA: Harvard University Press, 2010), 143. Muhammad Sa'id al-Muslim, *Sahil al-dhahab al-aswad: dirasah tarikhiyah insaniyah li-mintaqat al-Khalij al-'Arabi*, 2nd edn (Beirut: Dar Maktabat al-Hayat, 1962), 30–43, 218 (quotation).
31. Abdelrahman Munif (trans. Peter Theroux), *Cities of Salt* (New York: Vintage, 1987), 214, 217–18.
32. Thomas Lippman, *Inside the Mirage: America's Fragile Partnership with Saudi Arabia* (Boulder, CO: Westview Press, 2004), 21. See also Bird, *Crossing Mandelbaum Gate*, 99.
33. Fadia Basrawi, *Brownies and Kalashnikovs: A Saudi Woman's Memoir of American Arabia and Wartime Beirut* (Reading: South Street Press, 2009), 11.
34. Lippman, *Inside the Mirage*, 125; Barnes, *Looking Back Over My Shoulder*, 152 (see also 46–7, 'Automobiles').
35. Grant Butler, *Kings and Camels: An American in Saudi Arabia* (Reading: Garnet, 2008), reprint edn, 24–7, 85; and Michael Sheldon Cheney, *Big Oil Man from Arabia* (New York: Ballantine Books, 1958), 6, 118. Aramco also used the DC-3 and other aircraft for desert exploration. See memo for the files by Mulligan [1954, date obscured], folder 37, box 2, William E. Mulligan Papers [WEM], Special Collections, Lauinger Library, Georgetown University, Washington, DC.
36. Roy Lebkicher, George Rentz, Max Steineke et al., *Aramco Handbook* (Dhahran: Arabian American Oil Company, 1960), 193. See also Butler, *Kings and Camels*, 39; and 'Those Big Tires', *Arabian Sun & Flare* 46 (23 November 1949), 3.

37. Lippman, *Inside the Mirage*, 125–9.
38. Cheney, *Big Oil Man*, 40, 160–1.
39. Quoted in Basrawi, *Brownies and Kalashnikovs*, 11.
40. Ali al-Naimi, *Out of the Desert: My Journey from Nomadic Bedouin to the Heart of Global Oil* (London: Portfolio Penguin, 2016), 9, 34.
41. Cheney, *Big Oil Man*, 25, 165.
42. Lippman, *Inside the Mirage*, 129.
43. Bird, *Crossing Mandelbaum Gate*, 101.
44. Cheney, *Big Oil Man*, 2, 19.
45. Basrawi, *Brownies and Kalashnikovs*, 12.
46. Lippman, *Inside the Mirage*, 55; Bird, *Crossing Mandelbaum Gate*, 90.
47. Cheney, *Big Oil Man*, 123.
48. Solon T. Kimball, 'American Culture in Saudi Arabia', *Transactions of the New York Academy of Sciences* 18, no. 5 (1956), 477–8.
49. Butler, *Kings and Camels*, 28.
50. Hartzell to mother, 8 August 1952, folder 8; and Hartzell to mother, 16 November 1953, folder 9, box 11, WEM. On the 300 'bachelorettes', see Kimball, 'American Culture in Saudi Arabia', 480.
51. Cheney, *Big Oil Man*, 123.
52. Kimball, 'American Culture in Saudi Arabia', 477. See also Benjamin Jones, 'Women on the Oil Frontier: Gender and Power in Aramco's Arabia', *Rice Historical Review* 2 (Spring 2017): 55–69.
53. Nora Johnson, *You* Can *Go Home Again* (Garden City, NY: Doubleday, 1982), 48, 71, 78.
54. Lippman, *Inside the Mirage*, 64.
55. Cheney, *Big Oil Man*, 100, 121.
56. Hartzell to mother, 8 August 1952, folder 8; Hartzell to mother, 28 January 1953, folder 9; Hartzell to mother, 14 August 1953, folder 9, box 11, WEM.
57. Quoted in Nathan J. Citino, *Envisioning the Arab Future: Modernization in U.S.–Arab Relations, 1945–1967* (New York: Cambridge University Press, 2017), 119.
58. Malcolm Quint, 'Home and Family in the Qatif Oasis', November 1960, p. 7, folder 5, box 3, WEM.
59. Marr to Mulligan, 'Home Ownership Program – Rahimah', 22 December 1960, p. 3, folder 5, box 3, WEM.
60. Quint, 'Home and Family in the Qatif Oasis', November 1960, p. 7, folder 5, box 3, WEM.

61. Mueller to Cypher, 'Definition of the Saudi Employee Family', 29 December 1954, folder 44, box 2, WEM.
62. Quoted in Citino, *Envisioning the Arab Future*, 120.
63. Bird, *Crossing Mandelbaum Gate*, 107, 114.
64. Kimball, 'American Culture in Saudi Arabia', 280.
65. Basrawi, *Brownies and Kalashnikovs*, 3.
66. See Najib Alamuddin, *The Flying Sheikh* (London: Quartet, 1987), 111; and Barnes, *Looking Back Over My Shoulder*, 133.
67. Cheney, *Big Oil Man*, 64, 92.
68. Quoted in Citino, *Envisioning the Arab Future*, 119.
69. Johnson, *You* Can *Go Home Again*, 79.
70. Ibid., 71, 72. See also Hartzell to mother, 12 September 1952, folder 8, box 11, WEM.
71. Lippman, *Inside the Mirage*, 37, 38.
72. Cheney, *Big Oil Man*, 187.
73. Dhahran to State Department, 4 November 1953 [pp. 6, 13], 886A.062/11-453, State Department Central Files, Record Group [RG] 59, National Archives and Records Administration [hereafter NARA].
74. Sultan al-Jumayri, *'An al-tajriba al-nidaliyya al-'ummaliyya fi al-Sa'udiyya*, https://docs.google.com/document/d/1IwuBr-E4sM6-Nk4Wdt__pdZ_ZTfozQCTXh91iu7opcg/edit, p. 4 (last accessed 14 February 2020).
75. NARA: RG 59, 886A.06/11-2453. Memo of conversation by Fritzlan, 24 November 1953.
76. Nasir al-Sa'id, *Risala ila Su'ud min Nasir al-Sa'id* (Cairo: al-Dar al-Qawmiya, 1958), 16, 36.
77. See Giuliano Garavini, *The Rise and Fall of OPEC in the 20th Century* (New York: Oxford University Press, 2019), Chs 4 and 5; and David M. Wight, *Oil Money: Middle East Petrodollars and the Transformation of US Empire, 1967–1988* (Ithaca: Cornell University Press, 2021).
78. See Natasha Zaretsky, *No Direction Home: The American Family and the Fear of National Decline, 1968–1980* (Chapel Hill: University of North Carolina Press, 2007), 86.
79. Meg Jacobs, *Panic at the Pump: The Energy Crisis and the Transformation of American Politics in the 1970s* (New York: Hill & Wang, 2016), 74–9; quotation on p. 75. See also Shane Hamilton, *Trucking Country: The Road to America's Wal-Mart Economy* (Princeton: Princeton University Press, 2008), 223.

80. Toby Craig Jones, 'Rebellion on the Saudi Periphery: Modernity, Marginalization, and the Shi'a Uprising of 1979', *International Journal of Middle East Studies* 38, no. 2 (May 2006), 216 (quotation), 223. See also Toby Matthiesen, *The Other Saudis: Shiism, Dissent and Sectarianism* (New York: Cambridge University Press, 2015), 103–10.
81. See Eleanor A. Doumato, 'Gender, Monarchy, and National Identity in Saudi Arabia', *British Journal of Middle Eastern Studies* 19, no.1 (1992), 31–47.
82. Pascal Menoret, *Joyriding in Riyadh: Oil, Urbanism, and Road Revolt* (New York: Cambridge University Press, 2014), 'Urban Unrest and Non-Religious Radicalization in Saudi Arabia', in Madawi al-Rasheed and Marat Shterin (eds), *Dying for Faith: Religiously Motivated Violence in the Contemporary World* (New York: Palgrave Macmillan, 2009), 131.
83. Pascal Menoret, 'Learning from Riyadh: Automobility, Joyriding, and Politics', *Comparative Studies of South Asia, Africa, and the Middle East* 39, no.1 (May 2019), 133, 136, 139.

13

OIL, MOBILITY AND TERRITORIALITY IN THE TRUCIAL STATES/ UNITED ARAB EMIRATES, MUSCAT AND OMAN

Matthew MacLean

This chapter examines the impact of oil on the twin processes of state formation and space-making in the Trucial States and United Arab Emirates and Sultanate of Oman in the mid-twentieth century.[1] In much of the literature on the history of the Gulf and Arabian Peninsula, these processes are linked through oil concessions. Concessions necessitated the demarcation of domestic and international boundaries in the Arabian Peninsula, a key part of the state formation process.[2] This chapter looks instead at state formation through a new means of oil-fuelled mobility – automobility. Beginning in the early 1950s and surging dramatically at the end of the 1960s, the automobile rapidly displaced older modes of transportation, in the process becoming synonymous with modernisation and state-building. The automobile's speed and power sparked violence, necessitated new modes of regulation as well as a new road network, and made the state visible and tangible in even the most remote areas of the region. New boundaries between states were demarcated, with different rules for travel by car and by foot or animal. In the process, new understandings of space emerged, and state control over territory dramatically intensified. Eventually, it became both physically possible and morally permissible for UAE and Omani citizens (and others) to travel to places that had not been open to them before, while other patterns of circulation were closed off by a new international border; automobility and roads created both new freedoms and new restrictions. Through the lens of automobility, oil's role in state formation becomes more complex and contested, as various actors ranging from British Political Agents to local sheikhs wrestled with how new forms of movement ought to be governed.

Two spatial imaginaries frame the chapter's analysis – the pre-oil *dirah,* rooted in seasonal migrations and kinship relations, and the nascent *dawla* (state), which required free movement within demarcated boundaries. The shift from the *dirah* to the *dawla* is traced through several episodes involving automobile travel. The potential of automobility to undermine the existing political and spatial order is seen in the 1938 Majlis Movement in Dubai and in a 1950 conflict in Sha'am, in northern Ras al-Khaimah. When the British began to demarcate boundaries between the Trucial States to prepare the way for oil concessions in the mid-1950s, an incident in Masafi led to the understanding that ruling sheikhs could not restrict automobile travel, nor could their own movements be restricted within the Trucial States. In the next few years, conflicts in Madha and Asimah and Tayyibah showed how automobility and boundary-making both sparked violence and enabled the 'peacekeeping' efforts that furthered the growth of state power in rural hinterlands, while attempting to respect the older patterns of movement of the *dirah*. Demarcation of the Sultanate of Muscat's boundaries with the Trucial States, however, led to new state restrictions on automobile travel. These restrictions were contested by people ranging from ruling sheikhs of the Trucial States, and notables of the Omani interior such as Suleiman bin Himyar, to ordinary women. Finally, the new UAE state's road-building projects reshaped migration and settlement along its border with Oman. Taken together, they show how oil and state formation were manifested in everyday life rather than only in high-level political negotiations or international commerce.

Space, Boundary-making and Mobility in the Mid-century Trucial States – the *Dirah* and the *Dawla*

In the Trucial States and northern Oman of the early twentieth century, control over territory was constituted through property, kinship, migration patterns and various other informal arrangements. A group's territory was *al-dirah*, literally, the area in which one circulates. Seasonal migrations – *al-muqeet* – were a major part of life until the late 1960s. In the summer, families travelled from coastal towns to cooler inland oases to take advantage of the date harvest, exchange gifts, trade, and renew kinship and other ties. Because summer migrations coincided with the pearl-diving season, the migrants were mostly women and children. In some cases, these migrations were quite short, from the coast to nearby oases that have long since been subsumed into urban sprawl, but other families moved several dozen kilometres inland. The 'carrying trade', which provided an escort on the way to the oases, was an important source of income to Bedouin tribes in the deserts between the coast and the mountains; they provided camels and donkeys to the migrating families as well as dates and hospitality.[3] A regional code of social conventions revolved around travel, visiting and exchanging gifts.[4] In practice, the system of seasonal migrations through a *dirah* meant that local people knew who had the right to use a specific piece of land or travel route; contrary to an oft-imagined pre-state world of free mobility, overland travel was quite strictly controlled through the aforementioned social conventions and traditions.

However, those mechanisms of control were often illegible to British officials and incompatible with the legal necessities of oil concessions. A patchwork of kinship and other informal relations meant that no clear boundaries could have been drawn between territories. Ruling sheikhs were somewhat distant figures who arbitrated disputes and provided for common defence, but those functions were more frequently exercised by local sheikhs in a given area.[5] By contrast, British officials understood sovereignty as exercised by recognised ruling sheikhs in coastal towns, who in turn exercised authority over various tribes, who in turn had the rights to use certain lands. J. G. Lorimer's 1908 *Gazetteer* is replete with references to (often undefined) borders and boundaries between tribes. Likewise, a British official's 1951 list of the Sultanate of Muscat's tribes mentions each tribe's 'capital', suggesting a rather centralised understanding of how regional political authorities functioned.[6]

These hierarchical and territorial understandings of sovereignty would have to be reconciled with the realities of the *dirah*. The *dirah* would have to be made legible to the British and to oil companies; the efforts to do so were a key element in the rise of the *dawla,* the modern state. The British had to create a new map of the Trucial States and Oman in which a uniform sovereignty would be exercised by the seven British-recognised ruling sheikhs and the Sultan of Muscat. Such a map would be legible to oil companies operating under concession agreements, and by extension other enterprises used to the norms of global capitalism. By drawing a line around tribal lands, one could theoretically determine the extent of a ruling sheikh's legitimate authority, and thus the boundaries of the state and/or the relevant oil concessions. Determining where those boundaries were was the task of Julian Walker and other employees of the British Political Agency in Dubai.[7] Walker travelled throughout the northern Trucial States and into the lands then controlled by the Sultan of Muscat, inquiring to which ruling sheikh the local people owed their allegiance, to whom they paid *zakat*, who owned property there, and who exercised de facto control over a place; on the basis of Walker's and subsequent reports on these matters and past agreements made with ruling sheikhs, the British assigned territory to one recognised ruling sheikh or another.[8] The process of boundary demarcation began in 1955 but was fraught with disputes and ambiguities, many of which would take decades to resolve.[9] The difficulties in the demarcation process stemmed

Figure 13.1 Julian Walker (second from right) and his Land Rover in the sand near Al ʿAin, 1954. © National Archives, UAE.

from the fact that many tribes and villages were effectively independent and others divided their allegiance between two (or more) of the British-recognised ruling sheikhs.

Automobility and State Formation in the Mid-twentieth-century Trucial States

This task would be futile, however, without the means to actually exercise control over the territories in between the lines on a map. The means in question would be the automobile, but the automobile was not merely a top-down effect of the emerging *dawla*. Rather, its presence produced and gave shape to the spatial imaginaries that made the state visible in daily life. In this, the automobile's political impacts foreshadowed by a few decades the large-scale rapid change usually attributed to oil. The automobile preceded the discovery of oil and the accompanying large-scale transformation of daily life in the region; its use sparked several local conflicts and violent incidents in the early 1950s that gave shape to the future state. Three will be examined here: the Majlis Movement in Dubai in 1938, a local feud in northern Ras al-Khaimah in 1950–1, and the first episode in a long-running conflict in Masafi, a mountain village on the border between Ras al-Khaimah and Fujairah. Collectively, the three examples illustrate the automobile's economic and political potency. A fourth episode, an investigation by British authorities into a murder in Wadi Madha in 1956, hints at how and why the automobile provoked such conflicts.

By challenging the older customs and habits of mobility of the *dirah*, the automobile put great stress upon the existing mid-century political order. The British Political Agent imported the first automobile to the Trucial States in 1928, a time when the region's economy was under severe strain. The next decades were a time of relative poverty; the pearling trade had been decimated by the invention of cultured pearls, the Great Depression and World War II. A drought from 1947 to 1955 was exacerbated by the outmigration of a large percentage of the Trucial States labour force to work in the oil and construction industries of Kuwait, Bahrain and Saudi Arabia.[10] Long-standing agreements and understandings regarding territory and sovereignty could often not be upheld because nobody had enough funds to pay the tributes and *zakat* required. For example, the Ruler of Ras al-Khaimah had, since at least the 1920s, been paid tribute by the sheikhs of various towns and tribes, in exchange for the Ruler's acknowledgement of their local suzerainty. With the end of pearling and the decline of agriculture, these local sheikhs had less income.[11] In this context, the automobile provided a chance for various actors to challenge existing forms of authority over space, and likewise for ruling sheikhs to assert their authority in new ways.[12] Likewise, the British used the automobile (or more precisely the Land Rover) to extend their influence and the influence of ruling sheikhs into the interior of the Trucial States and northern Oman to realise a new organisation of territory.

It was in an atmosphere of economic uncertainty that the so-called 'Majlis Movement' for reform of Dubai's government took place in 1938. Several merchant families and a dissident branch of the ruling al-Maktoum family demanded a say in the governance of the sheikhdom and attempted to create an administrative state structure led by merchant notables that would have a significant voice in the sheikhdom's finances, education and trade policies. The Majlis Movement was crushed in short time, but is still sometimes understood as one of the only significant attempts to check the power of the ruling al-Maktoum family.[13] It was inspired in part by similar efforts in Kuwait and Bahrain.[14] But perhaps most tellingly, the movement's immediate origins lay not only in abstract notions of constitutional governance and general economic distress, but in a protest over the Ruler's son's attempt to monopolise a newly-lucrative taxi business in Dubai, and between Dubai and Sharjah. With the collapse of the pearling industry, the economic potential of the automobile was obvious. After an attack on the Ruler's son's car on 26 May 1938, and the subsequent imprisonment of the attackers, merchant notables and some members of the dissident al-Maktoum faction presented a list of demands to Sheikh Saeed, including the abolition of monopolies held by the Ruler and his immediate family. This eventually evolved into demands for a more consultative form of government, which were quashed only by the armed force – although the Ruler did give up some of his monopolies over land and sea transportation.[15]

Twelve years later in the villages of Sha'am and al-Jeer, an automobile brought another underlying political shift to its breaking point, again demonstrating the power of the automobile to disrupt the life world of the *dirah*. Sha'am and al-Jeer were the northernmost settlements under the nominal

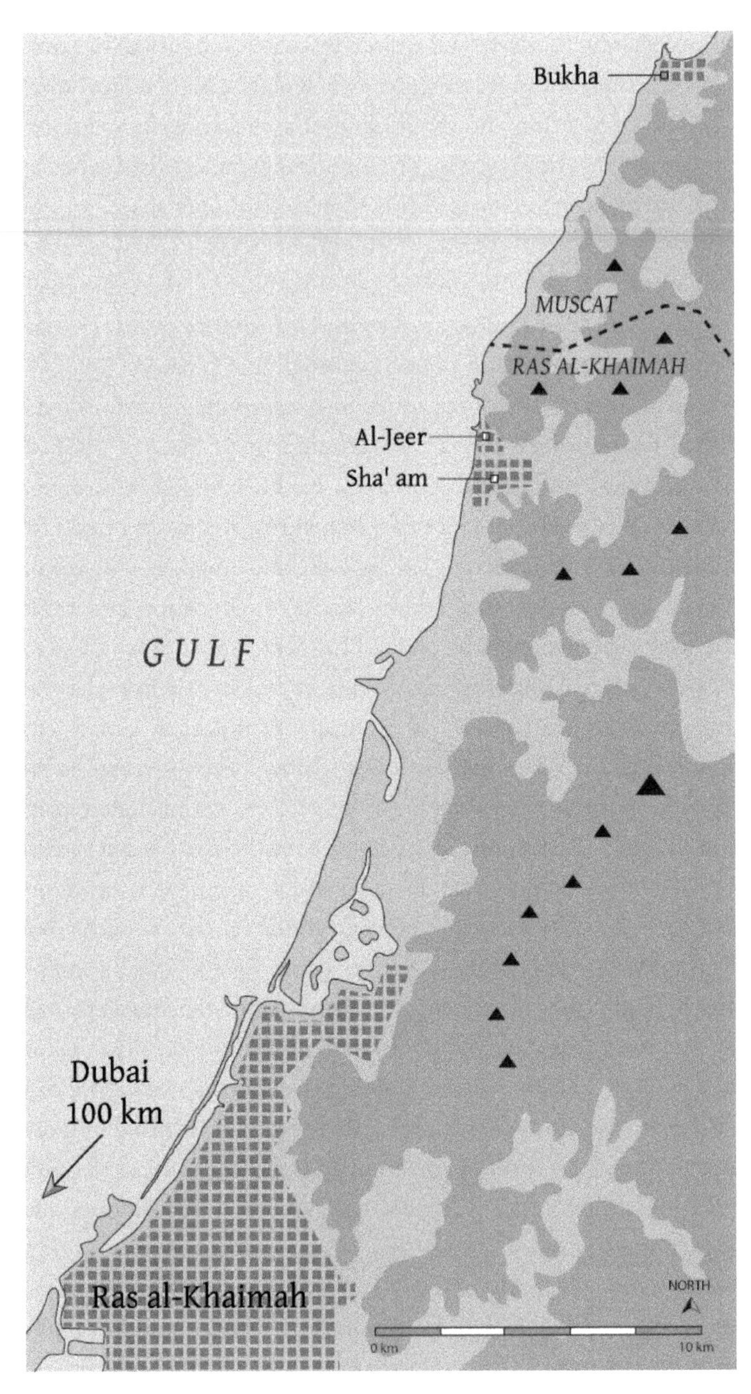

Figure 13.2 Map of Bukha and northern Ras al-Khaimah.

control of Ras al-Khaimah, near what today is the border with the Omani exclave of Musandam. As in other outlying locales, the Ruler of Ras al-Khaimah (Sheikh Saqr bin Mohammed al-Qasimi) appointed a Wali to resolve disputes and collect taxes in his name, but it was only rarely that Sheikh Saqr exercised any sort of direct control – an arrangement typical of the world of the *dirah*. Only large disputes necessitated the involvement of outside authorities. It was precisely this sort of dispute that began in 1950 between Sheikh Saqr and Muhammad Ahmad al-Shehhi, the semi-independent sheikh of the small coastal village of Bukha in what today is Musandam. British officials in the early 1950s considered Muhammad Ahmad, as he was commonly known, to be a subject of the Sultan of Muscat. Although Muscat exercised almost no influence in Bukha at the time (and Muhammad Ahmad considered himself independent) the British made this judgement on the basis of J. G. Lorimer's 1908 *Gazetteer*. Lorimer assigned Bukha to the Sultanate by virtue of Bukha's location (geologically, a part of the Ru'us al-Jibal mountains rather than the coastal plain that characterised Ras al-Khaimah) and tribal affiliation. He described Sha'am, where the mountains meet the waters of the Gulf, as the limits of Ras al-Khaimah's jurisdiction.[16]

Whatever Lorimer's understanding of Bukha's status, the political situation slowly changed over the ensuing decades. Muhammad Ahmad became wealthier, owning ten dhows that sailed as far as India and East Africa.[17] In the early 1940s, he bought a Land Rover to use to travel to his stores in Dubai. Because it was impossible to drive to Bukha overland, he kept the Land Rover in a garage in al-Jeer, just over the imagined 'border' with Ras al-Khaimah, and began to make loans to local residents, threatening the primacy of al-Qawasim authority in the area.[18] By 1950, the British estimated that approximately 300–400 of the 1,400 residents of al-Jeer and Sha'am were followers of Muhammad Ahmad, while the rest looked to Sheikh Saqr's appointed Wali. In the summer of 1950 residents of al-Jeer objected to Muhammad Ahmad's driving through their village and fired at his car. He escaped, but the car broke down, and he got into an accompanying vehicle to make his way to Dubai.[19] In retaliation, he sent a raiding party to Sha'am that set four houses on fire, cut down date palms (just before the annual harvest) and stole jewellery. Four people were wounded, with one dying soon after. Over the next year, the British Navy had to intervene twice to keep the peace in the region and reinforce Qawasim authority in Sha'am; Muhammad Ahmad was forced to return to Bukha, stop driving his car in Sha'am and recognise the entire area as Ras al-Khaimah territory.[20]

Muhammad Ahmad's automobile provoked violence for economic, political and social reasons. Like the Dubai taxi service monopoly held by the Ruler in 1938, the automobile was a form of obvious prosperity in decades of relative poverty.[21] His Land Rover solidified his commercial connections to Dubai, furthering his political influence in al-Jeer and Sha'am by drawing a portion of its residents away from their previous loyalties to Ras al-Khaimah. But perhaps equally important is the impression a loud, powerful motor vehicle would have made. The incident happened in July at the height of the summer migrations, when al-Jeer and Sha'am would have been at their most crowded with people and animals. An automobile would have been a major intrusion into the extraordinarily quiet soundscape of an agricultural village, and no doubt many of the residents and summer visitors might have taken fright at the unfamiliar sights and sounds. At a time of year when hospitality was at its peak during summer migrations to the area, the automobile would have insulated Muhammad Ahmad from his surroundings and been an obstacle to the social interactions usually expected at the time. On a larger scale, it reminded the British of the necessity (from their perspective) of establishing and maintaining the fixed borders characterising a modern state. In earlier times, the entire episode may have been safely ignored as an example of the slow, almost imperceptible evolution of 'sovereignty' and influence from one sheikh to another. But by 1950, with oil concessions being granted and Julian Walker's boundary-drawing missions not far off, the British saw a need to clearly define where Ras al-Khaimah territory ended and Bukha's territory (under nominal Muscat sovereignty) began. To do so, they enforced Lorimer's 1908 definition of the 'boundary' between Muscat and Ras al-Khaimah, which remains the UAE–Oman border today.

The rules for travel by automobile were further clarified a few years later following violence in the mountain town of Masafi, where the influence of Ras al-Khaimah met that of the newly-recognised sheikhdom of Fujairah. The Masafi incident demonstrates how the *dawla* became synonymous with automobility. Compared to impoverished coastal towns in the mid-twentieth century, the mountains were both remote and comparatively abundant in water, dates and other resources.[22] They were largely self-governing, although they might have connections to the coast through trade, seasonal migrations and the payment of *zakat*. Masafi's strategic location at the intersection of major routes across the Hajar Mountains gave it added importance. In this case, the two major tribes were the Mahariza (theoretically owing allegiance to Ras al-Khaimah) and the Sharqiiyiin (belonging to Fujairah), but both paid *zakat* to Ras al-Khaimah, presenting the British with a confusing situation. The town's geography did not lend itself to Walker's project of boundary-making, and in fact he left the region's fate to his successors. Disputes would continue well past the founding of the UAE federation in 1971.[23] A boundary could theoretically remove any ambiguity by assigning all land and people to Ras al-Khaimah or Fujairah, but boundary-making created more problems than it resolved.

In an unstable political situation, automobility again provided the spark for violence. As Walker started his work in 1954, Fujairah Ruler Sheikh Mohammed al-Sharqi constructed a wall across a dirt track used by the occasional Land Rover passing through Masafi. While he would have been within his rights to do so in his own territory, the wall lay several kilometres within territory claimed by Ras al-Khaimah. Ras al-Khaimah Ruler Sheikh Saqr bin Mohammed al-Qasimi attempted to visit Masafi in March 1954. Both the wall construction and the visit were provocative acts in the context of the boundary demarcation process. Fujairah guards opened fire on Sheikh Saqr's vehicle, riddling it with bullet holes. The British Political Agent forced Sheikh Mohammed to tear down his wall and pay a fine for his aggression. More importantly, the ensuing negotiations recognised a hierarchy of different forms of travel and means of controlling mobility. Only *ad-dawla* – the Political Agent, the Trucial Oman Scouts and the ruling sheikhs – and motor vehicles could move freely on the road.[24] Travel by foot or animal continued to be subject to the customs and restrictions of pre-oil movement and went unmentioned in the archive. The *dawla* was thus synonymous with travel by automobile, and with a freedom and power of movement that was not necessarily enjoyed by others.

The task of making the *dawla* real would fall to the Trucial Oman Scouts (founded in 1951 as the Trucial Oman Levies), the first modern standing army in the region. Drawing its officers from the UK and Jordan, it was ultimately accountable to the Political Resident in the Gulf, not to the individual ruling sheikhs. The TOS was the core of the *dawla,* distinguished by its use of regular drills, timekeeping, modern weaponry and its transportation – the Land Rover. The TOS served two functions: first, to 'keep the peace' in the hinterlands, and second, to extend the power of recognised coastal ruling sheikhs throughout the bounded territories designated as under their sovereignty. In practice, 'keeping the peace' meant extending the power of British-recognised sheikhs over all of the land within the boundaries drawn by Walker and other British officials, and enforcing the Political Agent's judgements in cases like that of Masafi. This required the removal or submission of tribal leaders and village headmen in the interior, as well as mediating conflicts between the nominal subjects of different British-recognised sheikhs.

The state-building combination of boundary demarcation and automobility was a potent instigator of violence. With oil concessions at stake, territory took on an outsize importance; automobiles were still rare, and few if any social conventions regarding automobile travel had emerged. These factors would again come together in a series of incidents in the village of Madha, located a few kilometres from the Gulf of Oman between Khor Fakkan and Fujairah. Madha was generally understood to belong to the Sultan of Muscat – an island of Muscati territory surrounded by lands belonging to Sharjah and Fujairah – although no borders had been officially defined or demarcated. As in the above example of Bukha, local residents were not used to a Land Rover's noise; and as in the Masafi incident, travel by automobile exacerbated disputes over territory. But automobility also enabled the British to project the power of the nascent *dawla* into the hinterlands in ways that were not previously possible.

On 10 April 1956 in Madha, a man named Ahmed bin Hilal, the local representative of the Ruler of Sharjah, had coffee at the house of Khamis bin Hassan.[25] Khamis bin Hassan lived next door to Mohammed bin Salem, the Sultan's Wali in Madha. Ahmed bin Hilal departed, handed the harness of his donkey to his nephew and 'knelt down to attend to nature'. When he stood up he was shot and killed.[26] Two days later, the Wali of Shinas (another Muscat representative) arrived to 'investigate' the murder, accompanied by several armed men who crossed through Sharjah territory on their way to Madha.

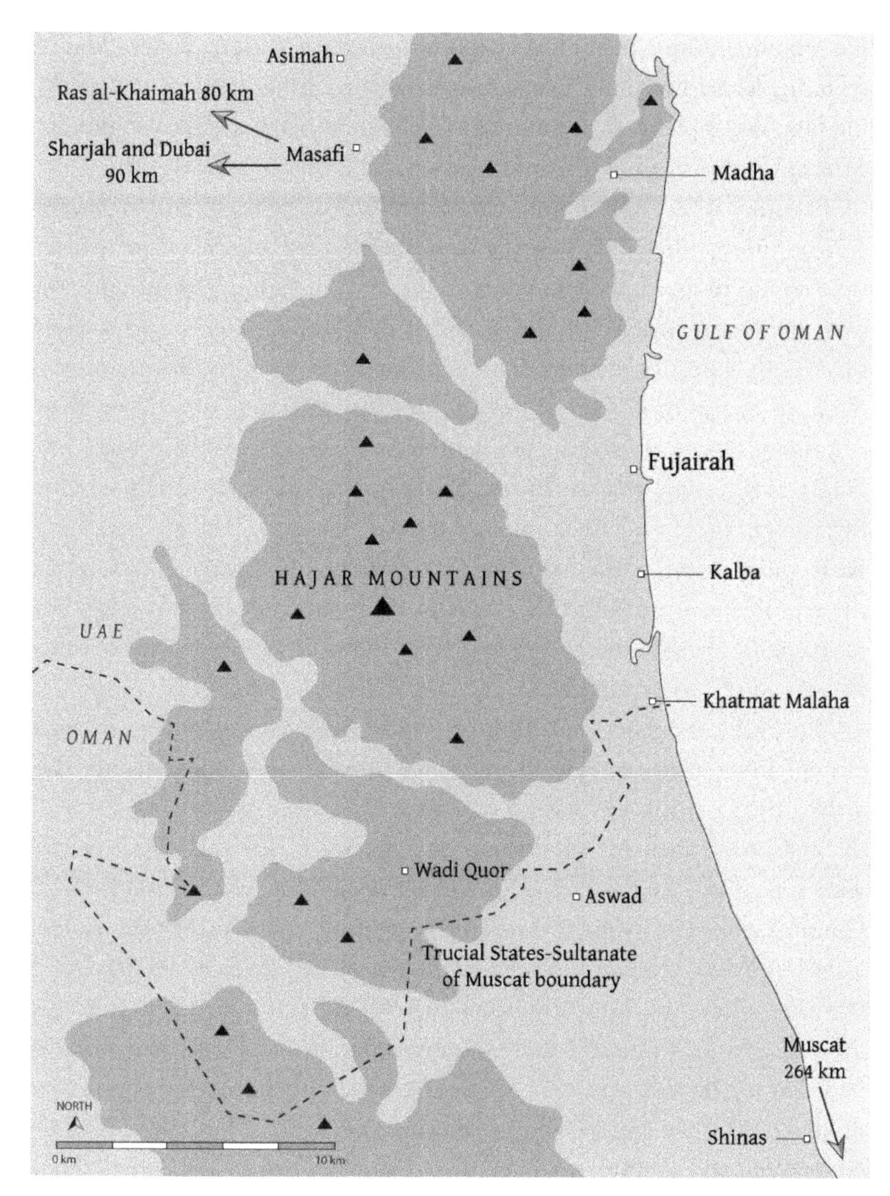

Figure 13.3 Map of Kalba–Aswad–Rafaq–Khatmat Malaha region.

He appointed Ahmed bin Hilal's nephew as the new Muscati Wali in Madha, replacing Mohammed bin Salem and quickly and rather conveniently absolving him of complicity in the murder. In doing so, his goal was to consolidate Muscat's authority in Madha and weaken Sharjah's territorial claims.

In earlier times, this murder would probably have been resolved locally. But in an era when British-recognised sheikhs and the Sultan of Muscat were trying to assert their direct authority in the region – amid questions over the precise extent of the territory over which they could exercise sovereignty – the dispute took on outsize importance and the British Political Agent in Sharjah and Consul General in Muscat became directly involved. Automobility exacerbated the problem while being a key to the ability of the British to intervene quickly. In fact, British intervention was prompted more by the Wali of Shinas's unannounced passage through the Sharjah exclave of Kalba on his way to Madha, which was protested by Sharjah.[27] This posed three problems for the British – how to maintain Muscat's authority, how to maintain Sharjah's authority, and what rights of passage existed between Muscat and Sharjah territories.

The Political Agency sent Ali Bustani (the Agency Clerk and a Jordanian citizen) and a small detachment of Trucial Oman Levies to investigate. They arrived on 17 April, but had to retreat in the face of gunfire. As the Wali of Shinas described it, Ali Bustani and his party 'attacked Wadi Madha with their cars', so the people of Madha responded with gunshots.[28] The Consul General in Muscat noted that 'it seems possible that the sudden arrival of motor vehicles appeared to the people of Madha as an attack'.[29] It was later alleged by Muscat officials that Ali Bustani's party had fired two shots at gazelles while still some distance away from Madha, and that this had alarmed the villagers, who immediately took up defensive positions and fired upon Bustani and his party when they appeared.[30] The confusion of the automobile and gunfire is telling, and may help explain the earlier shooting at Muhammad Ahmad of Bukha as well. In the extraordinarily quiet soundscape of the rural pre-oil Arabian Peninsula, the sound of an automobile would have been a loud and near-violent interruption; the sound of a gunshot was perhaps the only familiar thing comparable to a backfiring Land Rover. The British show of force led to a long-running arbitration process to determine the exact location of the border.

Why was the automobile so critical to state formation? Why did its use lead to violence in situations where older methods of dispute resolution would normally have succeeded? In a region where seasonal migrations of the *dirah* were a critical means of survival, mobility was a highly charged issue. Hidden inside a vehicle, passengers and drivers had new possibilities for evading and supplanting existing social conventions of mobility. The 'carrying trade' quickly became superfluous. In short, automobiles were not subject to older informal spatial regimes and had the power to produce a new spatial order.[31] This spatial order was synonymous with the state. The *dawla*, then, was the institution that was imagined as standing largely above and outside the kinship and other informal relations that formed the basis for spatial control in the pre-oil era. The *dawla* was the institution that did not have to abide by the social and economic customs that governed movement, and the automobile enabled the *dawla* to do this.

Figure 13.4 Sheikh Rashid bin Saeed al-Maktoum (centre) visits the Rover Car Company factory, Birmingham, UK, June 1959, as a guest of the British government. © The National Archives, UK.

The Trucial Oman Scouts, Automobility and Road Construction

Realising a new spatial order based on oil production, boundaries and the governance of automobility required the creation of a road network. And while the TOS is usually remembered for its military and 'peacekeeping' capabilities, the road network it built was perhaps equally as important a contribution to the emerging state.[32] The importance of graded roads – later upgraded to paved roads – became clear in a 1959 confrontation between two fertile villages north of Masafi, Asimah (assigned to RAK) and Tayyibah (assigned to Fujairah). In May 1959 the Ruler of Ras al-Khaimah gave rifles to a group of men from Tayyibah, an act which the Ruler of Fujairah's deputy deemed unlawful; this was followed by a series of raids and counter-attacks until the Trucial Oman Scouts arrived. As in Madha and Bukha, a simple act that in previous generations would not have concerned outsiders and would have been resolved locally now escalated into a conflict between two sheikhdoms in which hundreds of men were mobilised.

Political Agent Donald Hawley used the TOS to enforce an agreement between Ras al-Khaimah and Fujairah that clarified the spatial vision of the *dawla* in two ways: first, no people shall switch allegiances, thus fixing subjecthood and citizenship as well as territory; second, 'traditional freedom of movement' (*hurriyat al-muroor al-taqlid*) must be preserved in the area.[33] This term is left undefined, but surely refers to the patterns of circulation of the *dirah* which crossed the newly-imposed border. Given that this movement was governed by any number of informal conventions, 'freedom' is perhaps not the most accurate word to use. Most likely, Hawley meant that this movement on foot or by animal would be ungoverned by the *dawla*. Following the earlier Masafi agreement that allowed cars and representatives of the *dawla* to pass freely, all traffic was now allowed to cross borders between the Trucial States.

Another legacy of the Asimah-Tayyibah incident was that it made clear the need for a road network to make the long arm of the *dawla* visible in the most remote hinterlands. Although Asimah and Tayyibah were only twelve kilometres from the TOS base in Masafi, it took over eighteen hours for the TOS to reach the villages.[34] The need for a road network was obvious. Over the next decade, the TOS was engaged in an almost constant road construction campaign; at first clear tracks, then graded gravel roads, and eventually paved highways were built, in many cases by the soldiers themselves. Graded roads extended from Sharjah to Masafi and thence to Dibba and Fujairah, the construction of which required the dynamiting of mountain passes and narrow road-beds along the east coast of the Trucial States. This project took on added importance as plans for British withdrawal and a future federation of all seven Trucial States took shape in the late 1960s – a time when automobile ownership dramatically expanded as oil revenues began to flow to the populace. By the mid-1970s, paved highways linked the UAE to Oman and Saudi Arabia and connected all the major cities. Over 50,000 cars were on the road, just below one per family (and above that for UAE citizens), a figure which did not include commercial vehicles or taxis,[35] and probably represents an approximately 500 per cent increase in just over a decade.[36] This was a massive change, enabling the suburbanisation of the citizen population in new housing projects, Emiratis from the north to come to Abu Dhabi and Dubai for work, and much more. The new road network was trumpeted in the UAE's Arabic and English-language press as a sign of progress and national integration.[37] The freedom of movement and new patterns of movement facilitated by roads and automobiles had their origins in the mountainous hinterlands of Ras al-Khaimah and Fujairah. The *dawla* had supplanted the *dirah*.

Automobility and the Borders of the Sultanate of Muscat

Within the Trucial States, automobility brought increased freedom of movement. In contrast, the demarcation of the border between the Trucial States and the Sultanate of Muscat and Oman – the future United Arab Emirates and Sultanate of Oman – brought restrictions on automobile travel in the form of customs posts and tariffs. In the 1950s and 1960s, the border, the automobile and the *dawla* were co-constitutive. Travel by foot and animal – following the conventions of the *dirah* – continued as before. However, the outcome of two separate states was neither inevitable nor obvious to all involved, and the regulatory practices of the border were contested by a range of actors. Automobility and its governance was the key ground upon which these contestations took place. The Sultan of Muscat continually sought to restrict automobile travel from the Trucial States, while interior notables such as Suleiman bin Himyar and ruling sheikhs of the Trucial States sought greater freedom of mobility across the new border. After the independence of both the UAE and Oman, the UAE government used road construction to shift settlement patterns along the border in its favour.

The mid-twentieth century witnessed both the integration of what today is the Sultanate of Oman and its separation from the Trucial States/United Arab Emirates. In the early 1950s, the authority of the Sultan of Muscat extended primarily along the Batinah Coast and into some regions of the interior, but not to the mountains, which were governed by the Ibadi Imamate. Sultan Saeed bin Taimur recognised very early that the key to unifying Oman would be a road network that would extend his direct authority south to Dhofar and north-west to Buraimi, and thus surround his rivals in the Imamate.[38] Since 1920, the Treaty of Seeb had governed relations between the Sultan and the Imam, in which the Sultan (and thus the British) was responsible for external relations but the Imamate remained autonomous. As in the Trucial States, there were various tribes in the interior with tenuous ties to Muscat or the Imamate. Likewise, the Trucial States had deep and long-standing ties to northern Oman, and were culturally and economically part of a larger Oman; the sheikhdoms were sometimes called Trucial Oman or in Arabic *Sahil Oman*, the Oman Coast.

The foundation of not one but two states in historic 'Oman' was not inevitable, but an outcome of a confusing two-decade process beginning with the Buraimi Crisis of 1952–5. The British, on behalf of Abu Dhabi and Muscat, evicted a small Saudi force that had attempted to enforce a Saudi claim to the nine villages in the Buraimi region. The brief Saudi occupation of a strategic (and possibly oil-rich) oasis and ensuing negotiations gave urgency to the projects of extending the Sultan's power and defining the Sultanate's borders with the Trucial States to counter Saudi power. Saudi influence reached deep into the interior of Oman in the form of support for the last Imam, Ghalib bin Ali al-Khalili, who assumed power in 1954. After the ejection of the Saudis from Buraimi, Ghalib was deposed by a combination of the Trucial Oman Scouts and tribesmen loyal to the Sultan; but the Sultan's authority over the Imamate's territory would not be complete until the conclusion of the Jebel Akhdar War in 1959.[39]

Beginning with the Buraimi Crisis, the Sultanate and Trucial States imposed new controls on travellers. Soon after the Saudis occupied Buraimi, the Trucial States Council (all seven ruling sheikhs and the Political Agent) announced a new passport regulation that would be used to deport any 'undesirable alien' from the Trucial Coast. 'Normal traffic', however, could continue unhindered.[40] The Sultan of Muscat had required visas for all visitors other than British and Trucial States subjects since the previous year.[41] This was primarily directed first at Saudis but later at suspected Arab nationalist subversives. These regulations were mainly enforced at airports, but an increasing number of customs posts appeared near the Sultanate's land borders as well.

The complications posed by these new checkpoints can be seen especially on the east coast between Kalba and Sohar, where new borders drawn by Julian Walker and the *dawla* cut through long-standing circulation patterns of the *dirah*. As in the Trucial States, the regulation of automobility was dependent on personal status. In 1954 the long-standing Muscat customs post at Sohar, deep inside Sultanate territory, was replaced by two posts around ninety kilometres north in Khatmat Malaha and Aswad.[42] At the time, most travellers from Sharjah, Dubai and the rest of the towns of the Trucial Coast came through the Wadi Quor if they were going to Kalba, Fujairah and the east coast. The rulers of Sharjah, Ras al-Khaimah and Fujairah complained that their merchants faced onerous inspections and taxes at the Aswad customs post, even if they were only transiting through Sultanate territory.[43] After much debate the Sultan relented, allowing transiting merchants to proceed duty-free from Aswad to Khatmat Malaha with an armed escort to prevent smuggling.[44] Yet only a month later, Muscat soldiers blocked the Ruler of Sharjah's car as he passed through the Khatmat Malaha post on his way to Kalba, with his retainers knocking down a wall the Muscat government had constructed for good measure.[45] The British insisted, and the Sultan agreed, that all Trucial States ruling sheikhs could pass through Aswad and Khatmat Malaha unhindered and without prior notice through Muscat territory.[46] The agreement that representatives of *ad-dawla* – in this case the ruling sheikhs – had the right to free movement mirrored understandings within the Trucial States developed a few months earlier in the wake of the Masafi incident described above. But in the case of the 1956 murder in the Omani exclave of Madha, the ruler of Sharjah objected when the Sultan's Wali transited through Sharjah territory on his way to Madha. As British diplomats pointed out, free transit could hardly be granted to the Trucial States rulers in the Sultanate, but not to the Sultan's representatives in the Trucial States. This question was never clearly answered; perhaps Madha's remote location and small population meant that it was no longer an important problem after tensions faded.

Contestation and Control along the Sultanate Border

The increasing control of mobility by the *dawla* – both of the Trucial States and Muscat – did not go unquestioned or unchallenged. One alternative vision was articulated by Suleiman bin Himyar al-Nabhani, the self-styled ruler of Jebel Akhdar, a leader of the Ghaffiri tribal confederation that stretched across the Trucial States and Oman and was one of the major players in the struggle for the Omani interior.[47] He allied himself variously with the Imamate (and hoped to become Imam himself in 1954), the Saudis (as a leader of their 'faction' within the Imamate during the Buraimi Crisis) and Nasser (in resisting the Sultan's attempts to control Jebel Akhdar in the late 1950s). His ultimate ambition, like that of Muhammad Ahmed of Bukha, was to become a ruling sheikh recognised by the British; he would then be willing to grant a concession for oil exploration in the Omani interior.[48] Suleiman asked for permission to pass through customs posts in his car unchallenged, as did other recognised ruling sheikhs. At one point Suleiman's political opponents attempted to burn his car, an act the explorer Wilfred Thesiger attributed to a fanatical religious opposition to technology (which mirrored Thesiger's personal views) but which may have had a host of other causes.[49] Suleiman was hardly intolerant of outside influences, telling a visiting Omani employee of the British Consulate in Muscat:

> We have now opened our eyes by travelling abroad and have met people and learned their ideas: the way they enjoy their freedom in their country peacefully. We should now try to cope with them. Our aim is to make Oman into one nation and everyone should have equal right. There should be no Ghaffiri or Hinawi and an Omani should be able to walk through Oman freely ...[50]

In saying this, Suleiman gave voice to an alternative spatial imaginary for a Greater Oman in which Muscat, Oman and the Trucial States were one *dawla,* with one national identity. Naming the two major tribal confederations in the area (Ghaffiri and Hinawi), he criticises both the tribally-based regulations of mobility in the *dirah* and the new borders of the *dawla* that divided Omanis from each other. As one British diplomat noted, the border between the Trucial States and Muscat divided regions in which mobility was 'an immemorial practice ... [the people on both sides of the borders] are related and have interests in common in agriculture and in local customs; in fetching firewood and in grazing their flocks and a hundred and one other things'.[51] Many Trucial States subjects, for example, owned date palm farms on the Batinah Coast which they visited during summer.[52] In the coming decades, though, these older patterns of movement would come under increasing scrutiny.

Over the 1950s and 1960s, the Sultan's efforts to tighten control of the borders with the Trucial States, resulted in a patchwork of regulations and enforcement, dependent on the personal status of the traveller, type of mobility, activity and place.[53] One of his central demands was that Trucial States subjects obtain visas to enter Muscat territory, a departure from his previous policy. But he eventually made the same critical distinction between 'traditional' and vehicular traffic that had been made in the Trucial States. People who lived near the border were allowed to move to graze animals, visit wells and so on in their traditional *dirah.*[54] However, other nominally 'traditional' activities, like Sheikh Zayed's hunting trips to Muscat territory, were considered by the Sultan to require special permission because hunting from a motor vehicle was a non-traditional activity.[55] In general, automobiles were required to register at checkpoints because of worries about smuggling and gun-running to rebels in the interior; automobility thus posed a real challenge to the Sultan's control of his territory. The requirement that those travelling by motor vehicle have some form of official permission led to a series of increasingly complex regulations. By 1957, four truckloads of goods could transit the Sultanate duty-free if they were bound for the Trucial States via Aswad and Khatmat Malaha, and the vehicles of *ad-dawla* and the ruling sheikhs were exempt from inspection.[56] But the Sultan also held that these unwritten exemptions at Aswad could not be extended to Dubai and Abu

Figure 13.5 A Trucial Oman Levy outside a checkpoint in Mahadha, Oman, near Kahil; this may have been where Suleiman bin Himyar's car was stopped. © National Archives, UAE.

Dhabi, and a customs post at Kahil between Dubai and Buraimi would continue to collect revenue even on traffic bound for the Abu Dhabi side of the Buraimi Oasis. Zayed's wish that special 'sheikhly number plates' would give him some immunity from being checked by the Sultan's customs officers was not granted.[57] Cars belonging to Abu Dhabi sheikhs were regularly stopped by the growing Muscat police presence in the Buraimi region in the early 1960s; they assumed that they had access to Muscat territory but did not extend the same favour to Omanis driving into Abu Dhabi territory.[58] The *dawla* gained strength as an ad hoc series of arrangements took shape along the future border.

The power of automobility also enabled ordinary people to contest the new border regime in more mundane ways. Two cases involving alleged cross-border 'kidnappings' of women stand out in this regard, since women rarely appear in the British archive. But since they involved automobiles, they came under the purview of the *dawla*. A Wali of Khor Fakkan (owing allegiance to Sharjah) on the east coast of the Trucial States repeatedly ran into trouble because he drove his vehicle too fast for the Sultan's officers at Khatmat Malaha to stop him. But he used a return trip to Khor Fakkan to pick up a woman in Sohar named Aisha bint Ali and smuggle her out of Sultanate territory, violating a law that banned Omani women from leaving the Sultanate without the Sultan's permission. As it happened, the woman's husband was a Sharjah subject and the head mechanic for the Ruler of Sharjah's cars in Khor Fakkan. Aisha bint Ali was returned to Sohar, a proper application was filed for her to go to Khor Fakkan and live with her husband, and permission was granted after she gave birth to their child in the Sultanate.[59]

The ending to a second 'kidnapping' in 1965 is murkier. A Muscati subject, Khalfan bin Salim al-Washashi, was accused of abducting Moaza bint Rashid al-Washashiyah while she was grazing goats along the Batinah coast and then taking her to Dubai in his motor vehicle, where the two stayed in the house of Dubai Ruler Rashid bin Saeed. Her husband appeared to request her release, but Sheikh Rashid refused, and Muscat authorities appealed to the British.[60] Moaza bint Rashid claimed that she had not consented to marry her 'husband' – her cousins had attempted to force her to do so, so that one of them could marry the 'husband's' sister – and that since she was still a virgin, he had no legal authority over her and she could not be turned over to him.[61] The Sultanate authorities then claimed that she should have taken her case to the sharia courts there, and thus, she should be returned to the Sultanate.[62] And there the paper trail ends; Moaza's fate is lost to us. Yet it is unlikely that British officials would have taken notice of either Moaza or Aisha bint Ali had they not been travelling by automobile in violation of the new border regulations of the *dawla*.

Oman and the United Arab Emirates

The increasing control of motor vehicle movement between the Sultanate and the Trucial States during the 1950s and 1960s was a sign that this deeply interconnected region would not become a single independent state. At a larger political level, this was evident in the separation between the Trucial Oman Scouts (increasingly limited to action in the Trucial States) and the Sultan's Armed Forces, and the separate British political posts in Dubai, Abu Dhabi and Muscat. Automobility made this division manifest on the ground. When the negotiations for an independent confederation between the seven Trucial States, Bahrain and Qatar began in 1968, Oman was not included even though some British diplomats recognised that the Trucial States resembled Oman much more than they did Bahrain or Qatar. They were linked by geography, tribal composition, labour migration, seasonal migrations, and political opposition to Saudi influence.[63] Trucial States rulers, intent on modernisation, were understandably critical of Sultan Saeed bin Taimur's conservative policies; while some held out hope that Oman might be included in a federation after a new Sultan took power, this did not happen.[64]

In 1970, Qabus bin Said overthrew his father and began his project of modernising the Omani state. In 1971, the Trucial States became the United Arab Emirates, and the still largely undemarcated border between the UAE and Oman became an international border. Both states began massive spending programmes that dramatically improved the standard of living for their citizens. Automobiles became commonplace and the seasonal migrations of the *dirah* came to an end. Houses, roads and modern utilities became the norm.

Post-independence road construction reconfigured the space of the Dhahirah, the region south of Buraimi, creating both new patterns of mobility within and between Oman and the UAE. As in other areas, local people had long moved back and forth across what became the Abu Dhabi–Oman border in the 1950s. That border remained undemarcated on the ground, and after the accession of Sheikh Zayed bin Sultan al-Nahyan, many moved to the Abu Dhabi side and built temporary new homes. Many received Abu Dhabi passports and state benefits as well.[65] The Abu Dhabi government then built permanent homes and the oil minister, Mani al-Otaiba, suggested

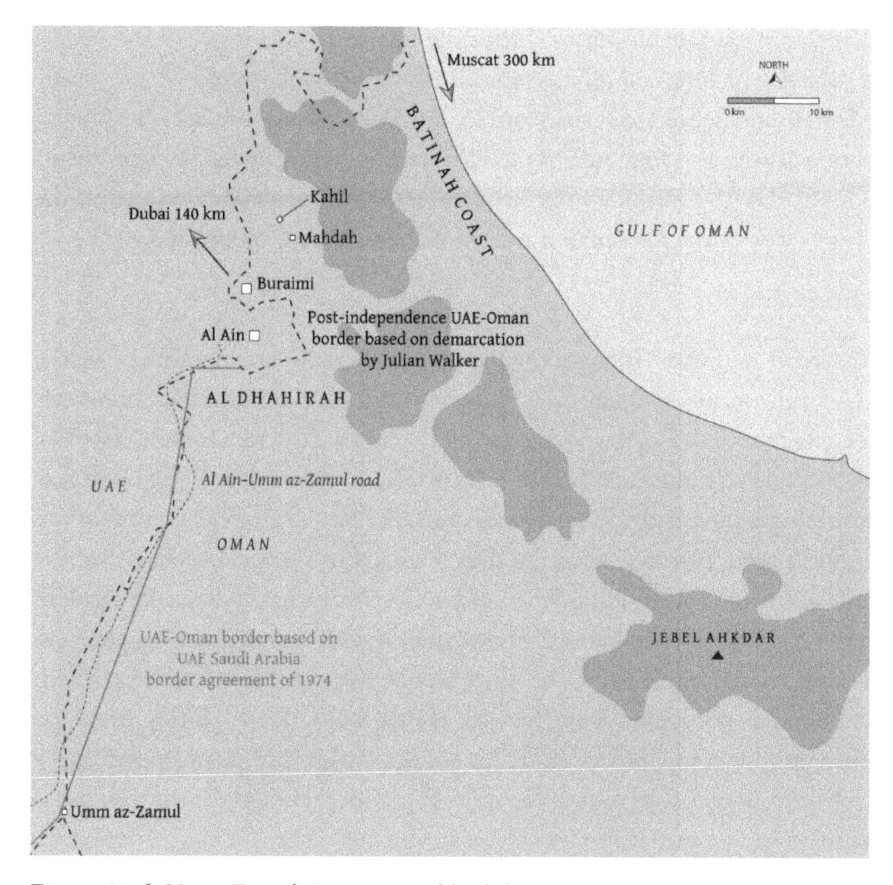

Figure 13.6 Umm Zamul–Buraimi road built by UAE government in Omani territory. NAUK, FCO 164/32, 'Saudi Arabia–Qatar–UAE–Oman Frontier Areas' map. The solid line represents the UAE border as defined by the UAE–Saudi Arabia border agreement of 1974, which had to include Buraimi as Omani and UAE territory. The dotted line represents the UAE–Oman border as demarcated by Julian Walker and other British officials.

building a tarmac road from Al Ain along the border.[66] The oil ministry did so, but without the benefit of the maps drawn up by Julian Walker and Martin Buckmaster over a decade before, and the road intruded well into Omani territory. Not only that, but an Abu Dhabi police post was stationed along the road on the Omani side of the border, and the road made it easier for Omanis in the area to access Al Ain and Abu Dhabi rather than Ibri and Muscat. Furthermore, the Omanis objected that the road went to Al Ain and thence to Abu Dhabi, not to Ibri or the interior where many of the people it served had long-standing connections.[67] But while the UAE eventually relented and rebuilt the road on its own side of the border, the Omani nationals who had migrated to the UAE largely stayed there. The settlements built by the Abu Dhabi government along the road remain as well, while the Omani side of the Dhahirah is today much less densely populated.

Conclusion

The great transformation occurring in the mid-twentieth century in the UAE and northern Oman saw the emergence of two states and large-scale reconfiguration of patterns and mobility in an area sometimes described as borderless in the time before oil. In fact, as the carrying trade and seasonal migrations of the *dirah* demonstrate, the pre-oil Arabian Peninsula was hardly a region with 'freedom of movement'. But since it largely lacked the official customs posts, boundaries and police forces that characterise modern states, historians – and many Emiratis and Omanis – sometimes imagine it as a time of freedom. It would be more accurate to say that the mid-twentieth century transformation brought with it new ways of disciplining mobility. Automobility and borders, and the political negotiations and occasionally violent resistance they encountered, were a critical component of that transformation. A Land Rover could easily disrupt older life worlds based on seasonal migrations, moral relations and informal local understandings and traditions. While British diplomats drew boundaries, it was motor power and a road network that brought those boundaries to life. The noise, speed and power of the automobile became synonymous with the violence and power of the modern state.

Two life worlds – the *dirah* and the *dawla* – coexisted uneasily during the 1950s and 1960s. Yet while the *dawla* incorporated and subordinated the older spatial and political order of the *dirah*, this transformation and the forms it took were hardly foregone conclusions. In Bukha, Masafi, Asimah and Tayyibah, Madha, and all along the UAE–Oman boundary, new borders and new forms of mobility were contested by actors ranging from local headmen to ruling sheikhs, the Sultan of Muscat, British diplomats, and ordinary people, including women. Nor was the emergence of two states, each with freedom of movement within its British-drawn borders, the only alternative to the 'traditional' spatial imaginaries of the mid-century world of *al-dirah*. Suleiman bin Himyar proclaimed the right of every Omani to travel freely in all of historic Oman – while asking to be recognised as a ruling sheikh and offering oil concessions in the interior. These episodes and vignettes serve as an important reminder that oil and state formation were not only manifested in the formal politics of concessions and boundary-making, but in automobility, road construction and everyday life.

Notes

1. Before independence and confederation in 1971, the seven emirates that make up today's UAE were known as the Trucial States (or 'Trucial Coast', 'Trucial Oman' or 'Trucial Sheikhdoms'). The name derives from the truces that were signed to prevent 'piracy' in the early 1800s, which were made 'perpetual' in 1853. The various sheikhdoms became semi-independent states under British protection, with their foreign affairs managed by the Government of India until 1947. Bahrain, Kuwait and Qatar had similar relations with the British Empire. Until 1970, the territory of today's Sultanate of Oman was known as the Sultanate of Muscat ('Oman' referring exclusively to the mountainous interior ruled from 1920 by the Ibadi Imamate based in Nizwa) and, later, the Sultanate of Muscat and Oman. The Sultanate's position in the British Empire was not unlike that of the other Gulf states, although the British were more directly involved in internal affairs than elsewhere, and the Sultanate's relations with Britain were not governed by 'truces' but by a series of treaties signed over the course of the 1800s.
2. Rosemarie Said Zahlan, *The Origins of the United Arab Emirates: A Political and Social History of the Trucial States* (London: Macmillan, 1978), 107–24; Muhammad Morsy Abdullah, *The United Arab Emirates: A Modern History* (London: Croom Helm, 1978), 291–315; Frauke Heard-Bey, *From Trucial States to United Arab Emirates* (London: Longman, 1982), 300.
3. William Lancaster and Fidelity Lancaster, *Honour is in Contentment: Life Before Oil in Ras al-Khaimah and Some Neighboring Regions* (Berlin: de Gruyter, 2011), 257–8.
4. The escort was in effect a sponsor whose protection could be counted on in alien territory; there are occasional accounts of similar practices at sea. Salem bin Ibrahim al-Saman, *My Life: Part I* (Abu Dhabi, 2008), 66.
5. Lancaster and Lancaster, *Honour is in Contentment*, 300.

6. NAUK: Lane to Bird, 19 November 1951. A later description of the Bani Kaab, a tribe north-west of Buraimi, mentions 'their capital, Mahadha'. NAUK: FCO 164/12, Howes to Prior, 1 April 1940.
7. Julian Walker, *Tyro on the Trucial Coast* (Durham: The Memoir Club, 1999), 109–21; NAUK: FCO 164/29.
8. NAUK: FCO 164/26, Pirie-Gordon to Residency, 9 April 1955.
9. For example, the UAE signed a final border agreement with Oman in 2008. WAM, 'Historic UAE–Oman Accord Involves 272 km of Border', *Gulf News*, 22 July 2008 http://gulfnews.com/news/uae/government/historic-uae-oman-accord-involves-272km-of-border-1.119592 (last accessed 29 January 2022).
10. NAUK: FO 371/114689, Eyre, 'Agriculture in the Persian Gulf: Report on a Visit to Bahrain, Trucial Oman, and Kuwait, February 28, 1955 to March 20, 1955'. Lancaster and Lancaster, *Honour is in Contentment*, 436–7.
11. These included, among others, the coastal towns of Rams and al-Jazira al-Hamra, and the interior tribe of the al-Khawatir. See Matthew MacLean, 'Spatial Transformations and the Emergence of "the National": the Formation of the United Arab Emirates, 1950–1980' (Ph.D. diss., New York University, 2017), 111–33, 268–86. See NAUK: FO 1016/44, 'Precis of Dispute between Rams and Ras al-Khaimah'.
12. See also Citino in this volume for the role of motorised driving in organising spatial relations in the Eastern province of Saudi Arabia.
13. Christopher Davidson, *Dubai: The Vulnerability of Success* (New York: Columbia University Press, 2008), 33–5.
14. Michael Herb, *The Wages of Oil: Parliaments and Economic Development in Kuwait and the UAE* (Ithaca: Cornell University Press, 2014), 80–2.
15. Rosemarie J. Said, 'The 1938 Reform Movement in Dubai', *Al-Abhath* 23, nos 1–4 (December 1970).
16. J. G. Lorimer, *Gazetteer of the Persian Gulf, Oman, and Central Arabia*, Vol. II A (Bombay: Superintendent Printing, 1908), 1,004. NAUK: ADM 1/21901, Hay to Foreign Secretary, 5 December 1950.
17. NAUK: FCO 164/15, Political Agent to Hay, 25 January 1947.
18. NAUK: FO 1016/44, Wilton to Finhey, 8 June 1951.
19. NAUK: FO 1016/32, Chauncey to Resident, 14 August 1950.
20. NAUK: ADM 1/21901, D. J. Godden, 'Enclosure "B(3)" to Flamingo's Letter', 23 July 1950.
21. Coincidentally (or not?), Muhammad Ahmad's wife was a member of Dubai's ruling family.

22. Lancaster and Lancaster, *Honour is in Contentment*, 156, 185, 226, 453.
23. Hesam al-Ulama, 'The Federal Boundaries of the United Arab Emirates' (Ph.D. diss., Durham University, UK, 1994), 270–1.
24. NAUK: FO 1016/375, 'Masafi', account of visit on 20 March 1954.
25. Today Madha is an exclave of Oman entirely surrounded by UAE territory; in turn the village of Nahwa is an exclave of Sharjah territory entirely surrounded by Madha.
26. NAUK: FO 1016/543, 'Report of Visit to Wadi Madha 15/16 April, 1956'.
27. NAUK: FO 1016/543, Tripp to Bahrain Residency, Telegram #81, 14 April 1956.
28. NAUK: FO 1016/542, Wali of Shinas to Minister of the Interior Sayyid Ahmad Ibrahim, 18 April 1956. Sadly, the original Arabic has been lost.
29. NAUK: FO 1016/543, Chauncy to Burrows, 26 April 1956.
30. NAUK: FO 1016/543, Chauncy to Burrows, 9 May 1956.
31. Mandana Limbert, *In the Time of Oil: Piety, Memory, and Social Life in an Omani Town* (Stanford: Stanford University Press, 2010), 33.
32. Donald Hawley, *The Trucial States* (London: Allen & Unwin, 1970), 176.
33. NAUK: FO 1016/669, Hawley to the Rulers of Ras al-Khaimah and Fujairah, 26 June 1959.
34. NAUK: FO 1016/669, 'Report on Incident between Ras al-Khaimah and Fujairah States in the Asimah/Tayyibah Area 18–20 May, 1959'. NAUK: WO 337/3, Trucial Oman Scouts Intelligence Report No 98, 3 October 1959.
35. Jamal al-Mehairi, 'The Role of Transportation Networks in the Development and Integration of the Seven Emirates Forming the United Arab Emirates, with Special Reference to Dubai' (Ph.D. diss., Durham University, UK, 1993), 250–2.
36. For example, in 1967 there were only 91 motor vehicles in Ras al-Khaimah, excluding TOS vehicles. NAUK: AY 4/2942, 'Report on a Visit to the Trucial States by Mr. Paul Cheshire, January/March 1967'.
37. 'Roads Play Role in Fostering Integration', *Emirates News*, 3 April 1978, 'Sharayeen al-Haya al-Hadeetha' (Arteries of Modern Life), *Al-Ittihad*, 2 December 1978.
38. NAUK: FO 371/104303, Bahrain to Foreign Office, 9 July 1953. This road network would not be realised until Said had been deposed by his son Qabus in 1970.
39. Jeremy Jones and Nicholas Ridout, *A History of Modern Oman* (Cambridge: Cambridge University Press, 2015), 113–20, 125–31.

40. NAUK: FO 371/104292, D. A. Greenhill, 'Saudi Arabian Frontier Dispute: Problem of Isolating Members of Turki's Party', 25 April 1953.
41. NAUK: FO 1016/198, Le Quesne to Muscat, 25 March 1952, and Chauncy to Bahrain, 5 August 1952.
42. NAUK: FO 1016/380, Walker to Sheikh Saqr bin Mohammed al-Qasimi, 3 August 1954, Pirie-Gordon to Muscat, 19 August 1954.
43. NAUK: FO 1016/380, Minute, 'Aswad Post'.
44. NAUK: FO 1016/380, Tristram to Political Agency Dubai, 19 September 1954.
45. NAUK: FO 1016/344, Pirie-Gordon to Muscat, 23 October 1954.
46. NAUK: FO 1016/344, Tristram to Sayyid Ahmad Ibrahim, 4 November 1954.
47. See for example his letter to al-Ghaylani, NAUK: FO 1016/355, 4 April 1954.
48. Wilfred Thesiger, *Arabian Sands* (New York: Penguin, 1991), 324; NAUK: FO 1016/354, 'Report by Mr. H. A. Ghaylany of Muscat Consulate', 10 October 1954.
49. NAUK: FO 371/104279, 'W. Thesiger's Notes – February 1953'.
50. NAUK: FO 1016/354, 'Report by Mr. H. A. Ghaylany of Muscat Consulate', 10 October 1954.
51. NAUK: FO 1016/817, Boustead to Phillips, 29 January 1963, Boustead to Phillips, 6 February 1963.
52. NAUK: FO 1016/817, Burton to Duncan, 1 July 1963, lists dozens of such subjects.
53. NAUK: FO 1016/818, Phillips to Brown, 25 April 1963.
54. NAUK: FO 371/148960, Middleton to Sheikh Shakhbut bin Sultan al-Nahyan, 10 September 1960.
55. NAUK: FO 371/148960, Phillips to Middleton, 10 September 1960.
56. NAUK: FO 1016/593, Chauncy to Burrows, 17 September 1957, and 12 December 1957.
57. NAUK: FO 1016/817, Phillips to Boustead, 8 February 1963.
58. NAUK: FO 1016/818, 'Extracts from Major Budd's Buraimi Report No S/1 dated 9 Jan 1963'.
59. NAUK: FO 1016/816, Balfour-Paul to Ruler of Sharjah, 4 May 1965. Tait to Pragnell, 28 July 1965, Pragnell to Tait, 26 October 1965.
60. NAUK: FO 1016/773, Secretary for External Affairs to Carden, 21 August 1965.
61. NAUK: FO 1016/773, Carden to Secretary for External Affairs, 28 October 1965.

62. NAUK: FO 1016/773, Secretary for External Affairs to Carden, 11 November 1965.
63. NAUK: FCO 8/849, Carden to Crawford, 20 April 1968.
64. NAUK: FCO 8/1077, Bullard to Crawford, 8 July 1969.
65. NAUK: FCO 8/2662, McCarthy to Allen, 19 June 1976; FCO 8/2900, McCarthy to Treadwell, 30 April 1977; FCO 8/2662, Treadwell to Day, 24 May 1976.
66. NAUK: FCO 8/3145, Harris to Walker et al, 17 November 1978.
67. NAUK: FCO 8/2891, Minute by Treadwell to Tatham, n.d. FCO 8/3145, Treadwell, 18 October 1978; FCO 8/3145, Lucas to Weir, 20 October 1978.

14

CONTESTATION AND CO-OPTATION IN THE DESERT LANDSCAPES OF OMAN

Dawn Chatty

Mobile peoples in the Middle East and elsewhere in the world have faced enormous pressure throughout the twentieth and twentieth-first centuries to change their way of life, to settle down and remain in one place. The notion that a settled existence is more modern than a mobile one continues to dominate expert thinking as a continuation of late nineteenth-century social evolutionist theories of the progress of civilisation.[1] Most of the modern nation-states of the Middle East have approached their mobile pastoral peoples with a determined view to making them stay put in one place and give up their pastoral subsistence livelihoods. Settlement schemes, it was assumed, would assure political and economic control over these difficult-to-emplace-and-control peoples. The development aid efforts, both bi-lateral and international, throughout the twentieth century followed these same biases and were designed to make mobile or nomadic peoples 'modern', using principles developed during the colonial era such as *terra nullius*, which declared all land not held privately as empty, and thus belonging to the state so that it could be disposed of or developed as the state wished. By the end of the twentieth century, most pastoral peoples' grazing lands had been expropriated and sedentarisation schemes of one sort or another were the mechanisms of choice. Pastoral peoples in Oman, however, had some success in challenging the notion of *terra nullius* in the deserts of the country. A younger generation of 'citizen' herders have been able to parlay further multinational oil industry intervention to support their continued mobility in the deserts of Oman and subsistence pastoral livelihoods.

I begin the chapter with a brief examination of the ways in which mobile pastoral communities in the Middle East have faced and then navigated around government land expropriation and sedentarisation efforts to create multi-resource livelihood successes without always being forced to settle. I then examine the situation in Oman, where a more 'enlightened' state policy regarding settlement was enacted and where oil concerns have been paramount. Determined to provide social benefits to its mobile pastoral communities without forcing them to settle, the government of Oman extended basic services to these communities late into the twentieth century. These provisions, I argue, gave the isolated and remote pastoralists in the country a breather, a space in which to catch up with the rest of the rural population of the country and become 'citizen' herders. However, in spite of this radical policy, it has become clear four decades on that the pastoral community has continued to be generally neglected and poorly understood by government bureaucrats, as well as some oil company employees, in comparison with the rest of the Oman's citizens. Furthermore, it is the general lack of a meaningful relationship with some of the oil companies whose concession areas extend over most of their traditional tribal lands which has highlighted the fundamental disadvantage of this 'stakeholder' group.

The lack of interaction with the state and with the oil companies is best understood, at the most basic level, by the legal-geographic doctrine of *terra nullius* which emerged during the colonial era and first played out in the Middle East during the mandate period between World Wars I and II and then continued into the modern nation-state period. This doctrine was used to justify and legalise European seizure of territory across the world. Where land was not held privately, as in cities or agricultural regions, it was often declared *terra nullius* or *territorium nullius* (territory belonging to no one) and thus served as justification for colonial dispossession of traditional lands, territory and sovereignty.[2] These principles were particularly important in the Gulf region in the 1970s when British advisors encouraged the newly recognised nation-states to declare all land not held privately as 'state land'. In general, *terra nullius*/state land was applied to all land not possessed by any person or used in ways recognised by European legal systems. Thus, pastoral tribal lands held in common for animal grazing were not recognised as 'occupied', and as pastoralists, nomadic herders were often regarded as trespassers on their territory, lands they had occupied for centuries, as the International Court of Justice Advisory Opinion on the Western Sahara case in 1975 set out.[3]

Mobile Pastoral Societies in the Middle East

The mobile pastoral tribes of the Arabian Peninsula are often referred to as Bedouin, a term derived from the Arabic word, *bedu*, meaning an inhabitant of the *Badia* – the large stretch of semi-arid land or desert that makes up nearly 80 per cent of the Arabian Peninsula. They have, for centuries, pushed their frontier regions into border areas of agricultural settlement and have, as often, been repulsed when central governments have had the strength to do so.

At the close of World War I, the semi-arid lands of the *Badia* of Northern Arabia were divided up and distributed, under League of Nations 'mandate status', to France and Great Britain. The southern wedge alone remained in the hands of ʿAbd al-ʿAziz Ibn Saʿud, the founding father of Saudi Arabia, and the Gulf emirates remained under political influence of the British government. This step, along with the subsequent establishment of British and French administrations and influence in their respective regions, was accompanied by a period of competitive oil exploration on the back of the principle of *terra nullius*.[4] Borders were drawn to support the oil pipelines between Kuwait and the Mediterranean often cutting through traditional Bedouin tribal grazing territory; the Haifa line, for example, was drawn by Mark Sykes in a straight line from Acre to Kirkuk and defined the national frontiers between the mandated states of Jordan, Syria and Iraq.[5] These new nation-state borders, many prophesied, would spell the death of the mobile pastoral herding Bedouin. However, these barriers to movement were quietly ignored and adjusted for by the Bedouin, who adapted and altered their migrations in order to maintain their pastoral subsistence livelihoods. In the second half of the twentieth century, development experts working in Northern Arabia came to regard mobile pastoral peoples with scorn, if not disdain.[6] Many of these experts assumed that the mobile pastoral peoples were 'trespassers', who were in the way of development efforts and were also ruining their physical

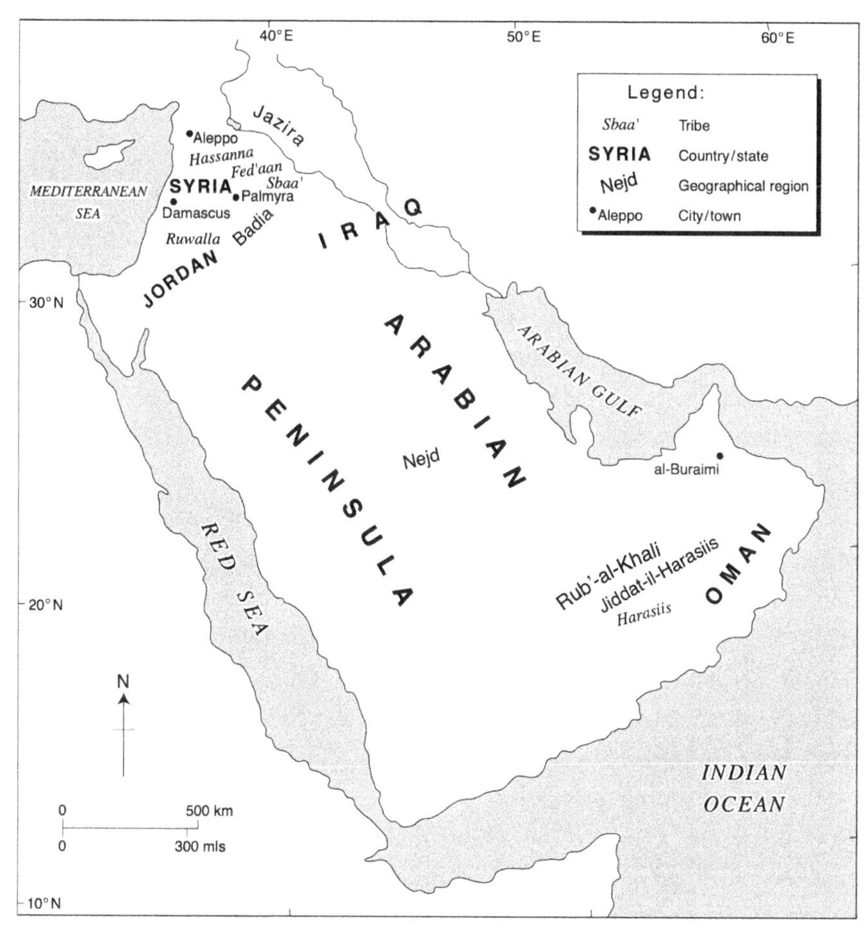

Figure 14.1 Map of the Arabian Peninsula and Fertile Crescent.

environment, thus providing further evidence for measures to get them off their traditional desert lands and settled in peri-urban landscapes.[7]

Governments in central and south-eastern Arabia, perhaps because they regarded their pastoral populations as 'weather-vanes' of internal political problems, sought local rather than international solutions to the issue of sovereignty over the desert regions. Accordingly, government policy in these regions has been less overt in expropriating mobile herders grazing lands and rather more directed at 'emplacing' these peoples in social housing allotments. Moving people out of the desert – where oil exploration was taking place – and settling them along the desert's edge came to be seen as the only way to control and integrate marginal and problematic populations who did not conform to the aspirations of the newly created modern nation-state.

Twentieth-century Settlement Efforts, Expropriation and their Contestation

In the northern part of the Arabian Peninsula, governments have attempted, for decades, to lure the mobile pastoralists out of the deserts and arid rangelands to housing settlements and agricultural pilot projects. These attempts, in large measure, failed. More forceful approaches have included revoking the traditional communal land holdings of these people by declaring their traditional grazing territory (*dar*) state land using the principle of *terra nullius* and creating effective barriers to access by livestock. In Lebanon, starting in the 1920s, French mandate policy of selling communal grazing lands in the Bekaa Valley as private holdings to local farmers has continued into the second half of the twentieth century, greatly restricting the movement of Bedouin herds and herding activity. This disregard for communal tribal land (*musha'*) was effectively expropriation of land held communally for centuries by the Bedouin in the Bekaa.[8]

The case of mobile pastoralists in Syria is instructive. With the discovery of oil in the region in the 1930s, the French Mandate power became concerned with safeguarding this potentially important international investment and ensuring there was no challenge to their sovereignty. Realising that inter-tribal Bedouin raiding and skirmishing were affecting the laying and protection of oil pipelines from the interior to the Mediterranean coastline, the French vigorously pacified the area, stripping the tribes of their semi-autonomous status and co-opting the Bedouin leadership into the urban elite of Damascus, Hama and Aleppo. This was accomplished largely through expropriation of the traditional tribal lands (*musha'*): grants of private ownership (*muluk*) of some of these large areas of communal land to tribal leaders; giving voting rights in Parliament to tribal leaders who co-operated with the French colonial power; granting privileged access to foreign education for the sons of these tribal leaders; and significant monetary compensation.[9]

In Saudi Arabia and the southern region of the Arabian Peninsula, the situation of mobile pastoral peoples has been complicated first by the discovery of oil, and more recently, the tremendous wealth that has come into the hands of these governments. Saudi Arabia has for decades tried to settle its large mobile pastoral peoples, beginning in the 1920s with schemes to control their movements and strip them of access to their traditional communal tribal lands. Issues of sovereignty over the desert landscape, which governments feared might potentially challenge eventual oil revenue, meant government decrees to convert these communal land tenure systems into state land, at the same time as the setting up of settlement projects. In later decades government settlement projects were even financed locally from oil revenues, but predictably the schemes failed. The large Wadi al-Sarhan Project and the King Faysal Settlement Project are two such examples. Here, mobile pastoralists, discouraged by these attempts to undermine their communal land holdings and turn them into settled tillers of the soil, simply flitted away having seen their traditional territories expropriated by the state; others adapted their livelihoods and turned to new endeavours more compatible with pastoralism, such as working in the transport industry or setting up trade ventures that capitalised on their mobility.

With the huge increase in the profit from petroleum extraction which the Arabian Gulf States, Saudi Arabia and the Sultanate of Oman experienced in the early 1970s came a new approach to the 'problem' of mobile pastoral tribes. Instead of mass settlement schemes and explicit land expropriation came 'enticements' to individual citizen pastoralists. Control, in a political sense, was attempted by encouraging the individual tribesman to come forward and register himself as a citizen. In return, these governments granted various privileges. In the wealthier states, with very few, settled populations, registration carried with it an entitlement to a plot of land, a house, an automobile and a subsidy for each head of livestock. In other states, registration meant a monthly stipend – generally in the region of the local equivalent of several thousand US dollars – often disguised as a salary for some form of national paramilitary service. The aim was the same, expropriation of land cleared largely for uncontested access to oil extraction and its revenues.

The Case of Oman: Co-optation, Adaptation and Contestation

The six largest mobile pastoral tribes of Oman divide themselves into two groups: those descended from Qahtan and are largely Hinawi supporters, and those whose origins go back to Nizar (Adnan) and are Ghafiri supporters. The main Qahtani tribes are the Mahra, the Jeneba, the Beit Kaythir and the Harasiis.[10] Increasingly over the past few decades, members of these mobile pastoral tribes have also begun to self-identify themselves as *'Arab* or Bedouin. Their livelihoods are based on domesticated livestock – mainly camel and goat – that can exist and sometimes thrive on the natural graze and browse of the desert, which regularly sees temperatures of 50 degrees Celsius and as little as 50 mm of rainfall a year. The well-being of these herds, and by association these peoples, depends upon their access to the widely spread-out grasses, trees and shrub cover throughout the desert region in response to infrequent and unpredictable rainfall and seasonal heavy occult precipitation. Livestock is key to this way of life, and mobility a basic feature of herd management and cultural identity. The daily routine, the seasonal routine and all migrations are determined by the needs of the herds. Whether these peoples live in a cement house, a house of plywood or palm fronds, or even dead tree branches and tarpaulin, the organising principle of their daily life is still livestock and movement to access natural graze and browse.

Throughout all the desert regions of Oman, life depends upon access to water. In the past, survival depended upon moving herds and households to water sources. Today, however, in most of Oman, water is moved to the herds and the households by truck or water bowser. The Harasiis tribe, which is the focus of this chapter, represents the community perhaps the most distant from urban centres, and hence the most extreme case. Up until the late 1950s, the rock and gravel desert plain (Jiddat-il-Harasiis) which composed of most of their traditional territory of about 40,000 square kilometres (about the size of Scotland in the UK or Delaware in the USA) contained no water whatsoever. The only source of water available to them was found along the Awta – the lowlands of the Huqf escarpment lying just along the coast of Oman from Duqm north towards Al-Hajj. During the winter months, the Harasiis moved along the Jiddat-il-Harasiis searching for pasture and browse for their herds of camels and goats. In the event of occult precipitation (heavy condensation), water would be collected by spreading blankets out and later wringing the water from them into drinking vessels. Their water consumption, however, was kept to a minimum, with families relying on their herds to provide them with enough milk for their nutritional and physiological requirements. During the summer the Harasiis abandoned the extremely hot Jiddat and moved down into the Awta where their herds had access to the brackish springs of the region.

In the late 1950s, the national oil company, which had been exploring for oil in the northern desert of Oman, moved into the Jiddat-il-Harasiis and drilled water wells as part of the exploration process to discover oil. One after another, these wells were capped as the oil company moved on. Finally, in 1958, the leader of the Harasiis tribe, Shaykh Shergi, insisted that these water wells be left open for the benefit of the local population. His demands made to the oil company were ignored. He then conveyed this potentially life-changing request to the Sultan, who ordered the oil company to leave at least two wells open: one at Al-Ajaiz and the other at Haima. Within a few years, several families began to remain on the Jiddat during the hot summer months, relying on their access to these water wells and some sporadic delivery of water by water tankers under the direction the oil company's 'Community Relations' personnel. For the Harasiis tribe, the oil company was the government, and a revolution in expectation was in the making – as had already happened in other tribal areas in the north of the country (e.g. the Duru' tribe near Ibri). The oil company had sunk wells for water, built animal troughs for the Duru' tribe and undertaken regular water delivery by bowsers trucks to key mobile households. Furthermore, and perhaps most importantly, Duru tribal leaders took on the role of selecting tribal labour for work with the oil company. The oil company recognised that Duru' grazing lands were very much still under the control of their tribal leaders, and not the Sultanate. The Harasiis wanted the same treatment. But oil had not been discovered (yet) in their territory – the first oil strike was made in 1964 at Fahud in Duru' territory. Discovery of oil precipitated some benefits to the mobile tribes traditionally located where oil discoveries had been made prior to the 1970s, but thereafter the thorny issue of pastoral communal land rights, private ownership and land expropriation by the state raised its head.

The 1970s were a period of tremendous transformation for Oman as a whole. In 1970, disillusionment with the enforced backwardness of the state and the medieval nature of the government under the reign of Sultan Sa'id reached such alarming proportions that rebellion was inevitable. In July of that year, the thirty-eight-year reign of Sultan Sa'id ended when he signed a formal abdication document in favour of his son, Qaboos bin Sa'id. Oman was kick-started into the twentieth century, with significant road building and other infrastructural development. The country's development budget in 1971, for example, during Sultan Qaboos bin Sa'id's first year as ruler was $60 million up from nearly 'nothing' in the years before.

By 1975, Oman's development budget had grown to $1,000 million. During this period British advisors strongly recommended in a white paper that all Omani land not held privately – that did not belong to anyone (*terra nullius*) – should be declared state land and under the control of the Sultan. This was accepted by the Sultan. Throughout this first decade of Sultan Qaboos' rule, modern roads and sea and air communications with the rest of the world were established. Telegraph and telephones were becoming widespread, as were schools, health clinics and hospitals. Social welfare services and the extension of water and electricity were also rapidly increasing in ever-expanding circles from Muscat and Salalah. And oil exploration in these 'empty lands' was rapidly expanding too.

By the late 1970s, with basic social services rapidly and methodically extended into the rural countryside of Oman, Sultan Qaboos issued a number of decrees of vital interest to the remote mobile pastoral communities. Reiterating this in a number of speeches, the Sultan determined that the desert regions of Oman were to receive the same care and attention as the villages and towns of the rest of the country. This mandate was interpreted by leading government ministers to mean that a way was to be found to extend the same social services to mobile pastoral nomads without forcing them to give up their traditional way of life. Plans were drawn up to create a number of tribal administrative centres throughout the desert where the basic social functions comprising healthcare, education facilities and welfare services would be available.

In 1981, the first United Nations (UNDP) project aimed at the development needs of a mobile pastoral population in the Arabian Peninsula was initiated at Haima in the central desert area of Oman. I was appointed 'Chief Technical Officer'. The first year of the project was devoted to conducting an anthropological study of the population to identify their felt needs, priorities and problems. The second and further years were focused on recommending and implementing practical programmes that would extend mobile and

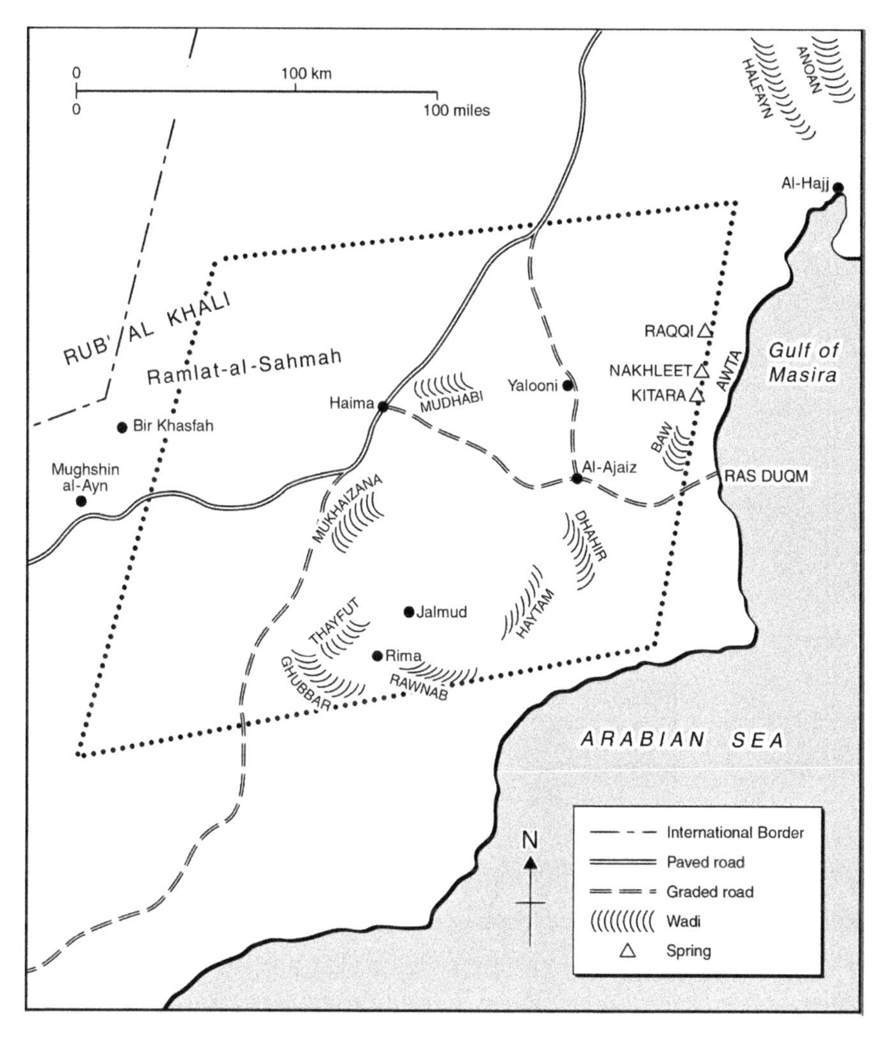

Figure 14.2 Map of the Jiddat-il-Harasiis.

fixed basic social services to this remote and marginal nomadic pastoral community. The mobile population associated with this region was the Harasiis tribe, a South Arabian-speaking people numbering around 3,000 occupying a large, nearly waterless gravel and rock plain – the Jiddat-il-Harasiis – about the size of Scotland.[11] Raising herds of camel and goat, mainly for the production of milk, these communities migrated across the vast arid expanse of the Jiddat. Their only water holes had been dug by the oil company in the

Figure 14.3 Aerial view of the Jiddat-il-Harasiis.

late 1950s and 1960s as the company had moved around the territory looking for oil.

In 1954, an exploratory party of the national oil company, Petroleum Development Oman (PDO), made its way from the southern coast of Oman at Duqm into the central Oman desert steppes. With the entry of active oil exploration in the region, the fortunes of the mobile pastoral tribes of Oman fell increasingly out of their own control. The first tribal agreement to allow exploration in their territory placed the mobile Duru' tribesmen in the position of pawns in a power struggle between the forces of the Sultan and the Ibadi Iman of the Jebel Akhdar. This confrontation, too, had international dimensions, drawing Saudi Arabia, Great Britain and the Arab League, as well as a special United Nations Investigative Committee, into the fray.

For the tribes and their leaders, the most significant aspect of these early years of contact with the oil company was in determining which leader would be recognised by the oil company, which needed to establish a system for meeting its needs for labour recruitment and labour supervision in order to effectively exploit the oil promise of the area. Such requests, however, were completely foreign to the world of the mobile pastoralists. Though the tribal leaders were accustomed to acting as political mediators on behalf of their tribesmen, the oil companies' mixed political and economic needs were novel to them. In some tribes this situation exacerbated the internal jealousies of the tribe, each sub-leader claiming the exclusive right to furnish labour for the oil company. In other tribes like the Harasiis, the entire tribal leadership structure was severely shaken when the oil company decided not to negotiate with the acknowledged leader of the tribe, but rather turned to a minor leader who was more amenable to its demands.[12]

By the mid-1970s, the Harasiis had engaged in a major technological change, shifting from camels to trucks for transport. By the early 1980s every household of the seventeen sampled for our study had a four-wheel-drive vehicle. Such a transformation required a fundamental shift in household economic organisation. Each household had to find means of keeping these vehicles running and 'local' employment for one member of each family became a significant survival strategy.

The Harasiis had first become exposed to opportunities for wage labour as early as the late 1950s when the oil company had been exploring for oil in their territory. Initially this contact had little impact on these mobile pastoral households other than to modify their migratory patterns in order that they could take advantage of possible access to water, food scraps for their herds and, later, petrol for their vehicles.[13] However, once motorised vehicles became widespread in the Jiddat, most household heads had to hold down regular employment in order to keep their vehicles running. When the joint United Nations–Omani Government effort to extend permanent and outreach (mobile) social services to these people commenced its work, it was met by the community with particular enthusiasm. Their logic was that this project would be able to offer schools which would transform their youth from potentially unskilled labourers into skilled, well-paid professionals. They saw this institution as a way out of the 'non-jobs' that were currently available to them as sentries, well-guards and installation watchmen. Instead, they looked to future employment with the oil company, the border police and the army – the only employers in this vast desert plain that they regarded as their home territory. But these forms of employment required literacy in Arabic, high-school diplomas and, in some case, fluency in English.

Within two years and with the full support of the community, the joint UN/Government project was able to set up a boarding school for boys (girls were admitted on a day basis), two mobile primary health care units, a welfare office for social affairs and a veterinary clinic with a mobile outreach programme. School enrolment, which began with forty-two boys and three girls in 1982, climbed yearly. By the mid-1990s the boarding school, which included a primary and a secondary school, had over 150 boys and girls in attendance. Some of the graduating boys had succeeded in getting jobs with the police, the army, or in the government civil service.

The Social Affairs office in Haima began to busy itself with determining what names might be removed from their welfare payroll (i.e. who had died), while the Harasiis, always opportunistic, looked for ways of increasing the government pay-outs to divorced, widowed and disabled individuals in their community. Government officials in Muscat soon began to notice an ever-increasing number of pastoral people coming forward to request monthly welfare assistance from the government. These officials came to regard such requests as proof of increased poverty among that sector of the country's population, evidence which required solutions in order to make these people 'economically productive'.

Even into the twentieth-first century, the mobile pastoral population of Oman has come to be increasingly classified by government administrators as a 'poor' people who are making no productive contribution to the national economy. They are seen as a drain on the country rather than as an asset. The expropriation of their traditional lands as *terra nullius* only served to confirm the notion of the Harasiis as poor, with no wealth/property of their own. This assessment was not derived from any particular set of facts from technical study, but rather from the long-standing ambiguous nature of the relations between the urban, settled societies of the towns and cities and the mobile, remote, pastoral peoples of the desert interiors. Hence government 'income-generating' schemes have been put forward which ignore the mobile livestock livelihood base of these communities. Instead, government focused on developing or teaching regionally acceptable craft skills to women (sewing, weaving and spinning) which might result in goods for sale at annual government-sponsored cultural events. But efforts to improve the quality of the livestock or systems of animal transport to towns and cities were not developed. Government schemes for pastoralists, generally an imitation of 'community development programmes' for oasis farmers, were regarded as a way of turning these 'poor' debt-ridden communities in the country into productive contributors to the country's gross national product. The negation of their rights to their lands and the profits that accrued from oil completed the expropriation of their centuries-old land holdings.

Despite this formal expropriation, the Harasiis continued to put forward initiatives to persuade government, and its most visible arm, the national oil company, to 'develop' their traditional homeland. Their 'rights' to these lands remained highly contested and their petitions were often 'invisible' – oral presentations by lineage leaders to ministers in Muscat. Tribal demands for road building, for passes under or over pipelines or, at the very least, regular road grading of desert tracks have been rejected or denied on the environmental grounds that road construction, grading and overpasses would be contrary to the country's conservation code, it being assumed that these 'state lands' were empty and the Harasiis were trespassers. Although not couched in these terms, it was clearly the principle of *terra nullius* that was in effect. This argument relates to the establishment of the UNESCO-recognised Arabian Oryx Sanctuary – a large area of the Jiddat-il-Harasiis claimed by government and recognised by international conservation agencies (the International Union for the Conservation of Nature [IUCN] and the World Wildlife Fund [WWF]) for the re-introduction of the Arabian oryx in the Sultanate of Oman.

In 1972, the oryx was declared extinct in Oman, the result of over-hunting by elites from Arabia and the Gulf. The relatively richest part of the desert, the north-eastern quadrangle of the Jiddat-il-Harasiis, was identified by IUCN specialists as the site for an oryx reintroduction project. The assumption here was that this was land that belonged to no one (*terra nullius*), was uninhabited, and could be set aside without consultation. The IUCN proposal was approved by government without consultation with the Harasiis who had occupied this land for centuries. The restrictions on infrastructural and other developments in the nature reserve have meant significant hardship and exclusion for the indigenous pastoral community.[14]

When I pursued the argument for road building on behalf of the Harasiis with the Office of the Environmental Conservation Advisor in the 1980s, I was told 'No Bedouin will keep his vehicle on a paved or graded road when a straight line across the desert floor is more direct'.[15] The suggestion here was clearly that there was something irrational in the way a Bedouin viewed mobility. Wear and tear on a vehicle would never come into consideration when driving from one point to another crossing the desert.

Tribal requests for government support for a system of water distribution were also regarded with deep scepticism and lack of understanding of the significance of mobility to their way of life. When the Harasiis tribal leader put forward a distribution plan which would rotate and change as households shifted their campsites around the desert, the proposal was denied by the government in favour of creating a system of 'fixed' facilities around brackish water wells where permanent water purification plants would be set up.

Even local Harasiis efforts to grow salt-resistant plants and fodder in the desert – following the example of the oil company Rahab Farm at Marmul established by royal decree to grow Rhodes grass – met with complete lack of interest on the part of Ministry of Agriculture personnel. Yet at the same time, the ministry was actively subsidising fodder farming and improving water-spreading systems in the agricultural regions of the country. It was as if

Figure 14.4 Oil company water bowser, 1982. Photo by Dawn Chatty.

these mobile herding communities in Oman's Central desert did not deserve assistance to adapt their way of life, but rather needed to be or become settled folk in order to be on the government's radar.

The Harasiis have been changed by these experiences. Their expectations of oil company and government assistance have grown, and their political consciousness also has been raised as they experience rejection or lack of interest relative to their settled neighbours in the north and the south. They have taken, and perhaps will always take, every opportunity to plead their case, to ask for assistance, to request help from the large society on the fringe of their universe. Their pragmatic position has been that sometimes they succeed and sometimes they do not. But at all costs they need to 'do the asking'. Although some of their material expectations – mainly centred on the motor vehicle and their increased mobility – have grown, their cultural integrity remains.

For the Harasiis, subsistence mobile animal husbandry livestock remains the central focus of their lives. A short study completed in 2001 – nearly two decades after the original study when motor vehicles were adopted – has revealed that household herds have remained remarkably stable, with a hundred head of goat and twenty-five head of camel remaining an average figure.[16] Employment of at least one male member of each household remains crucial for the well-being of the group. What has changed is the development of a growing sense of frustration at the kinds of employment available. Having engaged wholeheartedly with the government effort to provide education to them, the Harasiis now ask why they see so little return. They question why the oil company and its sub-contractors – the major employers in their locations – still employs so many unskilled labourers from abroad, when they themselves seek employment with them. And they remain puzzled and disconcerted as to why so few of their educated youth who have managed to gain training at oil company expense at local universities in the country still do not seem to be able to gain employment in the oil sector.

There is a perceptible sense of being 'left behind', of not benefiting from the wealth which oil – extracted from land expropriated from them – has generated for the rest of the country's citizens. Many of the Harasiis believe that the Mukhaizana Oil installation – an assessment they make on the basis of size alone – is a key producer of Oman's oil revenue. Few know that this oil field in the middle of their traditional grazing territory 'contributes around 13% to the Sultanate's entire oil output'.[17]

The Mukhaizana oil installation sits right in the middle of the Wadi Mukhaizana in the Jiddat. Adjacent to it is a large regional police station to protect the field. Harasiis families use this area for grazing when there has been good rain – an event which occurs every three to four years. So, although it may seem devoid of people for some years, it is still an important resource for the tribe. The oil field, the police station and the constant flow of tankers and other vehicle traffic have suggested to the Harasiis that this is a significant resource for the country. That the oil field contributes significantly to the entire oil output of the country only adds to the importance of the oil companies maintaining good relations with the local communities.[18]

Although the Harasiis do not use the vocabulary of the 'rights discourse' that has become so popular among indigenous groups around the world, particularly in Latin America and parts of Southeast Asia, the questions are basically the same: 'Why are the lands we regard as our homeland and which we have inhabited for centuries not being recognized as belonging to us?' As one young Harasiis woman asked at a workshop run by the oil company in 2002, 'Why do we not see any benefit from oil activities which are taking place on our lands, and which depend upon our goodwill?' And presciently, one Harasiis young man asked, 'Why are we being denied the benefits which all other Omani nationals have access to?'

In 2001, a meeting was set up for me at the oil company headquarters to review some recent Social Impact Assessments which challenged earlier Environmental Assessment claims that there were no people living in the desert area the oil company wished to exploit for its petroleum potential. The engineers attending the meeting stated that there were no indigenous people in the desert fields they were currently developing. They had been out there several times and had never seen anyone. The desert was empty of people, they said. Only when the oil company set up a camp, the engineers said, did people emerge out of nowhere to demand water and food. The oil company engineers did not see why such people should be given compensation for moving off these lands. This attitude was not unusual among oil company employees, and reflected a general lack of understanding of mobile

Figure 14.5 Government social housing in the Jiddat-il-Harasiis, 1989. Photo by Dawn Chatty.

pastoralists, their extensive, low-resource, traditional land holdings and contemporary claims to the land.

In part, these attitudes have surfaced because of a basic misunderstanding of the nature of mobile pastoralism on the part of government and some oil company personnel. Contestation has emerged from the unwillingness of some of these oil company employees to recognise the rights of these peoples to the extensive area of desert they have inhabited for centuries with their domesticated herds of camels and goats. When these herders are not physically present in particular locations, they are assumed by some oil company personnel to have no rights to them, to be trespassers. Hence, when Occidental had recently won the concession from Shell/PDO over the Mukhaizana area in the early 2000s, it came to the area not only in ignorance of traditional Harasiis grazing practices, but also with no interest in expanding its knowledge. When the company came across several Harasiis households camped in the Wadi Mukhaizana, it asked them to move out. As far as it was concerned, the Wadi was empty and these families were just opportunists/trespassers. The company gave each family 50 R.O. ($US 150)

Figure 14.6 Temporary camel camp (*azbah*) in the Jiddat-il-Harasiis (note oil company vehicle at campsite), 2005. Photo by Dawn Chatty.

to leave the area and considered it a generous gesture, as these peoples had no homes, and no fixed structures or land deeds to prove their rights to these particular spots. For the Harasiis, this was a payment to move away, temporarily, from one of their most important grazing lands for the season: nothing more, nothing less. For the oil company, this was an adequate permanent buy-off, or compensatory payment to establish its control over an 'empty' region which the Harasiis tribe considered their *dar* (heartland, country). Contestation in this part of the desert was inevitable.

As oil production increasingly focuses on fields in this central desert area, the questions the Harasiis have all asked are beginning to have national and international significance. In an age when multinational oil companies are increasingly being held accountable and pressure for transparency grows, the neglect of a significant stakeholder group – some would say 'rights holder' group – becomes problematic. The national oil consortium (including Shell International) holds concessions over a large portion of the Jiddat-il-Harasiis. Shell International and other oil companies in the consortium have had their concessions reviewed by the Sultan in 2020.[19] This review, as well as the growing international movement for sound social performance in concession areas, has meant that the oil company feels obliged to engage in some manner, even if only superficially, with this community. This effort to develop a relationship with the Harasiis comes not a moment too soon. Furthermore, news of disturbances in other parts of the country – regarding multinational gas companies, local employment and negotiated social investment policy – began to reach the Harasiis tribe in this, the remotest corner of the country. A large gas liquefaction project in the Sharqiyya region of Oman had recently run into trouble over the terms which the local community had been offered in compensation for loss of land holdings and fishing rights. The gas company's social investment package of 1.5 per cent of the profits, which was under discussion, has become common knowledge throughout the interior of Oman. Harasiis tribal elders have begun to lobby for a similar package from the oil company for the Harasiis – of 1–1.5 per cent of company profits – to be ploughed into a carefully designed and controlled social investment package for the Jiddat-il-Harasiis. The Harasiis began to make their concerns known over their past exclusion and to contest their lack of voice or representation. However, on their own and from their distant camps sites and shelters, their protestations were muted and unorganised and severely undermined by some of the bureaucratic decisions taken by the oil company and in government.

In 2001, the national oil company seemed to be responding to international pressure to integrate the local stakeholders, the nomadic pastoralists, into participatory social development. After commissioning a series of social impact assessments, the company hired a Shell Nigeria employee who had experience of social development. Her recruitment and terms of reference were managed at the highest level of the oil company, but her performance on the job was severely lacking. After less than two years in post she was dismissed, and all activity in the realm of social performance has come to a stop. Some of the more cynical Omanis have commented that her obvious lack of qualifications for the post in Oman, and her strident approach to management, worked in the interest of the 'old guard' in the oil company who had little time for contemporary ideas of social performance. Now that she has come and gone, they can go back to managing affairs as they did in the past.[20] The Harasiis tribe, young and old alike, could only voice their disquiet that after so many years of oil extraction from their traditional lands, they, as a people, saw so little benefit.

In 2009, on one of my annual visits to the Harasiis, I was told that the oil company had placed boulders on four of the graded roads in the Jiddat that connected oil camps to one another, and that were widely used by the Harasiis to traverse the desert. Several Harasiis had been hurt and a number of their vehicles had been 'totalled' crashing into these boulders unexpectedly while driving at night. I was asked to help them get these boulders removed as their complaints were not being acted on. I wrote to both oil companies operating in the area, PDO and Occidental, requesting that the boulders be removed. My first letter went unanswered; my second was replied to with the 'justification' that the boulders had been placed there by the company to stop local PDO employees from using the road. In my third letter I expressed my deep concern that 'placing such obstruction on these graded roads is a serious safety concern for the local community living in these desert areas. The desert is not empty. There are more than 5,000 Harasiis living in these areas …'[21] Finally, after this personal intervention, I received a letter from PDO dated 26 August 2009 informing me that, in response to my request, the boulders had been removed. Of course, the Harasiis were delighted with this news, and I received many telephone calls from them informing me of this road clearance. But at the back of my mind, I wondered for how much longer the Harasiis would accept being treated as invisible, irregular, if not illegal occupants of land considered *terra nullius* – a territory they had occupied for centuries – by the oil company.

In the wake of the Arab Uprising of 2011 and the disturbances reported in the media (most Harasiis have smartphones and are regularly on WhatApp and other social media), Harasiis youth decided to take matters into their own hands and contest the lack of interest in them by government and the oil company. The main oil company employer at the Mukhaizana oil field, Occidental, had been very slow to respond to any requests from the Harasiis, particularly for water distribution, job training and employment. Groups of young men decided to block access to the oil camp by tankers and other traffic until they were granted an opportunity to make their demands heard. After several weeks of blocking access into the camps, Occidental staff meet with these young people. The outcome in 2012 was that, yes, the oil company would provide ten scholarships a year to Harasiis youth to attend Dhofar University and that graduates would have an opportunity of employment in the Mukhaizana field installation. Those promises have not been entirely fulfilled. After a few years these scholarship programmes were dropped, and employment in the Mukhaizana installation is still difficult to come by for the local community of Harasiis. Occidental's website from 2020 says little of its Corporate Social Responsibility (CSR) programmes for the local community other than to indicate that it provides water to more than 1,000 inhabitants of the *villages* surrounding the Mukhaizana field. It also permits *local residents* access to its healthcare facility which services its personnel and contractors and their families. Clearly the grounds for contestation are there. Occidental's website refers to the local community in terms of settlement and sedentism. It has no understanding of the mobility of these herding families and can expect further contestation in the future.

Shell and the other partner companies of PDO, on the other hand, were more proactive and acted upon the recommendations made after a number of participatory workshops at Haima with the Harasiis. The company implemented some of the 'win–win' recommendations that emerged from those meetings with the local population and in that manner succeeded in co-opting some of its members. Their community relations office was active in efforts to preserve PDO' s social licence in its concession areas in the other parts of the Jiddat and throughout the country. Minimising impact, looking for constructive engagement and seeking to resolve community issues associated with the oil company were always high on its agenda. Water haulage to Harasiis household *and* to the camel *'azab*s (mobile camps), road grading, recruitment of thousands of local men on a temporary basis as watchmen and sentries, five-year scholarship programmes to study in colleges and universities in Oman, and constant evaluating of community requests through company grants, donations or social investment in their concession areas have become standard measures, judging from its website in 2020.

How individual tribal members will adapt to the recent developments in the middle of the Jiddat-il-Harasiis over the long durée is another matter. Here are a people caught in the modern postcolonial world of development and planning. In spite of the remoteness, the marginality and the isolation that characterise the Harasiis and their tribal lands, they are now beginning to face problems similar to those faced by the mobile pastoral tribes of the rest of Arabia. Their lands have been, in a sense, confiscated and are controlled *de facto* and *de jure* by the nation-state. Recent infrastructural growth has meant that they have access to and are affected by development hundreds of kilometres away. Motor transport and motorised water pumping facilities have revolutionised their lives, as have telecommunications. Some have had the opportunity to leave and take up new lives. But many have chosen to remain.

Although each country in the Arabian Peninsula has to face different sets of economic, political and social factors, one feature that is common throughout is the plight of the subsistence mobile pastoralists. Without exception, their territorial usufruct is no longer recognised by the central government; their struggle to subsist has required that they acquire modern motorised transport that can only be supported by some form of wage labour. This, in turn, has meant that male household-heads spend long periods of time away from the pastoral household, leaving the management of herds to women or to younger and less experienced male members of the family. By accident, mismanagement and force of circumstances, these peoples have been left to find their own solutions. Their dependence upon government has remained superficial, and these communities continue to adapt to changes in their environment, ever searching for a meaningful and viable existence for themselves and their herds. What remains to be seen is whether international pressure for accountability, transparency and respect for human rights among international non-government agencies and multinational extractive industries will continue to hold weight. Current indicators point to a movement in the direction of recognising as significant stakeholder groups the mobile pastoral communities which populate the concession areas of the major international oil and gas companies. As long as company profits hold, that movement should continue. How effective that recognition becomes depends on the continuing international demand for business accountability and transparency.[22] In the remote and largely arid lands inhabited by mobile pastoralists in the Middle East, indigenous human rights movements are non-existent. Good social performance policies by multinational corporations will depend upon continuing global efforts to promote sound social performance and prioritise local partnerships. Without such persistent external demand, it is unlikely that the pastoral communities of the Middle East could – in the near future – leverage support for their interests, or see investment in their traditional territory take the shape which they themselves feel would benefit them the most.

Notes

1. Lewis Henry Morgan, *Ancient Society* (Tucson: University of Arizona Press, 1985 [1877]).
2. The literature on *terra nullius* and its application as a principle of dispossession during the colonial era has begun to attract some attention. See Peter Novick, *That Noble Dream: The 'Objectivity Question' and the American Historical Profession* (Cambridge: Cambridge University Press, 1988).
3. The International Court of Justice (ICJ) advisory opinion ruled that land accommodation for 'tribes or people having a social and political organisation were not regarded as terra nullius'. *Advisory Opinion on Western Sahara*, International Courts of Justice Reports 39, 16 October 1975.
4. The first time troops of Western powers were used against pastoral groups in the Middle East to protect oil interests was in 1909. At that time Indian soldiers of the 18th Bengal Lancers were brought in to protect the drilling areas of the Anglo-Persian Oil Company in Iran from local pastoral tribesmen who were trying to defend their traditional grazing grounds. See George W. Stocking, *Middle East Oil: A Study in Political and Economic Controversy* (London: Penguin, 1971).
5. James Barr, *Line in the Sand: Britain, France, and the Struggle That Shaped the Middle East* (New York and London: Norton, 2011).
6. Riccardo Bocco, 'La Sédentarisation des Pasteurs Nomads: Les Experts Internationaux Face à la Question Bedouine dans le Moyen-Oriente Arab (1950–1970)', *Cahiers des Sciences Humaines* 26, no. 1–2 (1990), 97–117.
7. Jonathan Rae, 'Tribe and State: Management of the Syrian Steppe' (Ph.D. diss., University of Oxford, 1999).
8. Land tenure in Islam generally held four categories: *mulk*, land with full ownership (private land); *miri*, regarded as state land; *mawat*, unused but could be converted into private land tenure; and *musha'*, communal land and pasture land. For a full description of types of land tenure categories in Islam see *Islamic Principles and Land* (UN Habitat, 2011).
9. *Rapport sur la situation de la Syrie et du Liban soumis au conseil de la Société des Nations* (Paris: Minstère des Affaires Étrangères, 1923–38).

10. Dawn Chatty, 'The Bedouin of Central Oman'. *Journal of Oman Studies* 6, no. 1 (1984), 149–63.
11. For a full ethnographic description of the Harasiis tribe and their encounter with development planning see Dawn Chatty, *Mobile Pastoralists: Development Planning and Social Change in Oman* (New York: Columbia University Press, 1996).
12. For a full description of these events see Chatty, 'The Bedouin of Central Oman'.
13. Dawn Chatty, 'Petroleum Exploitation and the Displacement of Pastoral Nomadic Households in Oman', in Seteney Shami (ed.), *Population Displacement and Resettlement: Development and Conflict in the Middle East* (New York: Centre for Migration Studies, 1994), 89–106.
14. For more details see Dawn Chatty and Marcus Colchester (eds), *Conservation and Mobile Indigenous Peoples: Forced Settlement, Displacement and Sustainable Development* (Oxford: Berghahn, 2002).
15. Ralph Daly, personal communication, 1984.
16. Household figures for livestock numbers derived from 1981–2 and 1991–2 are generally comparable with figures derived from a social impact assessment which was carried out for the national oil company in the region in August 2001. See Chatty, *Mobile Pastoralists*, 94–100 and Jonathan Rae and Dawn Chatty, *Social Impact Assessment of Mukhaizana, Jiddat-il-Harasiis* (Muscat: Petroleum Development Oman, 2001). Visits to households and temporary camel camps in 2018 suggest that these animal numbers are fairly steady. These figures show that although human numbers are rising, livestock numbers have been maintained at reasonably steady numbers, in keeping with the Harasiis' assessments of what is a sustainable size of animal herd in the Jiddat.
17. Oxford Business Group, 'Discovery of Oil Reserves and New Concession in Oman Leverage Growth on Global Indices', 2019, p. 7, https://oxfordbusinessgroup.com/overview/pumped-reserve-discoveries-new-concessions-and-efficient-technology-leverage-steady-growth-global (last accessed 22 February 2022).
18. Titus Moser and Damian Miller, 'Multinational Corporations' Impacts on the Environment and Communities in the Developing World: A Synthesis of the Contemporary Debate', in Walter Wehrmeyer and Yacob Mulugetta (eds), *Growing Pains: Environmental Management in Developing Countries* (Sheffield: Greenleaf, 1999), 40–51.
19. A recent oil company document giving guidance on undertaking social performance reviews clearly sets out that the goals of the company 'minimizes negative

and optimises positive environmental and socio-economic impacts for local and other stakeholders'. It goes on to say that social performance is one of the key ways in which the company can contribute to sustainable development and is '*increasingly important for gaining and maintaining an operations' license to operate*' (italics mine).

20. Personal communications. This view is not far from that which can be gathered from the work of Judith Kimmerling in her assessment of oil company practices in Ecuadorian Amazon. Judith Kimmerling, 'Uncommon Grounds: Occidental's Land Access and Community Relations Standards and Practices in Quichua Communities in the Ecuadorian Amazon', *Law and Anthropology* 11 (2001), 179–247.
21. Correspondence, 25 May 2009.
22. Within the global conservation movement, for example, concern for the human rights of indigenous peoples has been translated into an international search for more participative forms of natural resource management so as to exhibit greater respect for indigenous livelihoods. At the World Parks Congress in Durban in 2003, a Plan of Action and recommendations devoted to Mobile Indigenous Peoples was agreed (Recommendation 5.27) which called for – among a number of issues – the restitution of land rights, recognition of traditional knowledges and practices and the adoption of the Dana Declaration on Mobile Peoples and Conservation (www.dana.declaration.org.). This achievement adds to the international body of instruments which can be called upon to protect mobile peoples in the context of conservation. Similar guidelines, recommendations and achievements for mobile peoples are emerging which tackle the issues that emerge from within the multi-resource extractive industry such as ILO 169, and OECD Guidelines.

BIBLIOGRAPHY

Archival Sources

British Petroleum Archive, University of Warwick, UK [BP]

Kuwait Oil Company (1934–1976)
Iraq Petroleum Company (1911–2001)
Anglo Iranian Oil Company (1909–1954)

ETH Zurich University Archives, Zurich

Arnold Heim Collection

Stanford University, Hoover Institute

Richard Sterling Finnie Papers (1945–1967)
Hizb al-Ba'th al-'Arabi al-Ishtiraki Records (Ba'ath Party Records) Kuwait Dataset [KDS]

Kuwait Oil Company Archive, Ahmadi, Kuwait [KOCA]

Photographic Collections, Miscellaneous

National Archives of India, New Delhi [NAI]

Ministry of External Affairs
Near East Branch

Middle East Branch
Emigration Branch

National Archives of the United Kingdom, London [NAUK]

Foreign and Commonwealth Office
FCO 8, Arabian Department and Middle Eastern Department (1967–1997)
FCO 164, Political Office, Dubai (1820–1979)
Foreign Office
FO 371, Political Departments, General Correspondence (1906–1966)
FO 1016, Political Residencies and Agencies, Persian Gulf (1917–1972)
Ministry of Labour, LAB 13, International Labour Division and Overseas Department (1923–2004)
War Office, WO 337, Headquarters British Forces Gulf Area (1957–1971)

Wellcome Trust Collection, London, UK

Sir Ernst Boris Chain Papers (1906–1980)

Georgetown University, Lauinger Library

William E. Mulligan Papers (1930–1992) [WEM]

US National Archives and Records Administration, College Park MD [NARA]

State Department Central Files (Record Group 59)

Princeton University, Seeley G. Mudd Library

William A. Eddy Papers (1859–1978)

Official Publications

British Government of India. Lorimer, J. G., *Gazetteer of the Persian Gulf, Oman, and Central Arabia, Volume II* (Bombay: Superintendent Printing, 1908).

France, Ministry of Foreign Affairs, *Rapport sur la situation de la Syrie et du Liban soumis au conseil de la Société des Nations* (Paris, 1923–38).

India, Ministry of External Affairs, 'History and Evolution of Non-Aligned Movement' (New Delhi: Government of India, 22 August 2012).

International Court of Justice, *Advisory Opinion on Western Sahara*, ICJ Report no. 39, 16 October 1975.

OAPEC, *Secretary General's Second Report*, 1975, 1977.

OAPEC Symposium on Petroprotein and Organization of Arab Petroleum Exporting Countries, *Proceedings of OAPEC Symposium on Petroprotein* (Kuwait: Organization of Arab Petroleum Exporting Countries, 1980).

UN Habitat, 'Islamic Principles and Land', 2011.

U.S. House of Representatives, Armed Services Committee, Subcommittee on Petroleum, 'Petroleum for National Defense', 80th Congress, 2nd Sess. (Washington: Government Printing Office, 1948).

Magazines and Periodicals

Al-Ittihad (The Union), 1978.

Ahl al-Naft (Oil People), 1958.

Al-'Amilun fil Naft (The Oil Workers), 1966.

Arabian Sun & Flare, 1945–9.

BP Magazine, 1953.

Emirates News, 1978.

Gulf News, 2008.

The Kuwaiti, 1954.

Iraq Petroleum, 1958.

The Lamp, 1944.

LIFE Magazine, 1945.

Mardom (People), 1947.

The Naft Magazine, 1943.

OAPEC News Bulletin (Organization of Arab Petroleum Exporting Countries), 1977–81.

Qafilah al-Zayt (The Oil Caravan), 1953–61.

Sabah, 2020.

Washington Post, 1991.

Films

al-Qadiri, Monira, dir., *Behind the Sun*, 2014.

Douglas, David, dir., *Fires of Kuwait*, IMAX, 1992.

Giesler, Rodney, dir., *Close-Up on Kuwait*, Kuwait Oil Company, 1961.

Guggenheim, David, dir., *An Inconvenient Truth*, 2006.

Herzog, Werner, dir., *Lessons of Darkness*, New Video Group, 1995.

Websites

al-Jumayri, Sultan, *'An al-tajriba al-nidaliyya al-'ummaliyya fi al-Sa'udiyya*, https://docs.google.com/document/d/1IwuBr-E4sM6-Nk4Wdt__pdZ_ZTfozQCTXh91iu7opcg/edit (last accessed 22 February 2022).

Bianet, 'Doğalgaz üretimine geçmek en az 4–5 yılı alacaktır', https://m.bianet.org/bianet/siyaset/229462-dogalgaz-uretimine-gecmek-en-az-4-5-yili-alacaktir (last accessed 28 October 2021).

Bianet, 'Karadeniz'deki doğalgazın değeri 60–70 milyar dolar', https://m.bianet.org/bianet/siyaset/229637-karadeniz-deki-dogalgazin-degeri-60-70-milyar-dolar (last accessed 25 August 2020).

CIA Freedom of Information Act Electronic Reading Room, https://www.cia.gov/library/readingroom (last accessed 22 February 2022).

Dana Declaration on Mobile Peoples and Conservation, 2002, https:www.danadeclaration.org (last accessed 22 February 2022).

Dubai Municipality, *Future of Dubai Gallery*, https://www.dubaiframe.ae/en/discover/discover-details?id=3 (last accessed 29 January 2022).

MAPEG, 'Orta Dönemli Petrol ve Doğal Gaz Arz-Talep Projeksiyonu', 2017, https://enerji.mmo.org.tr/wp-content/uploads/2019/06/PİGM-Orta_Donemli_Petrol_ve_Gaz_Arz-Talep_Projeksiyonu-2017.pdf (last accessed 23 July 2022).

Oxford Business Group, 'Discovery of oil reserves and new concession in Oman leverage growth on global indices', 2019, p. 7, https://oxfordbusinessgroup.com/overview/pumped-reserve-discoveries-new-concessions-and-efficient-technology-leverage-steady-growth-global (last accessed 22 February 2022).

Sabah, 'Enerjide denklem değişiyor! Doğal gaz keşfi ile hem devlet hem vatandaş kazanacak', https://www.sabah.com.tr/galeri/ekonomi/enerjide-denklem-degisiyor-dogal-gaz-kesfi-ile-hem-devlet-hem-vatandas-kazanacak (last accessed 19 October 2020).

US Energy Information Administration, *U.S. Natural Gas Markets: Midterm Prospects for Natural Gas Supply*, https://www.eia.gov/naturalgas/articles/storageindex.php (last accessed 23 February 2022).

Secondary Sources

Abdullah, Muhammad Morsy. *The United Arab Emirates: A Modern History* (London: Croom Helm, 1978).

Abrahamian, Ervand. *The Coup: 1953, the CIA, and the Roots of Modern U.S.–Iranian Relations* (New York: The New Press, 2013).

Abrahamian, Ervand. *Iran Between Two Revolutions* (Princeton: Princeton University Press, 1982).

Abrahamian, Ervand. 'The Strengths and Weaknesses of the Labor Movement in Iran, 1941–1953' in Michael E. Bonine and Nikki R. Keddie (eds), *Modern Iran: The Dialectics of Continuity and Change* (Albany: State University of New York Press, 1981), 211–32.

Abu-Ruwaida, A. S., Ibrahim M. Banat and Ibrahim Y. Hamdan. 'Chemostat Optimization of Biomass Production of a Mixed Bacterial Culture Utilizing Methanol', *Applied Microbiology and Biotechnology* 32, no. 5 (1990), 550–5.

Adas, Michael. *Machines as the Measure of Men: Science, Technology, and Ideologies of Western Dominance* (Ithaca: Cornell University Press, 1989).

Adelman, Morris. *Genie out of the Bottle: The World Oil Since 1970* (Cambridge, MA: MIT Press, 1995).

Adey, Peter. 'Security Atmospheres or the Crystallisation of Worlds', *Environment and Planning D: Society and Space* 32, no. 5 (2014), 834–51.

Agarwala, Rina. 'From Work to Welfare: A New Class Movement in India', *Critical Asian Studies* 38, no. 4 (December 2006), 419–44.

Alamuddin, Najib. *The Flying Sheikh* (London: Quartet, 1987).

Alani, Daham I. and Murry Moo-Young. *Perspectives in Biotechnology and Microbiology: Lectures and Papers from the First Arab Gulf Conference on Biotechnology and Applied Microbiology* (New York: Elsevier, 1986).

Al-Awadhi, Nader M., M. A. Razzaque, D. Jonker, Ibrahim M. Banat and Ibrahim Y. Hamdan. 'Nutritional and Toxicological Evaluation of Single-Cell Protein Produced from *Bacillus* Sp. KISRI-TM1A in Rats', *Journal of Food Quality* 18, no. 6 (1995), 495–509.

Alissa, Reem. 'The Oil Town of Ahmadi since 1946: From Colonial Town to Nostalgic City'. *Comparative Studies of South Asia, Africa and the Middle East*, 33, no. 1 (2013), 41–58.

al-Khalil, Samer. *The Monument, Art, Vulgarity and Responsibility in Iraq* (London: André Deutsch, 1991).

al-Mehairi, Jamal. 'The Role of Transportation Networks in the Development and Integration of the Seven Emirates Forming the United Arab Emirates, with Special Reference to Dubai' (Ph.D. diss., Durham University, UK, 1993).

al-Muslim, Muhammad Sa'id. *Sahil al-dhahab al-aswad: dirasah tarikhiyah insaniyah li-mintaqat al-Khalij al-'Arabi* (Beirut: Dar Maktabat al-Hayat, 1962), 2nd edn.

al-Naimi, Ali. *Out of the Desert: My Journey from Nomadic Bedouin to the Heart of Global Oil* (London: Portfolio Penguin, 2016).

Al-Nakib, Farah. *Kuwait Transformed: A History of Oil and Urban Life* (Stanford: Stanford University Press, 2016).

al-Rasheed, Madawi. *A Most Masculine State: Gender, Politics, and Religion in Saudi Arabia* (Cambridge: Cambridge University Press, 2013).

al-Rasheed, Madawi. *Muted Modernists: The Struggle over Divine Politics in Saudi Arabia* (London: Hurst, 2015).

al-Saman, Salem bin Ibrahim. *My Life: Part I* (Abu Dhabi, 2008).

al-Ulama, Hesam. 'The Federal Boundaries of the United Arab Emirates' (Ph.D. diss., Durham University, UK, 1994).

Akçam, Taner and Umit Kurt. *The Spirit of the Laws: The Plunder of Wealth in the Armenian Genocide* (New York: Berghahn, 2015).

Amklov, Petter and Vidar Hepsø. 'Between and Beyond Data', *Social Studies of Science* 41, no. 4 (2011), 539–61.

Appel, Hannah. *The Licit Life of Capitalism: U.S. Oil in Equatorial Guinea* (Durham, NC: Duke University Press, 2019).

Appel, Hannah, Arthur Mason and Michael Watts. 'Introduction: Oil Talk', in Hannah Appel, Arthur Mason and Michael Watts (eds), *Subterranean Estates: Life Worlds of Oil and Gas* (Ithaca: Cornell University Press, 2015), 1–26.

Appel, Hannah, Arthur Mason and Michael Watts (eds). *Subterranean Estates: Life Worlds of Oil and Gas* (Ithaca: Cornell University Press, 2015).

Arasteh, Reza. *Education and Social Awakening in Iran* (Leiden: Brill, 1962).

Arendt, Hannah. *The Origins of Totalitarianism* (New York: Meridian, 1958).

Atabaki, Touraj. 'Chronicles of a Calamitous Strike Foretold: Abadan, July 1946', in Karl Heinz Roth (ed.), *On the Road to Global Labour History* (Leiden: Brill, 2017), 93–128.

Atabaki, Touraj. 'Far from Home, but at Home: Indian Migrant Workers in the Iranian Oil Industry', *Studies in History* 31, no. 1 (1 February 2015), 85–114.

Atabaki, Touraj. 'Indian Migrant Workers in the Iranian Oil Industry, 1908–1951'. in Touraj Atabaki, Elisbetta Bini and Kaveh Ehsani (eds), *Working for Oil: Comparative Histories of Labor in the Global Oil Industry* (London: Palgrave Macmillan, 2018), 189–226.

Atabaki, Touraj. 'Writing the Social History of Labor in the Iranian Oil Industry', *International Labor and Working-Class History* 84 (2013), 154–8.

Atabaki, Touraj, Elisabetta Bini and Kaveh Ehsani (eds) *Working for Oil: Comparative Social Histories of Labor in the Global Oil Industry* (Cham: Springer, 2018).

Azoulay, Ariella. *The Civil Contract of Photography* (New York: Zone Books, 2008).

Bailey, Beth L. *From Front Porch to Back Seat: Courtship in Twentieth-Century America* (Baltimore: Johns Hopkins University Press, 1988).

Bakhtin, Mikhail (ed. Michael Holquist). *The Dialogic Imagination* (Austin: University of Texas Press, 1994).

Bakturtāsh, Nosratallah. *Chand Yādemān Az San'at Melli Shodan-e Naft Dar Ābādān va Qeireh* (Tehran: Enteshārāt-e ārtemis, 1396/2017).

Balslev, Sivan. *Iranian Masculinities: Gender and Sexuality in Late Qajar and Early Pahlavi Iran* (Cambridge: Cambridge University Press, 2019).

Bamberg, James H. *British Petroleum and Global Oil, 1950–1975: The Challenge of Nationalism* (Cambridge: Cambridge University Press, 2000).

Bamberg, James H. *The History of The British Petroleum Company: Vol. 2, The Anglo-Iranian Years, 1928–1954* (Cambridge: Cambridge University Press, 1994).

Banat, Ibrahim M., Mustafa Murad and Ibrahim Y. Hamdan. 'A Novel Thermo-tolerant Methylotrophic *Bacillus* Sp. and Its Potential for Use in Single-cell Protein Production', *World Journal of Microbiology and Biotechnology* 8, no. 3 (1992), 290–5.

Banat, Ibrahim M., Nader Al-Alwadhi and Ibrahim Y. Hamdan. 'Physiological Characteristics of Four Methylotrophic Bacteria and Their Potential Use in Single-Cell Protein Production', *MIRCEN Journal* 5, no. 2 (1989), 149–59.

Banita, Georgiana. 'From Isfahan to Ingolstadt: Bernardo Bertolucci's "La Via del Petrolio" and the Global Culture of Neorealism', in Ross Barrett and Daniel Worden (eds), *Oil Culture* (Minneapolis: University of Minnesota Press, 2014), 145–68.

Barak, On. *On Time: Technology and Temporality in Modern Egypt* (Berkeley: University of California Press, 2013).

Barnes, Larry. *Looking Back Over My Shoulder* (n.p., 1979).

Barr, James. *Line in the Sand: Britain, France, and the Struggle That Shaped the Middle East* (New York and London: Norton, 2011).

Barrett, Ross and Daniel Worden (eds). *Oil Culture* (Minneapolis: University of Minnesota Press, 2014).

Barrett, Ross and Daniel Worden. 'Oil Culture: Guest Editors' Introduction', *Journal of American Studies* 46, no. 2 (2012), 269–72.

Barry, Andrew. *Material Politics: Disputes along the Pipeline* (Chichester: Wiley Blackwell, 2013).

Barry, Andrew. 'The Oil Archives', in Hannah Appel, Arthur Mason and Michael Watts (eds), *Subterranean Estates: Life Worlds of Oil and Gas* (Ithaca: Cornell University Press, 2015), 95–107.

Barry, Andrew. 'The Political Geology of Area', *Political Geography* 57 (2016), 94–104.

Barry, Andrew. *Political Machines: Governing a Technological Society* (London: Athlone Press, 2001).

Barry, Andrew. 'Technological Zones', *European Journal of Social Theory* 9, no. 2 (May 2006), 239–53.

Basrawi, Fadia. *Brownies and Kalashnikovs: A Saudi Woman's Memoir of American Arabia and Wartime Beirut* (Reading: South Street Press, 2009).

Beasley, Betsy A. 'Service Learning: Oil, International Education, and Texas's Corporate Cold War', *Diplomatic History* 42, no. 2 (1 April 2018), 177–203.

Beaulieu, Jill and Mary Roberts (eds). *Orientalism's Interlocutors: Painting, Architecture, Photography* (Durham, NC: Duke University Press, 2003).

Behdad, Ali. 'The Orientalist Photograph', in Ali Behdad and Luke Gartlan (eds), *Photography's Orientalism: New Essays on Colonial Representation* (Los Angeles: Getty Research Institute, 2013), 11–32.

Benjamin, Walter. 'The Work of Art in the Age of Mechanical Reproduction', in Hannah Arendt (ed.), *Illuminations* (New York: Schocken, 1969).

Bet-Shlimon, Arbella. *City of Black Gold: Oil, Ethnicity, and the Making of Modern Kirkuk* (Stanford: Stanford University Press, 2019).

Beverley, Eric Lewis. 'Colonial Urbanism and South Asian Cities', *Social History* 36, no. 4 (November 2011), 482–97.

Bhagavan, Manu. 'A New Hope: India, the United Nations and the Making of the Universal Declaration of Human Rights', *Modern Asian Studies* 44, no. 2 (2010), 311–47.

Biglari, Mattin. 'Refining Knowledge: Expertise, Labour and Everyday Life in the Iranian Oil Industry, c.1933–51' (Ph.D. diss., SOAS University of London, 2020).

Bini, Elisabetta. 'Building an Oil Empire: Labor and Gender Relations in American Company Towns in Libya, 1950s–1970s', in Touraj Atabaki, Elisabetta Bini and Kaveh Ehsani (eds), *Working for Oil: Comparative Social Histories of Labor in the Global Oil Industry* (Cham: Springer, 2018), 313–36.

Bini, Elisabetta. 'From Colony to Oil Producer: US Oil Companies and the Reshaping of Labor Relations in Libya during the Cold War', *Labor History* 60, no. 1 (2 January 2019), 44–56.

Bini, Elisabetta and Francesco Petrini. 'Labor Politics in the Oil Industry: New Historical Perspectives', *Labor History* 60, no. 1 (2019), 1–7.

Bird, Kai. *Crossing Mandelbaum Gate: Coming of Age Between the Arabs and Israelis, 1956–1978* (New York: Scribner, 2010).

Bocco, Riccardo. 'La Sédentarisation des Pasteurs Nomads: Les Experts Internationaux

Face à la Question Bedouine dans le Moyen-Oriente Arab (1950–1970)', *Cahiers des Sciences Humaines* 26, no. 1–2 (1990), 97–117.

Bourdieu, Pierre. *Distinction* (London: Routledge, 1984).

Bowker, Geoffrey C. *Science on the Run: Information Management and Industrial Geophysics at Schlumberger, 1920–1940* (Cambridge, MA: MIT Press, 1994).

Brain, Robert. *The Pulse of the Modern* (Seattle: University of Washington Press, 2015).

Brown, Michael E. 'The Nationalization of the Iraqi Petroleum Company', *International Journal of Middle East Studies* 10, no. 1 (1979), 107–24.

Bsheer, Rosie. 'A Counter-Revolutionary State: Popular Movements and the Making of Saudi Arabia', *Past & Present* 238, no. 1 (February 2018), 233–77.

Butler, Grant. *Kings and Camels: An American in Saudi Arabia* (Reading: Garnet, 2008).

Carr, Helen. 'Modernism and Travel (1880–1940)', in Peter Hulme and Tim Youngs (eds), *The Cambridge Companion to Travel Writing* (Cambridge: Cambridge University Press, 2002), 70–86.

Cepek, Michael. *A Future for Amazonia: Randy Borman and Cofán Environmental Politics* (Austin: University of Texas Press, 2012).

Cetina, Karin Knorr. 'Strong Constructivism – from a Sociologist's Point of View: A Personal Addendum to Sismondo's Paper', *Social Studies of Science* 23, no. 3 (1993), 555–63.

Chakrabarty, Dipesh. 'Conditions for Knowledge of Working-Class Conditions: Employers, Government and the Jute Workers of Calcutta, 1890–1940', in Ranajit Guha (ed.), *Selected Subaltern Studies* (New York: Oxford University Press, 1988), 179–230.

Chatty, Dawn. 'The Bedouin of Central Oman', *Journal of Oman Studies* 6, no. 1 (1984), 149–63.

Chatty, Dawn. *Mobile Pastoralists: Development Planning and Social Change in Oman* (New York: Columbia University Press, 1996).

Chatty, Dawn. 'Petroleum Exploitation and the Displacement of Pastoral Nomadic Households in Oman', in Seteney Shami (ed.), *Population Displacement and Resettlement: Development and Conflict in the Middle East* (New York: Centre for Migration Studies, 1994), 89–106.

Chatty, Dawn and Marcus Colchester (eds). *Conservation and Mobile Indigenous Peoples: Forced Settlement, Displacement and Sustainable Development* (Oxford: Berghahn, 2002).

Cheney, Michael Sheldon. *Big Oil Man from Arabia* (New York: Ballantine Books, 1958).

Christoph, Horst. *Max Reisch: Über alle Straßen hinaus* (Innsbruck: Tyrolia-Verlag, 2012).

Citino, Nathan J. *Envisioning the Arab Future: Modernization in U.S.–Arab Relations, 1945–1967* (New York: Cambridge University Press, 2017).

Citino, Nathan J. 'Suburbia and Modernization: Community Building and America's Post-World War II Encounter with the Middle East', *The Arab Studies Journal* 13/14, no. 2/1 (Fall 2005/Spring 2006), 39–64.

Clark, Nigel and Kathryn Yusoff. 'Geosocial Formations and the Anthropocene', *Theory, Culture & Society* 34, no. 2–3 (2017), 3–23.

Clark, Ronald W. *The Life of Ernst Chain: Penicillin and Beyond* (New York: St Martin's Press, 1985).

Clarke, I. M. 'The Changing International Division of Labour within ICI', in Michael Taylor and Nigel Thrift (eds), *The Geography of Multinationals: Studies in the Spatial Development and Economic Consequences of Multinational Corporations* (New York: St Martin's Press, 1982), 90–116.

Cohen, Lizabeth. *A Consumers' Republic: The Politics of Mass Consumption in Postwar America* (New York: Knopf, 2003).

Corley, Thomas A. B. *A History of the Burmah Oil Company, 1886–1924* (London: Heinemann, 1983).

Coronil, Fernando. *The Magical State. Nature, Money, and Modernity in Venezuela* (Chicago: University of Chicago Press, 1997).

Cosgrove, Denis. *Apollo's Eye: A Cartographic Genealogy of the Earth in the Western Imagination* (Baltimore: Johns Hopkins University Press, 2001).

Crinson, Mark. 'Abadan: Planning and Architecture under the Anglo-Iranian Oil Company', *Planning Perspectives* 12, no. 3 (January 1997), 341–59.

Cronin, Stephanie. 'Popular Politics, the New State and the Birth of the Iranian Working Class: The 1929 Abadan Oil Refinery Strike', *Middle Eastern Studies* 46, no. 5 (September 2010), 699–732.

Damluji, Mona. 'The Image World of Middle Eastern Oil', in Hannah Appel, Arthur Mason and Michael Watts (eds), *Subterranean Estates: Life Worlds of Oil and Gas* (Ithaca: Cornell University Press, 2015), 147–64.

Damluji, Mona. 'The Oil City in Focus: The Cinematic Spaces of Abadan in the Anglo-Iranian Oil Company's *Persian Story*', *Comparative Studies in South Asia, Africa and the Middle East*, 33, no. 1 (2013), 75–88.

Damluji, Mona. 'Petroleum's Promise: The Neo-Colonial Imaginary of Oil Cities in the Modern Arabian Gulf' (Ph.D. diss., University of California at Berkeley, 2013).

Danby, Arthur. *Natural Rock: Asphalts and Bitumens, Their Geology, History, Properties and Industrial Application* (London: Constable, 1913).

DasGupta, Ranajit. *Labour and Working Class in Eastern India* (Calcutta: K. P. Bagchi & Co., 1940).

Davidson, Christopher. *Dubai: The Vulnerability of Success* (New York: Columbia University Press, 2008).

DeSouza, Wendy. *Unveiling Men: Modern Masculinities in Twentieth-Century Iran* (Syracuse, NY: Syracuse University Press, 2019).

Di-Capua, Yoav. 'Common Skies, Divided Horizons: Aviation, Class and Modernity in Early Twentieth Century Egypt', *Journal of Social History* 41, no. 4 (Summer 2008), 917–42.

Dietrich, Christopher R. W. *Oil Revolution: Anticolonial Elites, Sovereign Rights, and the Economic Culture of Decolonization* (Cambridge: Cambridge University Press, 2017).

Dobe, Michael E. 'A Long Slow Tutelage in Western Ways of Work: Industrial Education and the Containment of Nationalism in Anglo-Iranian and Aramco, 1923–1963' (Ph.D. diss., Rutgers, State University of New Jersey, 2008).

Doumato, Eleanor A. 'Gender, Monarchy, and National Identity in Saudi Arabia', *British Journal of Middle Eastern Studies* 19, no. 1 (1992), 31–47.

Duffy, Enda. *The Speed Handbook: Velocity, Pleasure, Modernism* (Durham, NC: Duke University Press, 2009).

Ehsani, Kaveh. 'Disappearing the Workers: How Labor in the Oil Complex Has Been Made Invisible', in Touraj Atabaki, Elisabetta Bini and Kaveh Ehsani (eds), *Working for Oil: Comparative Social Histories of Labor in the Global Oil Industry* (Cham: Palgrave Macmillan, 2018), 11–34.

Ehsani, Kaveh. 'Social Engineering and the Contradictions of Modernization in Khuzestan's Company Towns: A Look at Abadan and Masjed-Soleyman', *International Review of Social History* 48, no. 3 (December 2003), 361–99.

Ehsani, Kaveh. 'The Social History of Labor in the Iranian Oil Industry: The Built Environment and the Making of the Industrial Working Class (1908–1941)' (Ph.D. diss., Leiden University, 2015).

Eley, Geoff. *A Crooked Line: From Cultural History to the History of Society* (Ann Arbor: University of Michigan Press, 2005).

Eley, Geoff. 'Review of *Transnational Labour History: Explorations* by Marcel van der

Linden', *Labor: Studies in Working-Class History of the Americas* 3, no. 1 (2006), 164–6.

Elias, Norbert. *Involvement and Detachment* (Oxford: Blackwell, 1987).

Elling, Rasmus. 'On Lines and Fences: Labour, Community and Violence in an Oil City', in Ulrike Freitag, Nelida Fuccaro, Claudia Ghrawi and Nora Lafi (eds), *Urban Violence in the Middle East: Changing Cityscapes in the Transition from Empire to Nation State* (New York: Berghahn, 2015), 197–221.

El Shakry, Omnia '"History without Documents": The Vexed Archives of Decolonization in the Middle East', *The American Historical Review* 120, no. 3 (2015), 920–34.

Facey, William and Gillian Grant. *Kuwait by the First Photographers* (London: I. B. Tauris, 1999).

Fakhimi, Qobād. *Si sāl naft-e Iran: Az melli shodan-e naft tā enqelāb-e eslāmi* (Tehran: Mehrandish, 1387/2008).

Fanon, Frantz (trans. Constance Farrington). *The Wretched of the Earth* (London: Penguin, 2001 [1961]).

Farmanfarmaian, Manucher and Roxane Farmanfarmaian. *Blood and Oil: Inside the Shah's Iran* (New York: Modern Library, 1999).

Farrell, Alex and Adam Brandt. 'Risks of the Oil Transition', *Environmental Research Letters* 1, no. 1 (2006), 1–6.

Ferrier, Ronald W. *The History of the British Petroleum Company: Vol. 1, The Developing Years 1901–1932* (Cambridge: Cambridge University Press, 1982).

Finnie, David H. *Desert Enterprise: The Middle East Oil Industry in Its Local Environment* (Cambridge, MA: Harvard University Press, 1958).

Foucault, Michel. *The Order of Things* (New York: Vintage, 1970).

Foucault, Michel. *This Is Not a Pipe* (Berkeley: University of California Press, 2008).

Friedan, Betty (ed. Kirsten Fermaclich and Lisa M. Fine). *The Feminine Mystique* (New York: Norton, 2013).

Fuccaro, Nelida. 'Arab Oil Towns as Petro-Histories', in Carola Hein (ed.), *Oil Spaces: Exploring the Global Petroleumscape* (New York: Routledge, 2022), 129–44.

Fuccaro, Nelida (ed.). 'Introduction: Histories of Oil and Urban Modernity in the Middle East', Special Issue in *Comparative Studies in South Asia, Africa and the Middle East* 33, no. 1 (2013), 1–6.

Fuccaro, Nelida. 'Shaping the Urban Life of Oil in Bahrain: Consumerism, Leisure, and Public Communication in Manama and in the Oil Camps, 1932–1960s',

Comparative Studies of South Asia, Africa and the Middle East 33, no. 1 (2013), 59–74.

Gabrys, Jennifer. *Digital Rubbish* (Ann Arbor: University of Michigan Press, 2011).

Garavini, Giuliano. *The Rise and Fall of OPEC in the 20th Century* (New York: Oxford University Press, 2019).

Gisler, Monika. *'Swiss Gang': Pioniere der Erdölexploration* (Schweizer Pioniere der Wirtschaft und Technik) (Zurich: Verein für wirtschaftshistorische Studien, 2014).

Ghosn, Rania. 'Carbon Re-form', *Log: Overcoming Carbon Form*, 47 (Fall 2019), 106–17.

Ghosn, Rania (ed.). *Landscapes of Energy* (Cambridge MA: Harvard University Press, 2009).

Ghosn, Rania. 'Territories of Oil', in Amale Andraos and Nora Akawi (eds), *The Arab City: Architecture and Representation* (New York: Columbia Books on Architecture and the City, 2016), 164–75.

Ghosn, Rania. 'Where Are the Missing Spaces? The Geography of Some Uncommon Interests', *Perspecta* 45 (2012), 109–16.

Glover, William. *Making Lahore Modern: Constructing and Imagining a Colonial City* (Minneapolis: University of Minnesota Press, 2008).

Grosz, Elizabeth, Kathryn Yusoff and Nigel Clark. 'An Interview with Elizabeth Grosz: Geopower, Inhumanism and the Biopolitical', *Theory, Culture & Society* 34, no. 2–3 (2017), 129–46.

Guagnini, Anna. 'Worlds Apart: Academic Instruction and Professional Qualifications in the Training of Mechanical Engineers in England, 1850–1914', in Robert Fox and Anna Guagnini (eds), *Education, Technology, and Industrial Performance in Europe, 1850–1939* (Cambridge: Cambridge University Press, 1993), 16–41.

Günel, Gökçe. *Spaceship in the Desert: Energy, Climate Change, and Urban Design in Abu Dhabi* (London: Duke University Press, 2019).

Gustafsson, Bret. 'Fossil Knowledge Networks: Industry Strategy, Public Culture and the Challenge for Critical Research', in Owen J. Logan and John A. McNeish (eds), *Flammable Societies: Studies on the Socio-Economics of Oil and Gas* (London: Pluto Press, 2012), 315–16.

Gutfreund, Owen D. *20th-Century Sprawl: Highways and the Reshaping of the American Landscape* (New York: Oxford University Press, 2004).

Haber, Stephen and Victor Menaldo. 'Do Natural Resources Fuel Authoritarianism? A Reappraisal of the Resource Curse', *The American Political Science Review*, 105, no 1 (2011), 1–26.

Hackforth-Jones, Jocelyn and Mary Roberts (eds). *Edges of Empire: Orientalism and Visual Culture*, New Interventions in Art History (Oxford: Blackwell, 2005).

Halland, Ingrid. 'Being Plastic', *Log*, no. 47 (2019), 35–44.

Hajer, Maarten. 'Discourse Coalitions and the Institutionalization of Practice', in Frank Fischer and John Forester (eds), *The Argumentative Turn in Policy Analysis and Planning* (Durham, NC: Duke University Press, 1993), 43–76.

Hamer, G. and Ibrahim Y. Hamdan. 'The Transfer of Single Cell Protein Technology to the Petroleum-Exporting Arab States', *MIRCEN Journal* 1, no. 1 (1985), 23–32.

Hamilton, Shane. *Trucking Country: The Road to America's Wal-Mart Economy* (Princeton: Princeton University Press, 2008).

Hawley, Donald. *The Trucial States* (London: Allen & Unwin, 1970).

Heard-Bey, Frauke. *From Trucial States to United Arab Emirates* (London: Longman, 1982).

Helmreich, Stefan. *Alien Ocean: Anthropological Voyages on Microbial Seas* (Berkeley: University of California Press, 2009).

Henry, Todd A. *Assimilating Seoul: Japanese Rule and the Politics of Public Space in Colonial Korea, 1910–1945* (Berkeley: University of California Press, 2014).

Hepple, Peter (ed.). *Microbiology: Proceedings of a Conference Held in London, 19 and 20 September 1967* (London: Institute of Petroleum, 1968).

Herb, Michael. *The Wages of Oil: Parliaments and Economic Development in Kuwait and the UAE* (Ithaca: Cornell University Press, 2014).

Hindelang, Laura. *Iridescent Kuwait. Petro-Modernity and Urban Visual Culture since the Mid-Twentieth Century* (Berlin: De Gruyter, 2022).

Hindelang, Laura. 'Precious Property: Oil and Water in Twentieth-Century Kuwait', in Carola Hein (ed.), *Oil Spaces: Exploring the Global Petroleumscape* (New York: Routledge, 2021), 159–75.

Hitchcock, Peter. 'Velocity and Viscosity', in Hannah Appel, Arthur Mason and Michael Watts (eds), *Subterranean Estates: Life Worlds of Oil and Gas* (Ithaca: Cornell University Press, 2015), 45–60.

Holmberg, Eva J. 'The Middle East', in Carl Thompson (ed.), *The Routledge Companion to Travel Writing* (London, New York: Routledge, 2016), 372–84.

Houser, Heather. 'The Aesthetics of Environmental Visualizations: More than Information Ecstasy?', *Public Culture* 26, no. 2 (2015), 319–37.

Huber, Matthew T. *Lifeblood: Oil, Freedom, and the Forces of Capital* (Minneapolis: University of Minnesota Press, 2013).

Jacobs, Meg. *Panic at the Pump: The Energy Crisis and the Transformation of American Politics in the 1970s* (New York: Hill & Wang, 2016).

Javāheri'zādeh, Majid. *Pālāyeshgāh-e Ābādān dar 80 Sāl-e Tārikh-e Irān 1908-1988 (The Abadan Refinery in 80 Years of the History of Iran 1908–1988)* (Tehran: Nashr-e Shādegān, 1396/2017).

Jefroudi, Maral. 'Revisiting "the Long Night" of Iranian Workers: Labor Activism in the Iranian Oil Industry in the 1960s', *International Labor and Working-Class History* 84 (2013), 176–94.

Johnson, Nora. *You* Can *Go Home Again* (Garden City: Doubleday, 1982).

Jones, Benjamin. 'Women on the Oil Frontier: Gender and Power in Aramco's Arabia', *Rice Historical Review* 2 (Spring 2017), 55–69.

Jones, Jeremy and Nicholas Ridout. *A History of Modern Oman* (Cambridge: Cambridge University Press, 2015).

Jones, Toby Craig. *Desert Kingdom: How Oil and Water Forged Saudi Arabia* (Cambridge, MA: Harvard University Press, 2010).

Jones, Toby Craig. 'Rebellion on the Saudi Periphery: Modernity, Marginalization, and the Shi'a Uprising of 1979', *International Journal of Middle East Studies* 38, no. 2 (May 2006), 213–33.

Jongerden, Joost. 'The Spatial (Re)Production of the Kurdish Issue: Multiple and Contradicting Trajectories', *Journal of Balkan and Near Eastern Studies* 13, no. 4 (2011), 375–88.

Joyce, Patrick. *The State of Freedom: A Social History of the British State since 1800* (Cambridge: Cambridge University Press, 2013).

Kaplan, Amy. 'Manifest Domesticity', *American Literature* 70, no. 3 (September 1998), 581–606.

Karimi, Pamela. *Domesticity and Consumer Culture in Iran: Interior Revolutions of the Modern Era* (New York: Routledge, 2013).

Karuka, Manu. *Empire's Tracks: Indigenous Nations, Chinese Workers, and the Transcontinental Railroad* (Berkeley: University of California Press, 2019).

Karl, Terry Lynn. *The Paradox of Plenty: Oil Booms and Petro-States* (Berkeley: University of California Press, 1997).

Kaufman, Asher. 'Between Permeable and Sealed Borders: The Trans-Arabian Pipeline and the Arab–Israeli Conflict', *International Journal of Middle East Studies* 46, no. 1 (2014), 95–116.

Kennedy, Ryan and Lydia Tiede. 'Economic Development Assumptions and the Elusive Curse of Oil', *International Studies Quarterly* 57, no. 4 (2013), 760–71.

Khadduri, Walid(ed.). *'Abdullah al-Tariqi: al-A'mal al-Kamila* (Beirut: Markaz dirasat al-wahda al-'Arabiyya, 1999).

Khalili, Laleh. *Sinews of War and Trade: Shipping and Capitalism in the Arabian Peninsula* (New York: Verso, 2020).

Kimball, Solon T. 'American Culture in Saudi Arabia', *Transactions of the New York Academy of Sciences* 18, no. 5 (1956), 477–8.

Kimmerling, Judith. 'Uncommon Grounds: Occidental's Land Access and Community Relations Standards and Practices in Quichua Communities in the Ecuadorian Amazon', *Law and Anthropology* 11 (2001), 179–247.

Kinninmont, Jane. 'Bahrain', in Christopher Davidson (ed.), *Power and Politics in the Persian Gulf Monarchies* (New York: Columbia University Press, 2011), 31–62.

Kruse, Kevin M. *White Flight: Atlanta and the Making of Modern Conservatism* (Princeton: Princeton University Press, 2005).

Ladjevardi, Habib. *Labor Unions and Autocracy in Iran* (Syracuse, NY: Syracuse University Press, 1985).

Lakoff, Andrew and Stephen J. Collier. 'Ethics and the Anthropology of Modern Reason', *Anthropological Theory* 4, no. 4 (2004), 419–34.

Lancaster, William and Fidelity Lancaster. *Honour is in Contentment: Life Before Oil in Ras al-Khaimah and Some Neighboring Regions* (Berlin: de Gruyter, 2011).

Latour, Bruno. 'Drawing Things Together', in Michael E. Lynch and Steve Wooglar (eds), *Representation in Scientific Practice* (Cambridge, MA: MIT Press, 1990), 19–78.

Latour, Bruno. *Pandora's Hope: Essays on the Reality of Science Studies* (Cambridge, MA: Harvard University Press, 1999).

Latour, Bruno. *Science in Action: How to Follow Scientists and Engineers through Society* (Cambridge, MA: Harvard University Press, 1987).

Latour, Bruno. 'Zirkulierende Referenz', *ARCH+*, no. 238 (2020), 18–29.

Lebkicher, Roy, George Rentz, Max Steineke et al. *Aramco Handbook* (Dhahran: Arabian American Oil Company, 1960).

Leccese, Stephen R. 'John D. Rockefeller, Standard Oil, and the Rise of Corporate Public Relations in Progressive America, 1902–1908', *Journal of the Gilded Age and Progressive Era* 16 (2017), 245–63.

LeMenager, Stephanie. *Living Oil: Petroleum Culture in the American Century* (New York: Oxford University Press, 2014).

Levander, Caroline and Matthew Guterl. *Hotel Life* (Chapel Hill: University of North Carolina Press, 2015).

Limbert, Mandana E. *In the Time of Oil: Piety, Memory, and Social Life in an Omani Town* (Stanford: Stanford University Press, 2010).

Limbert, Mandana E. 'Reserves, Secrecy, and the Science of Oil Prognostication in Southern Arabia', in Hannah Appel, Arthur Mason and Michael Watts (eds), *Subterranean Estates: Life Worlds of Oil and Gas* (Ithaca: Cornell University Press, 2015), 340–53.

Lippman, Thomas. *Inside the Mirage: America's Fragile Partnership with Saudi Arabia* (Boulder: Westview Press, 2004).

Liu, Lydia H. 'Shadows of Universalism: The Untold Story of Human Rights around 1948', *Critical Inquiry* 40 (Summer 2014), 385–417.

Lumely, Laurence. 'The Invisible Bituminous Desert', *Log*, no. 47 (2019), 25–34.

Lynd, Robert S. and Helen Merrell Lynd. *Middletown: A Study in Contemporary American Culture* (New York: Harcourt, Brace & Co., 1929).

MacCabe, Colin. 'Preface', in Fredric Jameson. *The Geopolitical Aesthetic: Cinema and Space in the World System* (Bloomington: Indiana University Press, 1992).

MacLean, Matthew. 'Spatial Transformations and the Emergence of "the National": The Formation of the United Arab Emirates, 1950–1980' (Ph.D. diss., New York University, 2017).

Maier, Charles S. 'Between Taylorism and Technocracy: European Ideologies and the Vision of Industrial Productivity in the 1920s', *Journal of Contemporary History* 5, no. 2 (1970), 27–61.

Malallah, Hussain 'Isa. *The Iraqi War Criminals and Their Crimes during the Iraqi Occupation of Kuwait: A Legal Reading in the Documents of the Iraqi War Crimes against Kuwait and Its People* (Kuwait: Center for Research and Studies of Kuwait, 1998).

Malick, Emil A. 'Status of SCP in the USA and Provesta's Technologies', in *Proceedings of OAPEC Symposium on Petroprotein* (Kuwait: Organization of Arab Petroleum Producing Countries, 1980).

Mangan, J. A. *The Games Ethic and Imperialism: Aspects of the Diffusion of an Ideal* (Harmondsworth: Viking, 1986).

Manwaring, Tony and Stephen Wood. 'The Ghost in the Labour Process', in David Knights, Hugh Willmott and David L. Collinson (eds), *Job Redesign: Critical Perspectives on the Labour Process* (Aldershot: Gower, 1985), 171–96.

Mason, Arthur. 'Arctic Energy Image: Hydrocarbon Aesthetics of Progress and Form', *Polar Geography* 39, no. 2 (2016), 130–43.

Mason, Arthur. 'Consulting Virtue: From Judgment to Decision Making in Natural Gas Industry', *Journal of Royal Anthropological Institute* 25, no. 1 (2019), 1–19.

Mason, Arthur. 'Images of the Energy Future', *Environmental Research Letters* 1, no. 1 (2006), 1–8.

Mason, Arthur. 'The Rise of Consultant Forecasting in Liberalized Natural Gas Markets', *Public Culture* 19, no. 2 (2007), 367–79.

Matthee, Rudi. 'Transforming Dangerous Nomads into Useful Artisans, Technicians, Agriculturalists: Education in the Reza Shah Period', in Stephanie Cronin (ed.), *The Making of Modern Iran: State and Society under Riza Shah 1921–1941* (London: Routledge, 2003), 123–45.

Matthiesen, Toby. *The Other Saudis: Shiism, Dissent and Sectarianism* (New York: Cambridge University Press, 2015).

Maul, Daniel R. 'The International Labour Organization and the Globalization of Human Rights, 1944–1970', in Stefan-Ludwig Hoffman (ed.), *Human Rights in the Twentieth Century* (New York: Cambridge University Press, 2011), 301–20.

May, Elaine Tyler. *Homeward Bound: American Families in the Cold War Era* (New York: Basic Books, 1988).

Mazzarella, William. 'Culture, Globalization, Mediation', *Annual Review of Anthropology* 33 (2004), 345–67.

McClaren, Donald. 'The Great Protein Fiasco', *The Lancet*, 13 July 1974.

McFarland, Victor. *Oil Powers: A History of the U.S.–Saudi Alliance* (New York: Columbia University Press, 2020).

Melman, Billie. 'The Middle East/Arabia: "The Cradle of Islam"', in Peter Hulme and Tim Youngs (eds), *The Cambridge Companion to Travel Writing* (Cambridge: Cambridge University Press, 2002), 105–21.

Menashri, David. *Education and the Making of Modern Iran* (Ithaca: Cornell University Press, 1992).

Menoret, Pascal. 'Learning from Riyadh: Automobility, Joyriding, and Politics', *Comparative Studies of South Asia, Africa, and the Middle East* 39, no. 1 (May 2019), 131–42.

Menoret, Pascal. *Joyriding in Riyadh: Oil, Urbanism, and Road Revolt* (New York: Cambridge University Press, 2014).

Menoret, Pascal. 'Urban Unrest and Non-Religious Radicalization in Saudi Arabia', in Madawi al-Rasheed and Marat Shterin (eds), *Dying for Faith: Religiously Motivated Violence in the Contemporary World* (New York: Palgrave Macmillan, 2009), 123–37.

Michaels, David. *Doubt Is Their Product: How Industry's Assault on Science Threatens Your Health* (Oxford: Oxford University Press, 2008).

Miller, Karen S. *The Voice of Business: Hill & Knowlton and Postwar Public Relations* (Chapel Hill: University of North Carolina Press, 1999).

Mirzoeff, Nicholas. 'Visualizing the Anthropocene', *Public Culture* 26, no. 2 (2014), 213–32.

Mitchell, Timothy. *Carbon Democracy: Political Power in the Age of Oil* (London: Verso, 2011).

Mitchell, Timothy. *Rule of Experts: Egypt, Techno-Politics, Modernity* (Berkeley: University of California Press, 2002).

Monroe, Kristin V. 'Automobility and Citizenship in Interwar Lebanon', *Comparative Studies of South Asia, Africa, and the Middle East* 34, no. 3 (2014), 518–31.

Monteiro, Eric, Gasparas Jarulaitis and Vidar Hepsø. 'The Family Resemblance of Technologically Mediated Work Practice', *Information and Organization* 22, no. 3 (2012), 169–87.

Moore, Jason W. *Capitalism in the Web of Life: Ecology and the Accumulation of Capital* (New York: Verso, 2015).

Morgan, Lewis Henry. *Ancient Society* (Tucson: University of Arizona Press, 1985 [1877]).

Moser, Titus and Damian Miller. 'Multinational Corporations' Impacts on the Environment and Communities in the Developing World: A Synthesis of the Contemporary Debate', in Walter Wehrmeyer and Yacob Mulugetta (eds), *Growing Pains: Environmental Management in Developing Countries* (Sheffield: Greenleaf, 1999), 40–51.

Munif, Abdelrahman (trans. Peter Theroux). *Cities of Salt* (New York: Vintage, 1987).

Noble, David F. *America by Design: Science, Technology, and the Rise of Corporate Capitalism* (New York: Knopf, 1977).

Nye, David E. *Image Worlds: Corporate Identities at General Electric, 1890–1930* (Cambridge, MA: MIT Press, 1985).

Oldenziel, Ruth. *Making Technology Masculine: Men, Women and Modern Machines in America, 1870–1945* (Amsterdam: Amsterdam University Press, 1999).

Owen, Edgar W. *Trek of the Oil Finders: A History of Exploration for Petroleum* (Tulsa: American Association of Petroleum Geologists, 1975).

Özar, Şemsa, Nesrin Uçarlar and Osman Aytar (trans. Sedef Çakmak). *From Past to Present: A Paramilitary Organization in Turkey: Village Guard System* (Istanbul: DİSA, 2013).

Pálsson, Gisli and Heather A. Swanson. 'Down to Earth: Geosocialities and Geopolitics', *Environmental Humanities* 8, no. 2 (2016), 149–71.

Parkin, Katherine J. *Women at the Wheel: A Century of Buying, Driving, and Fixing Cars* (Philadelphia: University of Pennsylvania Press, 2017).

Painter, David S. *Oil and the American Century: The Political Economy of U.S. Foreign Oil Policy, 1941–1954* (Baltimore: Johns Hopkins University Press, 1986).

Pendakis, Andrew and Sheena Wilson. 'Sight, Site, Cite: Oil in the Field of Vision', *Imaginations: Journal of Cross-Cultural Image Studies* 3, no. 2 (2012), 4–5.

Philby, Harry S. J. B. *Arabia of the Wahhabis* (London: Constable, 1928).

Philipp, Hans-Jürgen. 'Arnold Heims erfolglose Erdölsuche und erfolgreiche Wassersuche 1924 im nordöstlichen Arabien', *Vierteljahrsschrift der Naturforschenden Gesellschaft in Zürich* 128, no. 1 (1983), 43–73.

Polanyi, Michael. *Personal Knowledge: Towards a Post-Critical Philosophy* (Chicago: University of Chicago Press, 1958).

Pollard, Lisa. *Nurturing the Nation: The Family Politics of Modernizing, Colonizing, and Liberating Egypt, 1805–1923* (Berkeley: University of California Press, 2005).

Povinelli, Elizabeth. *Geontologies: A Requiem to Late Liberalism* (Durham, NC: Duke University Press, 2016).

Prashad, Vijay. *The Darker Nations: A People's History of the Third World* (New York: The New Press, 2007).

Prokop, A., H. D. Ratcliffe, M. I. Fatayer, N. Al-Awadhi, A. Khamis, M. Murad, C. Bond and I. Y. Hamdan. 'Bacterial SCP from Methanol in Kuwait: Product Recovery and Composition', *Biotechnology and Bioengineering* 26, no. 9 (1984), 1,085–9.

Rae, Jonathan. 'Tribe and State: Management of the Syrian Steppe' (Ph.D. diss., University of Oxford, 1999).

Rae, Jonathan and Dawn Chatty. *Social Impact Assessment of Mukhaizana, Jiddat-il-Harasiis* (Muscat: Petroleum Development Oman, 2001).

Raman, Parvathi. 'Being Indian the South African Way', in Annie Coombes (ed.), *Rethinking Settler Colonialism: History and Memory in Australia, Canada, Aotearoa New Zealand and South Africa* (Manchester: Manchester University Press, 2006), 193–208.

Rancière, Jacques. *Politics and Aesthetics* (London: Continuum, 2004).

Reisch, Max. *Im Auto nach Koweit* (Vienna: Ullstein, 1953).

Reisz, Todd. 'Landscapes of Production: Filming Dubai and the Trucial States', *Journal of Urban History* 44, no. 2 (2018), 298–317.

Ricoeur, Paul. *The Rule of Metaphor* (London: Routledge, 1975).

Riley, Judith Merkle. *Management and Ideology: The Legacy of the International*

Scientific Management Movement (Berkeley: University of California Press, 1980).

Rogers, Douglas. 'Oil and Anthropology', *Annual Review of Anthropology* 44 (2015), 365–80.

Rosato, Donna. 'Worried about Corporate Numbers? How About the Charts?', *The New York Times*, 15 September 2002, Business Section.

Rose, Gillian. 'On the Need to Ask How, Exactly, Is Geography "Visual"?', *Antipode* 35, no. 2 (2003), 212–21.

Rose, Gillian and Divya P. Tolia-Kelly. 'Introduction: Visuality/Materiality. Introducing a Manifesto for Practice', in Gillian Rose and Divya P. Tolia-Kelly (eds), *Visuality/Materiality: Images, Objects and Practices* (Farnham: Ashgate, 2012), 1–11.

Rosenberg, Emily. 'Consuming Women: Images of Americanization in the "American Century"', *Diplomatic History* 23, no. 3 (1999), 479–97.

Ruxin, Josh. 'The United Nations Protein Advisory Group', in David F. Smith and Jim Phillips (eds), *Food, Science, and Regulation in the Twentieth Century: International and Comparative Perspectives* (London and New York: Routledge, 2016), 151–66.

Said, Edward W. *Orientalism* (New York: Vintage, 1979 [1978]).

Said, Rosemarie J. 'The 1938 Reform Movement in Dubai', *Al-Abhath* 23, nos 1–4 (December 1970), 247–318.

Saleh, Hassan Mohammad Abdulla. 'Labor, Nationalism and Imperialism in Eastern Arabia: Britain, the Shaikhs and the Gulf Oil Workers in Bahrain, Kuwait and Qatar, 1932–1956' (Ph.D. diss., University of Michigan, 1991).

Salgado, Sebastião. *Kuwait: A Desert on Fire* (Cologne: Taschen, 2016).

Salter, Arthur. *The Development of Iraq: A Plan of Action* (London: Caxton Press and Iraq Development Board, 1955).

Sampson, Anthony. *The Seven Sisters: The Great Oil Companies and the World They Shaped* (Toronto: Bantam Press, 1975).

Santiago, Myrna I. *The Ecology of Oil: Environment, Labor, and the Mexican Revolution, 1900–1938* (Cambridge: Cambridge University Press, 2006).

Santiago, Myrna. 'Women of the Mexican Oil Fields: Class, Nationality, Economy, Culture, 1900–1938', *Journal of Women's History* 21, no. 1 (2009), 87–110.

Sassoon, Joseph and Alissa Walter. 'The Iraqi Occupation of Kuwait: New Historical Perspectives', *The Middle East Journal* 71, no. 4 (2017), 607–28.

Saussure, Ferdinand de. *Course in General Linguistics* (London: Bloomsbury Revelations, 1983).

Schayegh, Cyrus. *Who Is Knowledgeable, Is Strong: Science, Class, and the Formation of Modern Iranian Society, 1900–1950* (Berkeley: University of California Press, 2009).

Scott, James C. *Seeing Like a State: How Certain Schemes to Improve the Human Condition Have Failed* (New Haven: Yale University Press, 1998).

Seccombe, Ian J. and Richard I. Lawless. 'Foreign Worker Dependence in the Gulf, and the International Oil Companies: 1910–50', *International Migration Review* 20, no. 3 (August 1986), 548–74.

Shafiee, Khatayun. *Machineries of Oil: An Infrastructural History of BP in Iran* (Cambridge, MA: MIT Press, 2018).

Sharara, Hayat. *When Days Dusked* (Beirut: Arab Institute for Research & Publishing, 2002).

Sharkey, Heather J. *Living with Colonialism: Nationalism and Culture in the Anglo-Egyptian Sudan* (Berkeley: University of California Press, 2003).

Shearer, Christine. *Kivalina: A Climate Change Story* (Chicago: Haymarket Books, 2011).

Sheller, Mimi. 'The Origins of Global Carbon Form', *Log*, no. 47 (2019), 57–68.

Sloterdijk, Peter. *Critique of Cynical Reason* (Minneapolis: University of Minnesota Press, 1987).

Smith, Chris and Peter Whalley. 'Engineers in Britain: A Study in Persistence', in Peter Meiksins, Chris Smith and Boel Berner (eds), *Engineering Labour: Technical Workers in Comparative Perspective* (London: Verso, 1996), 27–60.

Srivastava, Sanjay. *Constructing Post-Colonial India: National Character and the Doon School* (London: Routledge, 1998).

Stocking, George W. *Middle East Oil: A Study in Political and Economic Controversy* (London: Penguin, 1971).

Stoler, Ann Laura. 'Tense and Tender Ties: The Politics of Comparison in North American History and (Post) Colonial Studies', *Journal of American History* 88, no. 3 (December 2001), 829–65.

Strassler, Karen. *Demanding Images* (Durham, NC: Duke University Press, 2020).

Strassler, Karen. *Refracted Visions* (Durham, NC: Duke University Press, 2010).

Sugrue, Thomas J. *The Origins of the Urban Crisis: Race and Inequality in Postwar Detroit* (Princeton: Princeton University Press, 1996).

Szeman, Imre. 'The Cultural Politics of Oil: On *Lessons of Darkness* and *Black Sea Files*', *Polygraph* 22 (2010), 33–45.

Szeman, Imre. 'Literature and Energy Futures', *PMLA* 126, no. 2 (2011), 323–6.

Taylor, Keeanga-Yamahtta. *Race for Profit: How Banks and the Real Estate Industry*

Undermined Black Homeownership (Chapel Hill: University of North Carolina Press, 2019).

Tétreault, Mary Ann. *The Organization of Arab Petroleum Exporting Countries: History, Policies, and Prospects* (Westport, CT: Greenwood Press, 1981).

Thesiger, Wilfred. *Arabian Sands* (New York: Penguin, 1991).

Tinker Salas, Miguel. *The Enduring Legacy: Oil, Culture and Society in Venezuela* (Durham, NC: Duke University Press, 2009).

Tollefson, Hannah and Darin Barney. 'More Liquid than Liquid: Solid-Phase Bitumen and Its Forms', *Grey Room* 77 (2019), 38–57.

Tomeh, George. 'OAPEC: Its Growing Role in Arab and Oil Affairs', *The Journal of Energy and Development* 3, no. 1 (Autumn 1977), 26–36.

Toscano, Alberto and Jeff Kinkle. *Cartographies of the Absolute* (Alresford: Zero Books, 2015).

Tsing, Anna Lowenhaupt. *Friction: An Ethnography of Global Connection* (Princeton: Princeton University Press, 2005).

Tsing, Anna Lowenhaupt. *The Mushroom at the End of the World: On the Possibility of Life in Capitalist Ruins* (Princeton: Princeton University Press, 2015).

Tsing, Anna Lowenhaupt. 'On Nonscalability: The Living World is Not Amenable to Precision-Nested Scales', *Common Knowledge* 18, no. 3 (2012), 505–24.

Tufte, Edward R. *Visual Explanations: Images and Quantities, Evidence and Narrative* (Cheshire, CT: Graphics Press, 1997).

Vadén, Tere and Antti Salminen. *Energy and Experience: An Essay in Nafthology* (Chicago: MCM, 2015).

Vadén, Tere and Antti Salminen. 'Ethics, Nafthism, and the Fossil Subject', *Relations* 6, no. 1 (June 2018), 33–48.

Vali, Murtaza. *A Crude History of Modernity* (Dubai: Art Jameel, 2018).

Valizādeh, Iraj. *Anglo va Bangolo dar Abadan* (Tehran: Simiā Honar, 1390/2011).

Vico, Giambattista. *The New Science of Giambattista Vico* (Ithaca: Cornell University Press, 1988).

Vitalis, Robert. *America's Kingdom. Mythmaking on the Saudi Oil Frontier* (New York: Verso, 2009).

Vitalis, Robert. *Oilcraft: The Myths of Scarcity and Security That Haunt U.S. Energy Policy* (Stanford: Stanford University Press, 2020).

Von Bieberstein, Alice. 'Treasure/Fetish/Gift: Hunting for "Armenian Gold" in Post-Genocide Turkish Kurdistan', *Subjectivity* 10, no. 2 (2017), 170–89.

Walker, Julian. *Tyro on the Trucial Coast* (Durham: The Memoir Club, 1999).

Wapner, Paul. 'Politics Beyond the State', *World Politics* 47, no. 3 (1995), 311–40.

Watts, Michael. 'Righteous Oil? Human Rights, the Oil Complex and Corporate Social Responsibility', *Annual Review of Environment and Resources*, 30 (2005), 373–407.

Watts, Michael J. 'Specters of Oil: An Introduction to the Photographs of Ed Kashi', in Hannah Appel, Arthur Mason and Michael Watts (eds), *Subterranean Estates: Life Worlds of Oil and Gas* (Ithaca: Cornell University Press, 2015), 165–88.

Wight, David M. *Oil Money: Middle East Petrodollars and the Transformation of US Empire, 1967–1988* (Ithaca: Cornell University Press, 2021).

Wilmington, Michael. 'Beauty, Horror Emerge from IMAX's "Fires of Kuwait"', *Los Angeles Times*, 11 June 1993.

Wittgenstein, Ludwig. *Philosophical Investigations* (New York: Macmillan, 1956).

Wright, Andrea. 'From Slaves to Contract Workers: Genealogies of Consent and Security in Indian Labor Migration', *Journal of World History* 31, no. 2 (2020), 29–43.

Wright, Andrea. 'Imperial Labor: Strikes, Security, and the Depoliticization of Oil Production', in Neilesh Bose (ed.), *South Asian Migrations: A Global History: Labor, Law, and Wayward Lives* (New York: Bloomsbury, 2020), 63–84.

Yeğen, Mesut. 'The Turkish State Discourse and the Exclusion of Kurdish Identity', *Middle Eastern Studies* 32, no. 2 (1996), 216–22.

Yergin, Daniel. *The Prize: The Epic Quest for Oil, Money, and Power* (New York: Simon & Schuster, 1991).

Younis, Ala. 'Reparation of A Diluted Image', in Fawz Kabra (ed.), *No to the Invasion, Breakdowns and Side Effects* (Annandale-on-Hudson: CCS Bard and Barjeel Art Foundation, 2017), 54–60.

Zahlan, Rosemarie Said. *The Origins of the United Arab Emirates: A Political and Social History of the Trucial States* (London: Macmillan, 1978).

Zaloom, Caitlin. 'How to Read the Future: The Yield Curve, Affect, and Financial Prediction', *Public Culture* 21, no. 2 (2009), 246–68.

Zardini, Mirko. 'Homage to Asphalt', *Log*, no. 15 (2009), 8–15.

Zaretsky, Natasha. *No Direction Home: The American Family and the Fear of National Decline, 1968–1980* (Chapel Hill: University of North Carolina Press, 2007).

INDEX

Note: **bold** indicates illustrations